Clinical Applications for Next-Generation Sequencing

Clinical Applications for Next-Generation Sequencing

Edited by

Urszula Demkow

Rafał Płoski

AMSTERDAM • BOSTON • HEIDELBERG • LONDON
NEW YORK • OXFORD • PARIS • SAN DIEGO
SAN FRANCISCO • SINGAPORE • SYDNEY • TOKYO

Academic Press is an imprint of Elsevier

Academic Press is an imprint of Elsevier
125 London Wall, London EC2Y 5AS, UK
525 B Street, Suite 1800, San Diego, CA 92101-4495, USA
225 Wyman Street, Waltham, MA 02451, USA
The Boulevard, Langford Lane, Kidlington, Oxford OX5 1GB, UK

Notices
Knowledge and best practice in this field are constantly changing. As new research and experience broaden our understanding, changes in research methods, professional practices, or medical treatment may become necessary.

Practitioners and researchers must always rely on their own experience and knowledge in evaluating and using any information, methods, compounds, or experiments described herein. In using such information or methods they should be mindful of their own safety and the safety of others, including parties for whom they have a professional responsibility.

To the fullest extent of the law, neither the Publisher nor the authors, contributors, or editors, assume any liability for any injury and/or damage to persons or property as a matter of products liability, negligence or otherwise, or from any use or operation of any methods, products, instructions, or ideas contained in the material herein.

ISBN: 978-0-12-801739-5

British Library Cataloguing-in-Publication Data
A catalogue record for this book is available from the British Library

Library of Congress Cataloging-in-Publication Data
A catalog record for this book is available from the Library of Congress

For Information on all Academic Press publications
visit our website at http://store.elsevier.com/

Working together
to grow libraries in
developing countries

www.elsevier.com • www.bookaid.org

Typeset by TNQ Books and Journals
www.tnq.co.in

Printed and bound in the United States of America

Contents

v

CHAPTER 6 **Next Generation Sequencing in Neurology and Psychiatry.............97**
*Krystyna Szymańska, Krzysztof Szczałuba, Anna Kostera-Pruszczyk
and Tomasz Wolańczyk*

CHAPTER 12 The Role of Next Generation Sequencing in Genetic Counseling..241

Asude Durmaz and Burak Durmaz

CHAPTER 13 Next Generation Sequencing in Undiagnosed Diseases 259

Urszula Demkow

List of Contributors

Zofia T. Bilińska
Unit for Screening Studies in Inherited Cardiovascular Diseases, Institute of Cardiology, Alpejska, Warsaw, Poland

Izabela Chojnicka
Faculty of Psychology, University of Warsaw, Warsaw, Poland

Ozgur Cogulu
Department of Pediatric Genetics, Faculty of Medicine, Ege University, Izmir, Turkey; Department of Medical Genetics, Faculty of Medicine, Ege University, Izmir, Turkey

Silviene Fabiana de Oliveira
The Jackson Laboratory for Genomic Medicine, Farmington, CT, USA; Laboratório de Genética, Departamento de Genética e Morfologia, Instituto de Ciências Biológicas, Universidade de Brasília, Brasília, Brazil

Urszula Demkow
Department of Laboratory Diagnostics and Clinical Immunology of Developmental Age, Medical University of Warsaw, Warsaw, Poland

Asude Durmaz
Department of Medical Genetics, Ege University Faculty of Medicine, Izmir, Turkey

Burak Durmaz
Department of Medical Genetics, Ege University Faculty of Medicine, Izmir, Turkey

Eliza Glodkowska-Mrowka
Department of Laboratory Diagnostics and Clinical Immunology of Developmental Age, Medical University of Warsaw, Warsaw, Poland

Sławomir Gruca
Department of Immunology, Medical University of Warsaw, Warsaw, Poland; Bioinformatics Group, University of Leeds, Leeds, West Yorkshire, UK

Krystian Gulik
Department of Immunology, Medical University of Warsaw, Warsaw, Poland

Andrzej Kochański
Neuromuscular Unit, Mossakowski Medical Research Center, Polish Academy of Sciences, Warsaw, Poland

Anna Kostera-Pruszczyk
Department of Neurology, Medical University of Warsaw, Poland

John D. Lantos
Children's Mercy Hospital Bioethics Center, University of Missouri – Kansas City, Kansas City, MO, USA

Ankit Malhotra
The Jackson Laboratory for Genomic Medicine, Farmington, CT, USA

Iwona Malinowska
Department of Pediatrics, Hematology and Oncology, Medical University of Warsaw, Poland

Monika Ołdak
Department of Genetics, World Hearing Center, Institute of Physiology and Pathology of Hearing, Warsaw, Poland

Michal Okoniewski
Division Scientific IT Services, IT Services, ETH Zurich, Zurich, Switzerland

Jacub Owoc
Lubuski College of Public Health, Zielona Góra, Poland

Rafał Płoski
Department of Medical Genetics, Centre of Biostructure, Medical University of Warsaw, Warsaw, Poland

Dariusz Plewczynski
Centre of New Technologies, University of Warsaw, Warsaw, Poland; The Jackson Laboratory for Genomic Medicine, Farmington, CT, USA; Centre for Bioinformatics and Data Analysis, Medical University of Bialystok, Bialystok, Poland

Joanna Ponińska
Laboratory of Molecular Biology, Institute of Cardiology, Warsaw, Alpejska, Poland

Ravi Sachidanandam
Department of Oncological Sciences, Icahn School of Medicine at Mount Sinai, NY, USA

Robert Smigiel
Department of Social Pediatrics, Wroclaw Medical University, Wroclaw, Poland

Piotr Stawinski
Department of Immunology, Medical University of Warsaw, Warsaw, Poland

Tomasz Stoklosa
Department of Immunology, Medical University of Warsaw, Warsaw, Poland

Przemysław Szałaj
Centre for Bioinformatics and Data Analysis, Medical University of Bialystok, Bialystok, Poland; I-BioStat, Hasselt University, Hasselt, Belgium

Krzysztof Szczałuba
MEDGEN Medical Center, Warsaw, Poland; Medical Genetics Unit, Mastermed, Białystok, Poland

Krystyna Szymańska
Department of Experimental and Clinical Neuropathology, Mossakowski Medical Research Centre, Polish Academy of Sciences, Warsaw, Poland; Department of Child Psychiatry, Medical University of Warsaw, Poland

Marek Wiewiorka
Institute of Computer Science, Warsaw University of Technology, Warsaw, Poland

Tomasz Wolańczyk
Department of Child Psychiatry, Medical University of Warsaw, Poland

NEXT GENERATION SEQUENCING—GENERAL INFORMATION ABOUT THE TECHNOLOGY, POSSIBILITIES, AND LIMITATIONS

Rafał Płoski

Department of Medical Genetics, Centre of Biostructure, Medical University of Warsaw, Warsaw, Poland

CHAPTER OUTLINE

Next generation sequencing (NGS) is defined as technology allowing one to determine in a single experiment the sequence of a DNA molecule(s) with total size significantly larger than 1 million base pairs (1 million bp or 1 Mb). From a clinical perspective the important feature of NGS is the possibility of sequencing hundreds/thousands of genes or even a whole genome in one experiment.

The high-throughput characteristic of NGS is achieved by a massively parallel approach allowing one to sequence, depending on the platform used, from tens of thousands to more than a billion molecules in a single experiment (Figure 1). This massively parallel analysis is achieved by the miniaturization of the volume of individual sequencing reactions, which limits the size of the instruments and reduces the cost of reagents per reaction. In the case of some platforms (referred to as third generation sequencers) the miniaturization has reached an extreme and allows sequencing of single DNA molecules.

An important characteristic of main NGS platforms used today is the limited length of sequence generated in individual reactions, that is, limited read length. Despite constant improvements the read length for the majority of platforms has stayed in the range of hundreds of base pairs. To sequence DNA longer than the feasible read length, the material is fragmented prior to analysis. After the sequencing, the reads are reassembled in silico to provide the information on the sequence of the whole target molecule.

NGS VERSUS TRADITIONAL (SANGER SEQUENCING)

The term NGS emphasizes an increase in output relative to traditional DNA sequencing developed by Sanger in 1975 [1], which, despite the improvements introduced since then, still has an output limited to ~75,000 bp (75 kb). This increase in output translates to the possibility of genome-wide analyses, again contrasting with Sanger DNA sequencing, allowing in practice the analysis of single genes or parts thereof. Despite the spreading use of NGS, Sanger sequencing remains the method of choice for validation necessary for all clinically relevant NGS findings.

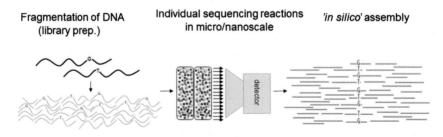

FIGURE 1

General principles of technical solutions for NGS. The central part of the process consists of a large number of sequencing reactions carried out in parallel on fragmented DNA in very small volumes (multicolored dots). The outcome of the individual reactions is read by an optical or electronic detector. The final step is the assembly of the thus-generated sequences (reads) allowing the determination of the sequence of the DNA molecule(s) before fragmentation.

COVERAGE

An important feature of NGS is multiple sequencing of each base of the target sequence. The number of times a given position has been sequenced in an NGS experiment (i.e., number of reads containing this position) is termed "coverage." On one hand, multiple coverage is a consequence of the abovementioned random target fragmentation necessitated by short read lengths. On the other hand, obtaining multiple reads covering the same target is necessary for eliminating random sequencing errors and, equally importantly, enabling the detection of individual components in DNA mixtures. The DNA mixtures that commonly need to be resolved in a clinical setting are those due to heterozygosity.

Sufficient coverage is important for good quality of an NGS experiment. Although the detection of heterozygosity may seem straightforward with a coverage of ~10, it should be realized that the probability of obtaining all reads from the chromosome without the variant is 1 in $2^{10} = 1/1024$, meaning that in whole-genome sequencing (WGS) hundreds of heterozygous variants can be missed. Even more challenging is the detection of a variant present in a proportion smaller than 50%, which often is the case for somatic mutations in neoplastic tissue, chimerism, and mosaicism, or heteroplasmy in mitochondrial DNA.

Coverage, sometimes also called "sequencing depth" or "depth or coverage," can be quantified by "mean coverage," that is, the sum of coverage for all nucleotides in the target sequence divided by the number of nucleotides. Mean coverage gives a general idea about experiment design but it may be misleadingly high if some few regions are covered excessively and others poorly or not at all. A more informative way to characterize coverage is to calculate what percentage of the target has been sequenced with a specified (or higher) depth that is deemed satisfactory. A reasonable result when looking for germ-line variants (e.g., disease-causing mutations, expected to be present in 50% or 100% of appropriately positioned reads) is to have more than 80% of the target covered a minimum of 20 times.

From an economical perspective it is also desirable to obtain coverage that is maximally smooth, that is, there are no discrete regions that are covered excessively or insufficiently. Excessive coverage is unnecessary and it generates cost since it uses up expensive sequencing reagents. A parameter describing smoothness of coverage is "fold 80 base penalty"—the fold overcoverage necessary to raise 80% of the bases in targets to the mean coverage in those targets. A value of 1 indicates a perfectly smooth coverage (unrealistic). If the mean coverage was satisfactory, a value of 2 indicates that the sequencing done so far should be repeated (doubled) to have 80% of targets satisfactorily covered. A fold 80 base penalty of <3 is regarded as satisfactory.

NGS LIBRARY PREPARATION

The steps needed to prepare DNA for NGS analysis are collectively called "library preparation." NGS libraries are platform specific, so that a library prepared for one platform cannot be used on another unless it is explicitly compatible (usually coming from the same manufacturer). NGS libraries can be prepared starting directly from target DNA (usually total genomic DNA) or from polymerase chain reaction (PCR) products. To undergo sequencing, DNA molecules in a library need short sequences called "adapters" to be present on both ends.

If a library is prepared by PCR the simplest approach is to incorporate adapter sequences into PCR primers so that they become part of the PCR products, which are then ready for sequencing. PCR usage for library preparation is particularly attractive when few genes/exons need to be analyzed. However, advanced approaches based on emulsion PCR have also been developed allowing large-scale analyses [2] (see also below).

If library preparation is not based on PCR the first stage is fragmentation of DNA. The next step after fragmentation is ligation of the adapters. Adapters may contain an "index"—4- to 10-bp sequences that provide tags allowing one to distinguish different samples sequenced together. The indexing is also known as "bar coding." Multiplexing of samples (pooling samples for a single sequencing experiment) is a common strategy allowing the use of high-throughput machines for analysis of samples that individually require less extensive and/or less deep coverage than that offered by a given platform. It is particularly efficient to use double indexing, usually in a strategy using a separate index for each of the two paired end reads (see below). The samples are then identified by the combination of two indices, which increases the multiplexing possibilities (10 indices when used in pairs allow multiplexing of 100 samples). Most of the commonly used NGS applications allow multiplexing of 24–96 samples, in some cases this number is 384; even higher numbers can be achieved with customized approaches.

Traditional methods of fragmentation are based on sonication: The Adaptive Focused Acoustics™ technology patented by Covaris (http://covarisinc.com) or Adaptive Cavitation Technology of Diagenode (http://www.diagenode.com/en/index.php). While sonication gives high-quality results in terms of randomness of break points and reproducible fragment size distribution, it introduces damage at the ends of DNA molecules necessitating an additional enzymatic step of repair.

An ingenious advancement in the preparation of NGS libraries relies on enzymatic reaction with transposase, which catalyzes simultaneously both DNA fragmentation and adaptor/tag incorporation [3], a process that has been nicknamed "tagmentation." Tagmentation greatly reduces the amount of material needed for library construction, allowing one to routinely process samples of 50 ng DNA or less (vs ~1 µg typically required if sonication was used). It also speeds up library preparation and allows easy automation. For example, using tagmentation a library for whole-exome sequencing (WES) can be finished within 3 h, whereas previous protocols required ~2 days.

Libraries made from genomic DNA, although not based on PCR in the initial stages, often use 10–20 PCR cycles at the final stage. Although PCR compensates for sample losses during library preparation and increases the yield of molecules with correctly ligated adapters, it has disadvantages: (1) during PCR fragments with extremely high or low GC content are less efficiently amplified. Since GC-rich sequences are often located in functionally important 5′ regions of genes (first exons, in particular), this leads to annoying gaps in coverage [4]. (2) PCR decreases the diversity of a library, generating "duplicates," that is, multiple fragments that are all copies of a single molecule. Duplicates decrease the quality of sequencing—they can falsely suggest homozygosity and/or amplify a random error to an such extent that it can be accepted as a true variant. A solution to these problems, increasingly often used for WGS, is provided by protocols and kits that allow one to make PCR-free libraries (http://www.illumina.com, http://www.biospace.com).

SEQUENCE ASSEMBLY: DE NOVO SEQUENCING VS RESEQUENCING

Owing to short reads generated by NGS platforms, the important step of analysis is the assembly of the sequence. Two basically different approaches exist: *de novo* assembly and resequencing. *De novo* assembly is performed whenever a completely unknown target is analyzed, as is typically the case when a genome/plasmid is sequenced for the first time. *De novo* sequencing requires high coverage to provide enough overlapping reads to guide assembly throughout the whole target. It is also computationally demanding since all reads need to be checked against one another for overlaps. Further challenge in *de novo* sequencing comes from abundant repetitive regions often present in genomes. Sequences of such regions are particularly difficult to infer from the short reads generated by sequencers.

In resequencing the assembly of reads is guided by an a priori knowledge of the target available as a reference sequence. The ideal reference sequence is a consensus sequence providing a general framework of the target with its most prevalent variants. When a reference sequence is available it is typically used as a target for alignment of the generated reads. Despite their short length the majority of reads can usually be mapped with high confidence, that is, defined as coming from a given part of the genome. After being mapped, the reads are scanned for mismatches with the reference sequence and these are interpreted as variants.

Given the high and constantly improving quality of the human reference genome [5–7], resequencing is the predominant approach in medical genetics. In comparison to *de novo* sequencing, resequencing requires less coverage and is simpler computationally. Resequencing is efficient at detecting variants much shorter than the length of the reads. Typically these are single nucleotide variants (SNVs) as well as small insertions/deletions. Conversely, detection of larger variants such as copy number variants (CNVs, which involve fragments >1000 bp) or even bigger structural chromosomal variants is more challenging or even impossible. Obviously, detection of variants in repetitive regions is also challenging as it is difficult to confidently map reads from such regions.

PAIRED-END AND MATE-PAIR LIBRARIES AND LONG FRAGMENT READ TECHNOLOGY [8]

Problems associated with short read lengths generated by NGS platforms can be to some extent alleviated by certain strategies of library preparation and/or sequencing. A common approach is to perform sequencing from both ends of the fragments contained in the library. This is known as paired-end sequencing and allows one to effectively double the length of the sequenced DNA molecule. Paired-end sequencing is typically performed on libraries of DNA fragments longer than the part that undergoes sequencing to ensure that the reads from both ends do not overlap. Usually paired-end sequencing is used for libraries of fragments <1 kb.

An approach allowing one to sequence fragments located much farther apart (up to 25 kb) is known as mate-pair sequencing [9]. After an initial gentle fragmentation that leaves appropriately long DNA fragments the ends of the DNA molecules are labeled with biotin and the molecules are circularized. Then, the second round of fragmentation yielding fragments <1 kb is performed, followed by enrichment for biotin-labeled molecules. Finally paired-end sequencing is carried out on the molecules, which effectively consist of terminal ends of the initial long DNA fragment joined together. Mate-pair sequencing allows one to overcome to some extent the limitations of NGS associated with long repetitive stretches commonly present in the human genome and is useful for the detection of chromosomal rearrangements.

An interesting approach to sequencing relatively long fragments (~10 kb) using the available short reads is offered by long fragment read (LFR) technology [8]. The first step of LFR is to dilute high-molecular-mass DNA (fragments ~10 kb) and physically separate it into aliquots, which are then processed in parallel: DNA in each well is fragmented, amplified, and ligated to uniquely indexed adapters, thus allowing them to be distinguished from the other DNA in all the wells. Next, DNA from all the wells is pooled and submitted to a standard NGS procedure. Three hundred eighty-four aliquots are commonly prepared on a microtiter plate and this number is regarded as sufficient for whole human genome analysis. The DNA concentration in each aliquot (well) is low enough to ensure that a given DNA fragment, with a reasonably high likelihood, is present in a number of wells as a *single* copy. Since information about the "well of origin" is kept owing to indexing, provided successful bioinformatics assembly, each individual ~10-kb sequence can be regarded as representing a continuous stretch of DNA from a single chromosome. The LFR approach has been implemented in a commercially available kit (TruSeq Synthetic Long-Read DNA Library Prep Kit, http://www.illumina.com).

The LFR approach is not fully equivalent to single-molecule sequencing since in some cases the ~10-kb fragments may be difficult to assemble owing to repetitive sequences. Notwithstanding this, LFR is a valuable tool to obtain phase information, that is, information on which variants are located together in a single maternal or paternal chromosome. Phase information can be of paramount importance in a number of settings [10], in medical genetics it is, for example, important in the search for compound heterozygous mutations in diagnosing autosomal recessive diseases. Phase information allows one to easily filter out variants found in *cis*, whereas without it each candidate pair of mutations has to be verified in a family study, which is laborious, or by inclusion of the parents in the initial study, which is expensive.

NGS PLATFORMS
ILLUMINA

Illumina platforms rely on fluorescence-based sequencing of single DNA molecules after a non-PCR-based clonal amplification on solid support. The approach was developed in 2006 by the company Solexa, which was subsequently acquired by Illumina.

Library preparation for Illumina platforms originally included DNA fragmentation by physical means and enzymatic repair of the ends of molecules with subsequent addition of a single adenine base to the 3′ end of the DNA fragments. The final step was ligation of adapters. The ligation is facilitated by a single thymine overhanging the 3′ end of each adapter, which complements the adenine overhang of the DNA fragments. Although this procedure is still used, alternative protocols based on transposase-catalyzed tagmentation are gaining increasing popularity [3]. Illumina adapters always include (1) so-called P5 and P7 binding regions, which are complementary to oligos on the surface of the flow cell (see below) and (2) sequencing primer binding regions. Whenever multiplexing is planned adapters should also have one or two indices.

On Illumina platforms sequencing takes place on the surface of a fluidic chamber (flow cell) designed to provide access to reagents and make optical imaging possible [11]. A flow cell can have up to eight channels called lanes, which can accommodate independent samples. The surface of a flow cell is coated with a lawn of oligonucleotides complementary to the P5 and P7 binding regions in the adapters.

The prerequisite for sequencing is binding of the DNA library to the flow cell. A denatured (single stranded) and appropriately diluted library is applied to the flow cell allowing hybridization (noncovalent binding) between DNA fragments and oligos at the flow cell's surface. The relatively weak noncovalent binding is subsequently converted to strong covalent bonds by synthesis of the complementary (reverse) strand followed by washing away of the originally bound DNA strand.

The next step is bridge amplification—a cyclic process that clonally replicates DNA molecules bound to the flow cell, creating so-called "clusters." During bridge amplification a single-stranded molecule flips over and forms a bridge by hybridizing to an adjacent, complementary primer. After extension by polymerase a double-stranded bridge is formed, which, after denaturation, yields a reverse copy of the original (forward) DNA fragment covalently bound to the flow cell surface. The process is cyclically repeated, and at the final step the reverse strands are cleaved away leaving a homogeneous cluster with ~1000 forward strands.

Appropriate density of clusters is a critical determinant of successful sequencing. Too few clusters decrease the sequencing yield; too many clusters result in overlaps, which negatively affect the quality of data and in extreme cases cause total failure of the experiment.

The DNA sequencing proper on Illumina platforms is performed by sequencing-by-synthesis (SBS) technology. The reaction is started by hybridization of a primer complementary to the part of the adapter adjacent to the sequenced insert followed by cycles of (1) addition of DNA polymerase with four nucleotides, (2) imaging, and (3) cleavage of the fluorophore and deblocking. The nucleotides are reversibly blocked (terminated) and individually labeled fluorescently, which ensures that during each cycle the primer is extended by a single base only and that this base can be identified by its fluorescence during appropriate excitation and scanning.

Depending on the application and the particular platform 36–301 cycles can performed, allowing one to sequence 35–300 bp (base calling at the nth cycle requires fluorescence data for this cycle as well as the $n-1$ and $n+1$ cycles; thus the number of cycles is always higher by 1 than the length of sequence obtained).

All Illumina platforms support paired-end sequencing. The sequencing of the other end of a DNA molecule is achieved by stripping off the strand synthesized during the first read and performing a single cycle of bridge amplification with the cleavage of the original forward strand. This converts single-stranded DNA molecules in each cluster into their reverses, which are then sequenced as described above.

The indices are sequenced in separate additional reads (one or two as required). Each index read starts by hybridization of a dedicated primer followed by a number of SBS cycles appropriate to the length of the index.

Illumina apparatuses

The first platform using the described technology was the Genome Analyzer (GA), initially offered by Solexa, a company acquired by Illumina in 2007. Although still used, GA is being largely replaced by HiSeq instruments (HiSeq 1000 and 2000), and their software and/or hardware upgraded versions (HiSeq 1500, HiSeq2500 and HiSeq 3000, HiSeq4000, respectively) as well as by MiSeq and MiSeqDx. The important upgrade of the HiSeq 3000/4000 is a patterned flow cell with nanowells directing cluster formation, which ensures optimal cluster density. The HiSeq instruments are as of this writing the dominant platforms for high-throughput NGS applications worldwide. The MiSeq machines belong to a category of benchtop sequencers—MiSeq is being developed toward low-scale research projects and MiSeqDx is focused on clinical applications. It is noteworthy that as of 2014 MiSeqDx is the only NGS instrument that has received FDA clearance. A comparison of the most popular Illumina NGS platforms is shown in Table 1.

Table 1 Comparison of Most Popular Illumina NGS Platforms

	Max. Output (Gb)	Max. Read Number (M, Paired-End Reads)	Max. Read Length (bp)	No. of Lanes
Genome Analyzer IIx	95	300	2×150	8
HiSeq 2500	1000	4000	2×125	2×8
HiSeq 2500 rapid mode/single flow cell	90	300	2×150	2
HiSeq 3000/4000	750/1500	2500/5000	2×150	8/16 (2×8)
NextSeq 500	120	400	2×150	1
MiSeq	15	25	2×300	1

For specialized centers Illumina also offers the HiSeq X Ten platform, which is a set of five or 10 machines sold together to enable human WGS at a population scale with a price of US $1000 per genome (www.illumina.com).

Sequencing on GA or HiSeq 1000/1500/2000/2500 (but not MiSeq, NextSeq 500) requires a separate machine (cBOT) for the clustering.

SEMICONDUCTOR-BASED PLATFORMS

Semiconductor-based platforms rely on detection of pH changes occurring during DNA synthesis [12]. Sequencing is performed after single-molecule amplification by a process known as emulsion PCR [13].

Library preparation starts from DNA fragmentation and adapter ligation similar to the Illumina approach, and then emulsion PCR is performed [13]. The library is mixed with microscopic beads in an environment of oil and an aqueous solution containing PCR reagents and the mixture is shaken to form an emulsion. The dilution of the library is low enough to ensure that only single DNA molecules have a chance to be encapsulated together with a bead in one emulsion micelle. The emulsion is then subjected to thermal changes allowing PCR. The micelles are separated from one another by oil so that PCR occurs independently in each without diffusion of products. As the beads are covalently coated with oligonucleotides complementary to adapter sequences, the PCR products generated within a micelle adhere to the bead surface. After PCR the emulsion is broken, and the beads are separated, enriched for those that contain PCR products, and primed for sequencing by annealing of an appropriate primer.

The primed beads are placed into wells of a specialized chip (Ion Chip). The size of the Ion Chip wells ensures that each can accommodate a single bead only. The chip is then cyclically flushed with four nucleotides (one after another, in a constant order) and reagents allowing DNA synthesis. Each time a nucleotide is incorporated there is a release of H^+ ions leading to a pH drop in the well, which is detected by a sensor located at the bottom of the well. If two or more identical nucleotides are present side by side in the sequenced fragment their number can be inferred from the stronger decrease in pH relative to what is observed for a single nucleotide.

Since signals from all wells are collected simultaneously without the need for sequential scanning the semiconductor-based sequencing is fast, with a single run taking ~2h. The use of unlabeled nucleotides simplifies sequencing chemistry, lowering the cost. The availability of chips with different numbers of wells makes the size of sequencing experiments easily scalable.

Table 2 Characteristics of Performance of Semiconductor-Based Apparatuses				
Machine	Chip	Max. Output (Gb)	Max. Read Number (M)	Read Length (bp)
Ion PGM™	Ion 314™ v2	0.03–0.1	0.4–0.55	~400
	Ion 316™ v2	0.3–1	2--3	~400
	Ion 318™ v2	0.6–2.0	4–5.5	~400
Ion Proton	Ion Proton I Chip	10	60–80	~200

Semiconductor-based apparatuses

The first semiconductor-based machine was the Ion PGM released in 2010 by Ion Torrent, a company later acquired by Life Technologies Corp. (now part of Thermo Fisher Scientific, Inc.). The Ion PGM can be used with three chips, allowing 0.03–2Gb of output. It is a low-throughput, low-cost benchtop sequencer dedicated mainly to amplicon sequencing. A considerably upgraded version of the PGM is the Ion Proton, which, although still in the benchtop class, is capable of WES and, according to company claims (www.lifetechnologies.com), in the near future should allow WGS as well. The characteristics of performance of semiconductor-based apparatuses from Life Technologies are shown in Table 2.

A recent development in the field of clinically oriented semiconductor-based NGS platforms comes from Vela Diagnostics (http://www.veladx.com/), who offer the Sentosa system, including both a sample preparation station and a sequencer. The sequencer is manufactured by Thermo Fisher according to Vela Diagnostics specifications and uses Ion Torrent technology. As of this writing this system is dedicated to running CE-IVD cancer panels developed by the company.

SEQUENCING BY THE OLIGO LIGATION DETECTION (SOLiD) PLATFORM

The SOLiD platform relies on ligation [14] with fluorescence-based detection. After the standard steps of fragmentation and adapter ligation, the library is amplified by emulsion PCR or, in the final upgrade, by an isothermal amplification on the surface of a flow cell (flow chip), called "template walking" or "Wildfire," which is a process with some similarities to the bridge amplification of Illumina [15].

Sequencing on the SOLiD [14] is based on ligation. It starts with annealing of a primer ending at the last base of the adapter. Next a mixture of 16 oligonucleotide octamer probes labeled with four different fluorochromes is added. Each probe at one end has an interrogation sequence representing one of the 16 combinations of a 2-base sequence followed by a 6-bp degenerate stretch and the fluorochrome label at the other end. After hybridization ligation is performed, which covalently links the primer and the adjacently annealed probe. Next, unbound probes are washed away, fluorescence is read, and the probe is cleaved, removing the label and three neighboring bases (leaving a 5-mer bound). The hybridization–ligation cycle is repeated six more times, the only difference being that ligation occurs with the previously bound probe instead of the primer. This completes the first, so-called, "round" of sequencing. Next the synthesized strand is removed and the second round is started by annealing a new primer, finishing one base before the end of the adapter (n − 1 primer). Five rounds are performed with successively more offset primers (to n − 4). Although four fluorochromes do not allow discrimination of 16 probes, the sequence can be determined using information from all the offset cycles. In addition, in some cycles knowledge of the adapter sequence is used to interpret the data (see [16] for a detailed explanation for the n − 1 cycle). The advantage of SOLiD sequencing is accuracy due to effective double interrogation of each position.

The first SOLiD platform was released by Applied Biosystems in 2007, followed by upgrades of which the 5500xl with the Wildfire chemistry was the most advanced. The SOLiD platform has good accuracy (up to 99.99%), moderate output (30 Gb), and rather short reads (from the initial 35 to 85 bp for SOLiD 5500xl). Although SOLiD is potentially attractive for high-throughput-dependent diagnostic applications such as WES or WGS, in 2013 Life Technologies announced that it has no plans for further development and in 2014 SOLiD was available only to existing customers.

PYROSEQUENCING ON ROCHE/454 PLATFORMS

Pyrosequencing relies on the detection of the pyrophosphate molecule released during DNA synthesis [17]. It was the first NGS platform available commercially and some concepts behind it were later used in semiconductor sequencing. Both processes share: (1) emulsion PCR on microbeads, (2) deposition of the beads on a microplate (PTP or picotiterplate) according to the one bead–one well principle, (3) sequencing by strand elongation after sequential flushing with four nucleotides, and (4) detection of a product released during strand elongation (pyrophosphate or H^+, respectively) whose concentration is proportional to the number of bases incorporated. A difference is that in pyrosequencing the signal ultimately comes from conversion of luciferin into oxyluciferin, which generates visible light [16].

Since the first release of the system by Roche in 2005 the platform has been upgraded, with its final high-throughput version being the 454 GS FLX Titanium system. This system used eight independent lanes each allowing ~100,000 reads and produced 14 Gb in an ~10-h run. In 2009 Roche, as the first company, introduced a benchtop NGS sequencer called Junior. Junior is a machine similar to FLX but it can accommodate a PTP with a single lane only.

The advantage of the Roche/454 system is the long read length (up to 800 bp); the disadvantages are low output, problems with homopolymers, and very costly reagents. As announced in 2013, both Roche platforms will be no longer developed.

COMPLETE GENOMICS ANALYSIS (CGA™) PLATFORM

The CGA platform employs sequencing by ligation with fluorescence-based detection. Sequencing is performed on self-assembling DNA nanoarrays or DNB™ arrays [18,19].

An unique feature of the library preparation for the CGA is amplification of fragmented DNA by rolling-circle replication, which produces covalently linked tandem copies of single-stranded DNA, called "DNA nanoballs" (DNBs). DNB formation allows very dense packaging of amplified library molecules—hundreds of fragments are effectively squeezed, forming a sphere with a diameter of approximately 200 nm. Next, the DNBs are immobilized on the surface of a chip manufactured to contain ~3 billion regularly patterned sticky spots, each binding only one DNB. The chip with bound nanoballs is called the DNB™ array. The dense and ordered pattern of the DNB™ array reduces the volume of sequencing reagents and maximizes the efficiency of the imaging by ensuring an optimal alignment with the camera, so that every two pixels are used to image a different DNB.

The sequencing by ligation on the CGA™ platform has some similarities to the SOLiD platform. The difference is that in the CGA protocol nucleotide positions are interrogated one at a time. Furthermore, the CGA approach is fully "unchained," that is, there is no need to determine the first base before reading the second one, etc. Thus, possible errors (in particular deletions/insertions) introduced at the beginning of sequence do not affect the quality of downstream bases as is the case with other methods.

Complete Genomics, Inc., was established in 2006, in Mountain View, California, USA, and in 2013 it was acquired by BGI-Shenzhen (www.completegenomics.com). The company has never commercialized its platform but offers DNA sequencing as a service with a focus on high-quality human WGS [19].

SINGLE-MOLECULE SEQUENCING

All NGS platforms described above rely on clonal amplification of a library prior to sequencing (bridge amplification, emulsion PCR, etc.), which is necessary to make the sequencing signal strong enough for detection but can lead to errors and biases (GC bias in PCR, preferential amplification of shorter fragments in bridge amplification). Single-molecule sequencing, also known as third generation sequencing (TGS), avoids these pitfalls.

Pacific biosciences single-molecule real-time sequencing

On the PacBio SMRT platform the sequencing is carried out by monitoring in real time the activity of a single DNA polymerase extending a primer annealed to the sequenced template. This is achieved by recording the fluorescence emitted each time a labeled nucleotide is bound by the enzyme [20,21]. The reactions are performed in "zero-mode waveguide" microwells—sophisticated ultra-small wells with a transparent bottom, which allow one to immobilize a single molecule of DNA polymerase and guide the light emitted by the nucleotides it binds in a way that facilitates detection [22].

The PacBio library is prepared by ligating SMRTbell™ adapters to both ends of double-stranded DNA fragments. The adapters have a hairpin structure so that after ligation a topologically circular single-stranded template is generated, called the SMRTbell. After annealing of a primer complementary to the adapter sequence the SMRTbell allows for multiple rounds of DNA synthesis so that the insert (especially a short one) can be sequenced many times.

The advantage of SMRT technology is long read length (up to 30 kb), whereas a high error rate (15%) and limited number of reads as well as high price, large size, and complicated maintenance of the instrument are disadvantages. In contrast to other platforms the errors of PacBio are essentially random [23] and can be corrected by multiple reads from SMRTbell templates (although at the expense of read length). A unique feature of the SMRT technology is direct detection of modified bases based on the variation of fluorescence duration reflecting differences in polymerase kinetics (although detection of 5-methylcytosine, the most common epigenetic modification in humans, is challenging) [24].

The first platform based on SMRT sequencing was commercialized in 2010. The latest upgrade, PacBio RS II with P5-C3 chemistry, yields, for each of 16 independent SMRT cells, 50,000 reads with 375 Mb of sequence, half of which comes from reads >10 kb (www.pacificbiosciences.com).

Helicos genetic analysis system (HeliScope)

The Helicos Genetic Analysis System (HeliScope) was the first TGS platform introduced in 2009 by Helicos BioSciences Corporation. The HeliScope library is prepared by adding a poly(A) tail to the molecules, which are then bound to a flow cell coated with poly(T) [25,26]. Sequencing is carried out by synthesis using fluorescently labeled nucleotide analogs [25–27]. Despite short reads (24 to 70 bp), low throughput (35 Gb/8 days), and a rather high error rate (~4%), the HeliScope has been shown to be capable of WGS [28]. The HeliScope is so far the only platform allowing direct sequencing of RNA [29] and it has been used in some high-profile studies of human gene expression [30,31] by the FANTOM consortium (http://fantom.gsc.riken.jp). The future of the Helicos technology is uncertain owing to the bankruptcy announced by the company in 2012.

TARGETED RESEQUENCING/ENRICHMENT STRATEGIES

The gold standard of medical NGS analysis is WGS, but owing to the size of the human genome (~3 Gb) this approach is too expensive for widespread use. A popular alternative to WGS is a selective analysis of regions of interest. The target may range from a single exon to a substantial part of the genome such as a whole exome. For small targets the selection (enrichment) is typically achieved by PCR, including long-range PCR used, for example, in whole mitochondrial DNA sequencing. The PCR enrichments usually rely on multiplex PCR, in some cases requiring sophisticated equipment such as the Fluidigm Access Array (allowing simultaneous amplification of up to ~500 loci in 48 samples) or ThunderStorm from RainDance (allowing PCR with up to 20,000 primer pairs to be performed in emulsion droplets). Large targets are usually enriched by hybridization with oligonucleotide probes (baits), which can be made as DNA or RNA molecules. There are also methods using ligation with or without circularization that can be used for intermediate/large enrichments [32]. PCR- and ligation-based enrichments start directly from DNA, whereas those using hybridization require preparation of a library appropriate for the platform used. PCR-based enrichment requires much less DNA than other strategies except for those based on transposons and pioneered by Illumina (Nextera).

An enrichment can be prepared locally but it is more common to order it from a commercial vendor (Table 3), either as a predesigned product or as a set of reagents prepared according to user specification (including genomic coordinates or gene symbols). Predesigned enrichments are usually well performing, whereas custom panels often need to be optimized. Typical optimization steps for hybridization enrichments include adjustment of library size and sequencing read lengths and, if necessary, redesign of the primers/probes set. As human exons have a median length of only ~120 bp, exon-targeted enrichments often perform better with relatively small library sizes and shorter read lengths. It is also desirable to reduce the number of PCR cycles throughout the procedure to decrease PCR-inherent biases.

An important issue with predesigned enrichments is a formal certificate for in vitro diagnostic use (IVD). Although there is a clear demand for such products at present (January 2015), there are only two enrichments with FDA clearance (Illumina's cystic fibrosis assays).

Enrichments, especially those that are customized, are expensive and they are seldom ordered if only a few samples are to be analyzed. Some enrichments are designed for up to 12 samples that need to be processed simultaneously, while others allow single-sample experiments. An example of a decision tree in enrichment selection has been proposed [33].

An important factor in considering an enrichment is the amount of sequencing necessary to obtain desired coverage. Whereas for enrichments of small targets by PCR the amount of sequencing reflects target size multiplied by desired coverage, for larger targets the situation is complicated by the relatively low enrichment efficiency possible, which implies that a considerable amount of sequencing is "wasted" on off-target regions. Figure 2 shows the relationship between target size and fraction of bases expected to be on target for an enrichment with a performance in the range of enrichments used in exome studies (~50) as well as the amount of sequencing necessary to achieve a mean coverage of 50. Note that relatively similar amounts of sequencing (~3 Gb vs ~8 Gb) are needed for disproportionately different targets (~1 Mb vs ~100 Mb, respectively). A detailed discussion of enrichment parameters vs amount of necessary sequencing can be found elsewhere [32].

Table 3 Commercially Available Enrichment Technologies for NGS				
Company	**Technology**	**Predesigned Panels**	**Custom Designs**	**Platforms**
Agilent Technologies www.agilent.com	120-nt RNA probes	Exome, kinome, X-chromosome, SureSelect Inherited Disease (2742 genes)	1 kb–24 Mb	Illumina, SOLiD
	Ligation based (up to 200,000 amplicons)	Exome, cancer, cardiomyopathy, Noonan syndrome, connective tissue disorder, X chromosome. ICCG (the International Collaboration for Clinical Genomics www.iccg.org/panel)	1 kb–5 Mb	Illumina, Ion Torrent (excl. whole exome enrichment)
Roche NimbleGen www.NimbleGen.com	60- to 90-nt DNA probes	Exome, cancer (575 genes), neurology (256 genes)	<7 Mb or <50 Mb	Illumina, Roche 454
Illumina www.illumina.com	DNA probes	Exome, tumor (somatic mutations), myeloid malignancies (somatic mutations), cancer (germ-line mutations), cardiomyopathy, inherited disease (recessive pediatric onset diseases), autism, HLA, TruSight One >4800 genes	Nextera rapid capture custom 0.5–15 Mb	Illumina
	Ligation based		Up to 1536 amplicons, covering ~650 kb	Illumina
Integrated DNA Technologies www.idtdna.com	Oligos individually synthesized on the IDT Ultramer® platform	Acute myeloid leukemia (>260 genes), cancer (127 genes), inherited disease (4503 genes)	Up to 2000 probes	Compatible with Agilent and NimbleGen libraries
Qiagen http://www.qiagen.com	PCR	Various cancer panels (up to 160 genes), cardiomyopathy (58 genes), carrier testing (157 genes)	A subset of 570 preverified gene primer sets	MiSeq/HiSeq, PGM/Proton Ion
Life Technologies	PCR	Cancer panels (>400 genes), fusion panel, inherited diseases (>300 genes), exome	Up to 6144 amplicons	PGM/Proton
	DNA probes	Exome (TargetSeq)		PGM/Proton
RainDance Technologies www.raindancetech.com	Microdoplet PCR (equipment required)	Cancer (142 genes), tumor (somatic), HLA, autism (62 genes), pharmacokinetic and pharmacology genes (36/242), X-linked disorders (800 genes)	Up to 20,000 amplicons	HiSeq/MiSeq, PGM, SOLiD, PacBio, FLX/Junior

Continued

Company	Technology	Predesigned Panels	Custom Designs	Platforms
Fluidigm Corporation www.fluidigm.com	Multiplex PCR in microfluidic chambers (equipment required)	ThunderBolts cancer panel (50 cancer genes)	Up to 480 amplicons per each of 48 samples	Illumina, Junior/ FLX, PGM/Ion
Multiplicom www.multiplicom.com	PCR-based MASTR™ (multiplex amplification of specific targets for resequencing) assays	Panels for >10 inherited diseases including CE-IVD validated panels for *BRCA1/2* and *CFTR* mutations, three onco-panels, an aneuploidy test	No	Junior/FLX, MiSeq

Table 3 Commercially Available Enrichment Technologies for NGS—cont'd

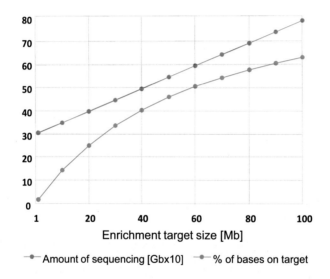

Enrichment target size [Mb]

━●━ Amount of sequencing [Gbx10]   ━●━ % of bases on target

FIGURE 2

Relationship between target size and fraction of bases expected to be on target as well as amount of sequencing required for mean coverage of 50, for an enrichment characterized by an enrichment factor or EF = 50. EF is defined as the mean coverage of the target (T) divided by the mean coverage of the rest of the genome. The fraction of bases on target F = EF × target size/[(EF − 1) × target size + genome size] [32]; the amount of sequencing = 50 × (T/F), genome size = 3 Gb.

WHOLE EXOME SEQUENCING AND WHOLE GENOME SEQUENCING

WES was described in 2009 [34] as a technique allowing one to sequence the exome, which is the portion of the genome including all of the protein-coding regions (exons). The attractiveness of WES comes from the fact that although it encompasses only ~1.5% of the genome it harbors the majority (~85%) of variants causing single-gene disorders. Indeed, in the same year the detection of congenital chloride-losing diarrhea in a patient misdiagnosed as having Bartter syndrome was reported as the first clinical application of

WES [35]. A report on spectacular WES-aided diagnosis and therapy of a case of severe inflammatory bowel disease illustrated the potential of this technology not only for diagnosis but also for treatment [36].

In 2011 Ambry Genetics, as the first CLIA-certified laboratory, offered WES together with medical interpretation for clinical purposes. As of October 2014 the GeneTests web site (www.genetests.org) listed 16 labs offering clinical WES. The turnaround time is in most cases 12–13 weeks, with the shortest being 4–8 weeks, and the price US $4000–5500. Notably, the cost of WES may be only two to four times as high as the cost of some much more selective panels. The majority of labs also offer interpretation of data, including in some cases interpretation of external data. The GeneTests web site (www.genetests.org) lists a single lab (Medical College of Wisconsin) offering WGS declaring a minimal coverage of 10 for over 90% of the genome with a turnaround time of 12–13 weeks. WGS can also be obtained from Illumina and BGI (www.bgiamericas.com).

WES is usually performed using hybridization-based enrichments using kits from Agilent Technologies, Roche NimbleGen, Illumina, or others. Although all the offered kits allow WES, differences exist in their design and performance [37]. In some cases the kits cover only protein-coding exons, whereas other kits include additional loci for noncoding RNAs (including micro RNAs) and 5′ as well 3′ untranslated regions. The resulting differences in target size (e.g., 37 Mb for Nextera Rapid Capture Exome from Illumina vs 96 Mb for SeqCap EZ Exome + UTR from NimbleGen) imply significant, up to threefold, differences in sequencing cost.

WES is particularly attractive when there is a suspicion that the disease is caused by an unknown locus—so far WES has allowed the linkage of more than 150 new genes to human Mendelian disorders [38]. However, in a clinical setting the majority of information supplied by WES remains difficult to interpret as it pertains to loci that have so far not been linked to any human disease. To focus on most clinically relevant loci, the so-called "clinical exome" panels have been developed, such as TruSight One from Illumina, which covers >4800 genes and has a cumulative target size of only ~12 Mb, or SureSelect Inherited Disease from Agilent (>2700 genes, 10.5 Mb).

The use of WGS, which in theory should offer a much more powerful diagnostic tool than WES, is at present limited by poor understanding of the role of variation in the noncoding part of the genome. Intriguingly, a study of 50 patients with severe intellectual disability in whom the causative molecular defect had not been found in extensive former studies, including WES and comparative genomic hybridization concluded that WGS has a high diagnostic yield of 42%, but mainly owing to findings of exonic variants apparently missed by earlier studies [39].

LIMITATIONS OF NGS IN CLINICAL MEDICINE

The major limitations of clinical NGS applications stem from technical shortcomings of current technologies (short read lengths, relatively high error rate, incomplete coverage) and from challenges in data interpretation (see later chapters).

The short reads support identification of SNVs and short deletions/insertions, whereas it is difficult or impossible to detect variants involving longer sequences. In particular, using NGS it is not possible to analyze stretches of short tandem repeats including those causing clinically important diseases such as fragile X, Huntington disease, and others [40]. Despite efforts (see "Analysis of Structural Chromosome Variants by NGS") it remains difficult to detect CNVs such as those causing DiGeorge syndrome (22q11.2 deletion syndrome) or Charcot–Marie–Tooth disease type 1A. Similarly chromosomal translocations and aneuploidy are likely to be missed by NGS analyses unless they are specifically designed for detection of such variants. Short read lengths cause problems in analyzing regions of the genome with segmental duplications containing stretches of sequences with high similarity.

The most commonly used NGS platforms (Illumina, PGM, Proton Ion) have error rates in the range of ~0.1–1%. These error rates translate into thousands of mistakes in extensive analyses such as WES or WGS. The problem is compounded by a generally nonrandom distribution of errors, which makes corrections based on increased coverage difficult [21]. It is noteworthy that a study of WGS in 12 subjects found only 66% concordance in variants detected by Illumina and CG platforms [41].

Relatively high error rates pose unique problems in the detection of somatic variants, for example, in searches of clinically actionable mutations in neoplastic tissue. Owing to tumor heterogeneity and contamination with normal cells, the rate of occurrence of such mutations may well be comparable to the NGS error rate, making their detection difficult/impossible. Ingenious strategies to minimize errors based on combining information from both strands of the DNA molecule have been proposed [42], but they are not routine.

Whereas small (<10 kb) targets can often be enriched (usually by PCR) and sequenced with 100% efficiency, this is not possible with longer targets, whose analysis as a rule has gaps from incomplete coverage. When considering such tests it should be borne in mind that adequate coverage can be generally expected only for 85–95% of the target. Although WGS can improve coverage by avoiding the shortcomings of enrichment techniques, it is still not perfect—in a 2014 study of 56 genes with a high clinical importance according to the American College of Medical Genetics and Genomics, a median of up to 17% were not appropriately covered by WGS [41].

Owing to the above-mentioned problems, positive results of NGS tests, in particular WES and WGS, should be confirmed with an independent technique (e.g., Sanger sequencing), whereas in the case of negative findings it should always be remembered that they may be a false negative. Importantly, even WES or WGS, despite their power, *cannot be* used to exclude a possibility of a genetic defect in a patient [41,43].

CONCLUSION

The development of NGS has revolutionized genetic research as well as the practice of clinical genetics. As will become apparent from the following chapters, NGS testing is heavily used in virtually all branches of medical science. An integral part of NGS implementation is related to the development of appropriate data analysis tools. Actually, even at present the challenges presented by analysis and interpretation of NGS results are bigger than those associated with sequencing itself. Thus, no discussion of NGS is complete without mentioning bioinformatics issues (see the chapters Basic Bio-informatic Analysis of NGS Data and Analysis of Structural Chromosome Variants by NGS). An important aspect of NGS is linked to ethical problems (see Ethical Issues), in particular the problem of incidental findings, that is, finding variants that, with a high probability, indicate the presence or high risk of a disease that was not searched for. The strategy of dealing with incidental findings is the most hotly debated issue that has emerged from NGS. Another aspect of NGS, relevant in particular to its clinical applications, is the necessity for development of efficient and balanced reimbursement strategies (see Organizational and Financial Challenges). Finally, it should be emphasized that the process of NGS development is by no means over. There is a constant progress in increasing the throughput of existing platforms; in parallel, as discussed in Future Directions, there is an ongoing search for completely new technical strategies, such as nanopore sequencing and novel massively parallel bioinformatical solutions.

REFERENCES

[1] Sanger F, Coulson AR. A rapid method for determining sequences in DNA by primed synthesis with DNA polymerase. J Mol Biol May 25, 1975;94(3):441–8.

[2] Tewhey R, Warner JB, Nakano M, Libby B, Medkova M, David PH, et al. Microdroplet-based PCR enrichment for large-scale targeted sequencing. Nat Biotechnol November 2009;27(11):1025–U94.

[3] Adey A, Morrison H, Asan, Xun X, Kitzman J, Turner E, et al. Rapid, low-input, low-bias construction of shotgun fragment libraries by high-density in vitro transposition. Genome Biol 2010;11(12):R119.

[4] Ross MG, Russ C, Costello M, Hollinger A, Lennon NJ, Hegarty R, et al. Characterizing and measuring bias in sequence data. Genome Biol 2013;14(5).

[5] Lander ES, Linton LM, Birren B, Nusbaum C, Zody MC, Baldwin J, et al. Initial sequencing and analysis of the human genome. Nature February 15, 2001;409(6822):860–921.

[6] Venter JC, Adams MD, Myers EW, Li PW, Mural RJ, Sutton GG, et al. The sequence of the human genome. Science February 16, 2001;291(5507):1304–51.

[7] Genome Reference Consortium. http://www.ncbi.nlm.nih.gov/projects/genome/assembly/grc/human/; 2014.

[8] Peters BA, Kermani BG, Sparks AB, Alferov O, Hong P, Alexeev A, et al. Accurate whole-genome sequencing and haplotyping from 10 to 20 human cells. Nature July 12, 2012;487(7406):190–5.

[9] Korbel JO, Urban AE, Affourtit JP, Godwin B, Grubert F, Simons JF, et al. Paired-end mapping reveals extensive structural variation in the human genome. Science October 19, 2007;318(5849):420–6.

[10] Tewhey R, Bansal V, Torkamani A, Topol EJ, Schork NJ. The importance of phase information for human genomics. Nat Rev Genet March 2011;12(3):215–23.

[11] Lebl M, Buermann D, Reed MT, Heiner DL, Triener A. Flow cells and manifolds having an electroosmotic pump. 08.02.12 [Google Patents].

[12] Rothberg JM, Hinz W, Rearick TM, Schultz J, Mileski W, Davey M, et al. An integrated semiconductor device enabling non-optical genome sequencing. Nature July 21, 2011;475(7356):348–52.

[13] Dressman D, Yan H, Traverso G, Kinzler KW, Vogelstein B. Transforming single DNA molecules into fluorescent magnetic particles for detection and enumeration of genetic variations. Proc Natl Acad Sci USA July 22, 2003;100(15):8817–22.

[14] Shendure J, Porreca GJ, Reppas NB, Lin XX, McCutcheon JP, Rosenbaum AM, et al. Accurate multiplex polony sequencing of an evolved bacterial genome. Science September 9, 2005;309(5741):1728–32.

[15] Ma ZC, Lee RW, Li B, Kenney P, Wang YF, Erikson J, et al. Isothermal amplification method for next-generation sequencing. Proc Natl Acad Sci USA August 27, 2013;110(35):14320–3.

[16] Voelkerding KV, Dames SA, Durtschi JD. Next-generation sequencing: from basic research to diagnostics. Clin Chem April 2009;55(4):641–58.

[17] Margulies M, Egholm M, Altman WE, Attiya S, Bader JS, Bemben LA, et al. Genome sequencing in micro-fabricated high-density picolitre reactors. Nature September 15, 2005;437(7057):376–80.

[18] Drmanac R, Sparks AB, Callow MJ, Halpern AL, Burns NL, Kermani BG, et al. Human genome sequencing using unchained base reads on self-assembling DNA nanoarrays. Science January 1, 2010;327(5961):78–81.

[19] Reid C, Complete Genomics Inc. Future Oncol 2011;7(2):219–21.

[20] Eid J, Fehr A, Gray J, Luong K, Lyle J, Otto G, et al. Real-time DNA sequencing from single polymerase molecules. Science January 2, 2009;323(5910):133–8.

[21] Mardis ER. Next-generation sequencing platforms. Annu Rev Anal Chem 2013;6(6):287–303.

[22] Levene MJ, Korlach J, Turner SW, Foquet M, Craighead HG, Webb WW. Zero-mode waveguides for single-molecule analysis at high concentrations. Science January 31, 2003;299(5607):682–6.

[23] Carneiro MO, Russ C, Ross MG, Gabriel SB, Nusbaum C, DePristo MA. Pacific biosciences sequencing technology for genotyping and variation discovery in human data. BMC Genomics August 5, 2012;13.

[24] Flusberg BA, Webster DR, Lee JH, Travers KJ, Olivares EC, Clark TA, et al. Direct detection of DNA methylation during single-molecule, real-time sequencing. Nat Methods June 2010;7(6):461–U72.

[25] Braslavsky I, Hebert B, Kartalov E, Quake SR. Sequence information can be obtained from single DNA molecules. Proc Natl Acad Sci USA April 1, 2003;100(7):3960–4.

[26] Ozsolak F. Third-generation sequencing techniques and applications to drug discovery. Expert Opin Drug Discov March 2012;7(3):231–43.

[27] Bowers J, Mitchell J, Beer E, Buzby PR, Causey M, Efcavitch JW, et al. Virtual terminator nucleotides for next-generation DNA sequencing. Nat Methods August 2009;6(8):593–5.

[28] Pushkarev D, Neff NF, Quake SR. Single-molecule sequencing of an individual human genome. Nat Biotechnol September 2009;27(9):847–50.

[29] Ozsolak F, Platt AR, Jones DR, Reifenberger JG, Sass LE, McInerney P, et al. Direct RNA sequencing. Nature October 8, 2009;461(7265):814–8.

[30] Forrest ARR, Kawaji H, Rehli M, Baillie JK, de Hoon MJL, Haberle V, et al. A promoter-level mammalian expression atlas. Nature March 27, 2014;507(7493):462–70.

[31] Andersson R, Gebhard C, Miguel-Escalada I, Hoof I, Bornholdt J, Boyd M, et al. An atlas of active enhancers across human cell types and tissues. Nature March 27, 2014;507(7493):455–61.

[32] Mertes F, ElSharawy A, Sauer S, van Helvoort JM, van der Zaag PJ, Franke A, et al. Targeted enrichment of genomic DNA regions for next-generation sequencing. Briefings Funct Genomics November 1, 2011;10(6):374–86.

[33] Altmuller J, Budde BS, Nurnberg P. Enrichment of target sequences for next-generation sequencing applications in research and diagnostics. Biol Chem February 2014;395(2):231–7.

[34] Ng SB, Turner EH, Robertson PD, Flygare SD, Bigham AW, Lee C, et al. Targeted capture and massively parallel sequencing of 12 human exomes. Nature September 10, 2009;461(7261):272–U153.

[35] Choi M, Scholl UI, Ji WZ, Liu TW, Tikhonova IR, Zumbo P, et al. Genetic diagnosis by whole exome capture and massively parallel DNA sequencing. Proc Natl Acad Sci USA November 10, 2009;106(45):19096–101.

[36] Worthey EA, Mayer AN, Syverson GD, Helbling D, Bonacci BB, Decker B, et al. Making a definitive diagnosis: successful clinical application of whole exome sequencing in a child with intractable inflammatory bowel disease. Genet Med March 2011;13(3):255–62.

[37] Chilamakuri CSR, Lorenz S, Madoui MA, Vodak D, Sun JC, Hovig E, et al. Performance comparison of four exome capture systems for deep sequencing. BMC Genomics June 9, 2014;15.

[38] Rabbani B, Tekin M, Mahdieh N. The promise of whole-exome sequencing in medical genetics. J Hum Genet January 2014;59(1):5–15.

[39] Gilissen C, Hehir-Kwa JY, Thung DT, van de Vorst M, van Bon BWM, Willemsen MH, et al. Genome sequencing identifies major causes of severe intellectual disability. Nature July 17, 2014;511(7509):344–7.

[40] Almeida B, Fernandes S, Abreu IA, Macedo-Ribeiro S. Trinucleotide repeats: a structural perspective. Front Neurol 2013;4.

[41] Feero WG. Clinical application of whole-genome sequencing proceed with care. JAMA March 12, 2014;311(10):1017–9.

[42] Kirsch S, Klein CA. Sequence error storms and the landscape of mutations in cancer. Proc Natl Acad Sci USA September 4, 2012;109(36):14289–90.

[43] Biesecker LG, Green RC. Diagnostic clinical genome and exome sequencing. N Engl J Med June 18, 2014;370(25):2418–25.

BASIC BIOINFORMATIC ANALYSES OF NGS DATA

2

Piotr Stawinski[1], Ravi Sachidanandam[2], Izabela Chojnicka[3], Rafał Płoski[4]

[1]*Department of Immunology, Medical University of Warsaw, Warsaw, Poland;* [2]*Department of Oncological Sciences, Icahn School of Medicine at Mount Sinai, NY, USA;* [3]*Faculty of Psychology, University of Warsaw, Warsaw, Poland;* [4]*Department of Medical Genetics, Centre of Biostructure, Medical University of Warsaw, Warsaw, Poland*

CHAPTER OUTLINE

Since 2005, the technological progress in medical genetics, particularly next generation sequencing (NGS) technology, has revolutionized the use of genomics in clinical applications. NGS enables rapid (in a few days), inexpensive (a few thousand dollars) analyses of the whole genome. NGS has allowed the use of a variety of molecular biological techniques in diagnostics and patient care, such as i) whole-genome sequencing (WGS), which involves sequencing the whole genome to study mutations and rearrangements; ii) mRNA-seq to study changes in expression profiles; iii) small RNA-seq to study the role of microRNAs and other small noncoding RNAs; iv) methyl-seq to study DNA modifications such as methylation; v) chromatin immunoprecipitation (ChIP) to study chromatin modifications such as histone marks and to map protein–DNA interactions; vi) targeted sequencing to study select regions of the genome (mitochondria, cancer panels, etc.); and vii) noncoding transcript profiling. These techniques can be used to identify differences in disease states and identify biomarkers, as well as to help identify therapeutic targets and help clinicians decide on the course of treatment [1–3].

NGS experiments require the cooperation of experts from various fields including molecular biology, clinical work, technology, instrumentation, and bioinformatics. After collecting genetic material and clinical information about the subject, preparing that material, and performing consecutive steps of the NGS experiment, the NGS sequencer produces large volumes of data that are impossible to interpret without bioinformatic analysis.

The main goal of bioinformatic analyses is to identify differences in the disease state compared to the normal. In the cases of genetic disorders, the aim is to identify the functional differences between the reference and the subject genomes. Of the various differences that show up in any differential analyses, the aim is the identification of specific variants that either partially or fully explain the observed clinical phenotype. Such analyses are often not definitive, providing only a statistical association, owing to the influence of variety of factors on disease etiology.

Current NGS technologies are limited to sequencing small fragments of DNA (up to several hundred bases) [4]. To get around this limitation, the input material (DNA or RNA) is fragmented and then processed for sequencing. This results in several millions of fragments that then need to be reassembled to ascertain the source. This is a difficult problem to solve ab initio, but the existence of the reference genome helps with this process immensely. But this also means that large-scale structural changes, such as rearrangements of large sections of the genome, are difficult to detect using NGS (other complementary techniques such as fluorescence in situ hybridization, involving visualizing hybridization of tagged probes, are better suited to this).

Most technologies of sequencing work by synthesis of the second strand using a single strand of a DNA fragment as a template [5]. This process often introduces errors, which are overcome in NGS through the generation of a large number of reads (>20) covering each position. Reliable calls of variants, remains an open problem, requiring understanding of the nature of the errors introduced by each instrument and the sample preparation techniques.

In this chapter we will focus on the detection of single nucleotide variants (SNVs) and deletions/insertions (called indels) that are significantly smaller than the read size. There are other types of variants, including structural variants, such as inversions, tandem duplications, long indels, and translocations. Most of the techniques and algorithms presented in this chapter are focused on targeted sequencing experiments but may be applied to the whole genome as well.

We will describe in detail various components of bioinformatic analysis, beginning with the data input, reads mapping, and processing of the mapping products, up to variant calling, with special attention paid to issues with annotations. We will close with a discussion of hardware-related matters, including requirements for data analysis and secure storage, with a description of various architectures of computing systems for NGS data processing.

SOFTWARE TOOLS

There are several software tools available to call SNVs and small indels from deep-sequencing data. Most tools have several analysis steps in common and create outputs in standard formats [6]. The common data flow and standard procedures are presented in Figure 1.

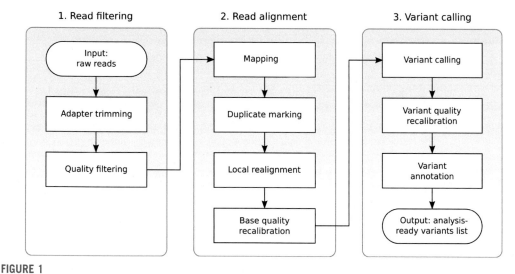

FIGURE 1

General NGS data processing pipeline [7].

INPUT SEQUENCE PREPROCESSING

Sequencing instruments from Illumina generate data files using the FASTQ format, which uses four lines per read, with names, sequences, and quality scores (Figure 2). This has become the de facto standard for NGS data, owing to the ubiquity of Illumina instruments. The read quality at each base is a phred score that can range from 0 to 60 on a logarithmic scale, for example, a score of 30 represents a 1 in 1000 chance of error, while a score of 10 represents a 1 in 10 chance of error. To compress the data, the read quality per position is encoded using ASCII characters (so it uses 1 byte or 8 bits per position); a common encoding format is the Phred+33 quality score, which is presented in Figure 3 [8].

Several NGS sequencers can read sequences from both ends of a single DNA, creating "paired-end reads" presented in Figure 4. In such cases, the sequences are placed in paired FASTQ files; each row holds one end of the molecule with the corresponding paired end on the same row in the paired file. The FASTQ files are compressed using gzip and most programs will accept these compressed files, usually with a fastq.gz or fq.gz extension, as input.

FASTQ files are always preprocessed, to apply various quality controls and remove any adapters that remain from the sample preparation process (although with long inserts, adapter trimming is less of an issue). Preprocessing is critical to remove potential sources of error that can propagate through to variant calling [9]. The quality of bases is not evenly distributed: it decreases with increasing position in the read (Figure 5(a)). In addition, the first few bases can be of lower quality owing to the method by which the clusters are recognized in several sequencing platforms. The reads are usually trimmed, to remove low-quality reads, either using a fixed trimming from both ends or using a phred score-based trimming, using a score of 20 as the cutoff. Reads can also contain contamination, such as adapter sequences, when the inserts are smaller than the read length (Figure 4(c)). In such cases, adapter trimming, using alignment to the adapter, is used.

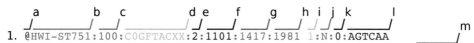

1. `@HWI-ST751:100:C0GFTACXX:2:1101:1417:1981 1:N:0:AGTCAA`
2. `CNCTGGACTTAGGTTTAGTGAGGGCTGACTCTGAAGCTGAGGGTCTGCAGAAGGGAGTNNNNNNN`
3. `+`
4. `D#4ADDDEHHHHHFHGGHJJJJJJJIJJJJJJJIIIIIIJJGHIIIIJJJGIHHIIJHGC######`

FIGURE 2

A read from a FASTQ file produced by Casava 1.8 software. In the FASTQ file each read sequence is described using four lines. 1. The sequence identifier and optional description: a) required @ symbol; b) the instrument name; c) run ID; d) flow cell ID; e) flow cell lane; f) tile number within the flow cell lane; g); h) the coordinate of the cluster within the tile; i) member of a pair; j) Y for reads filtered out, N for reads passing filter; k) 0 if the read is not identified as a control; l) index sequence. 2. The sequenced read: m) N denotes unidentified bases. 3. Line begins with +, optionally followed by a sequence identifier. 4. Encoded quality values.

$$Q = -10 \log_{10} p$$

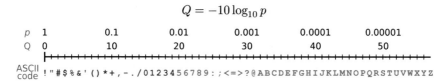

p	1	0.1	0.01	0.001	0.0001	0.00001
Q	0	10	20	30	40	50

ASCII code `!"#$%&'()*+,-./0123456789:;<=>?@ABCDEFGHIJKLMNOPQRSTUVWXYZ`

FIGURE 3

Phred quality score and Phred+33 encoding. p—probability of an error, Q—phred quality score, ASCII code—method of encoding the Q value in a FASTQ file, e.g., ASCII code I corresponds to Q score 40, which translates to a probability of 1×10^{-4} for the occurrence of an error.

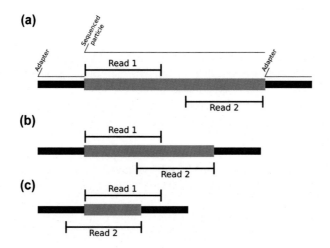

FIGURE 4

Paired-end reads and possible sequencing geometries; S—sequenced fragment length, R—read length. a) When S>2R, reads do not overlap and each base is read once. b) When 2R>S>R reads overlap, some bases are read twice. c) When R>S part of the read contains an adapter sequence, read ends contain bases not present in a sequenced fragment.

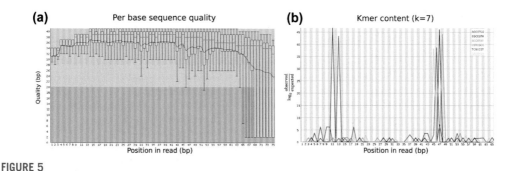

FIGURE 5

Examples of charts generated by FastQC for the FASTQ quality control. a) Quality chart; b) *k*-mer content chart.

Contamination can be detected by studying the distribution of *k*-mers, which are short sequences of length "*k*." The *k*-mer distribution charts are a great help for detecting contamination. When the contaminants share a common sequence pattern (Figure 5(b)), peaks indicate the positions where the observed level of short DNA sequence exceeds the number expected for random base distribution [10]. This kind of read contamination has a significant effect in the subsequent analysis and should be removed. Several applications exist allowing unsupervised quality and adapter trimming, such as the Trimmomatic [11] or Cutadapt programs [12]. It is also usually advisable to check the overall FASTQ quality statistics manually, for example, using the FastQC application (http://www.bioinformatics.bbsrc.ac.uk/projects/fastqc/), which may help in detecting potential problems in the early analysis step.

MAPPING

Mapping is the assignment of genomic location to each read. Correct mapping of the reads to the reference genome is one of the main challenges in the analysis of the NGS data. Mappers are used to rebuild the subject's sequence using short NGS reads from the experiment. While de novo assembly of the reads is possible and would allow re-creating the subject's sequence without the bias, it is time and resource consuming and will not be easy for regions with repetitive content. To simplify the task, most of the mapping tools use the reference sequence as a template. Mappers align reads to the reference and map each read to the most probable position. Such approach works well for finding small indels and SNVs, but may lead to omitting structural variations present and visible in the reads [13]. Examples of reads and their mapping positions to the reference sequence are shown in Figure 6.

It is not always possible to unambiguously map a read. Reads from duplicated or repetitive regions (such as ALU elements) pose problems (Figure 6(g) and (h)). Using paired-end reads and increasing the insert size as well as read lengths can reduce the ambiguity. However, in targeted sequencing experiments longer inserts are not favored as they increase the fraction of bases that are outside the region of interest. Most of the mappers also include information on the probability of the read being misplaced, usually reported on a phred scale [14]. This reports the similarity of the read to the reference sequence and also the uniqueness of the best mapping position.

There are two common algorithmic approaches to the problem of mapping [15]. In the first step both approaches try to find the best match in the genome for the each read using either:

1. A data structure implementing a substring index that enables finding a given sequence in the reference sequence in the shortest possible time (e.g., a suffix tree, a suffix array, or an FM index, a compressed data structure based on the Burrows–Wheeler transform) or
2. A hash-based algorithm, which allows fast finding of a given k-mer in the reference sequence, where k is set using a precomputed hash table.

Occasionally, to improve mapping of reads containing indels and SNVs, smaller parts of each read are mapped independently. After finding the potential positions of the parts, the algorithm performs a careful alignment of the whole read to the selected reference location along with a calculation of the mapping quality. Mapping algorithms can ascertain the best position of the read, where only a part of the read maps well while read ends do not match the reference sequence. Such read ends are usually marked as "soft clipped" by the mapper and are included in the result. Later they are usually omitted by the variant callers and not displayed by alignment browsers. Nevertheless, they may contain valuable information about the structural variations present in the sequence [13].

Whereas most mappers follow the general approaches described above, they differ substantially in the details. At the time of writing this chapter, there are more than 66 published mappers, each of them implementing a different mapping algorithm [15]. Some of the mappers specialize in aligning short paired-end reads of high quality, while others work best with long reads but handle reads of lower quality correctly. It is essential to choose a mapper suitable for the data that need to be analyzed.

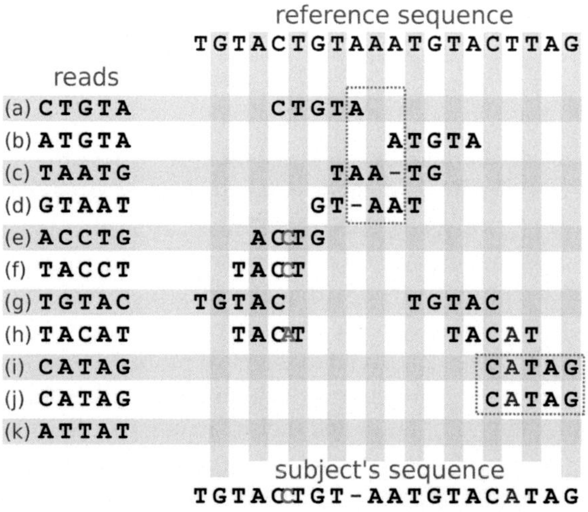

FIGURE 6

Example of reads (a–k) mapping to the reference sequence. The patient's sequence is shown at the bottom: (a, b) are reads that match exactly; (c, d) are reads with a small deletion; (e, f) are reads with an insertion; (g) is a read that maps to exactly two locations on the reference sequence; (h) is a read that has two near-matching positions in the reference, one with a small insert, one with an SNV; (i, j) are duplicated reads; (k) is an artificial read without a matching position.

The most popular way of representing alignment data is the BAM format, which is a binary version of the human-readable SAM—sequence alignment/map format [14]. An index file, with the extension .bai, that allows rapid selection of reads by position usually accompanies BAM files. There are also new formats emerging designed for specific purposes: a) the CRAM format, which offers a better compression ratio than a standard BAM file [16] and b) the ADAM file format, designed for storing alignment data in a distributed computing environment (http://www.eecs.berkeley.edu/Pubs/TechRpts/2013/EECS-2013-207.html).

A popular mapper using the first of the approaches described above is BWA [12], while an example of the second approach is the Mosaik mapper [17]. The results can differ significantly even when the underlying algorithms are similar owing to different implementations of heuristic algorithms. Most of the mappers align reads to the whole reference genome disregarding the possible use of targeted enrichment of the sample. However, there are also alternative approaches, such as MiST, that use the information about the target region to restrict the mapping space, increasing both sensitivity and specificity while improving the speed [18]. Although using information about the target region may result in better sensitivity, it should be used carefully. The procedure of target enrichment does not necessarily enrich only the region of interest but some of the reads may in fact be coming from regions sharing similar sequence. MiST does take this into account, but this might not always be computationally corrected.

Manual examination of the specific alignment regions is sometimes necessary to investigate reads in the regions of interest. There are several BAM file browsers that display the reads aligned to the reference sequence in a convenient way, such as the Integrative Genomics Viewer [19]. A window of the Integrative Genomics Viewer browser is shown as an example in Figure 7.

Another issue concerning mapping is that not every read maps to the reference sequence. In Figure 6 read (k) was an example of such a read. In the case of paired-end sequencing sometimes only one read from a pair maps, or reads from a single pair map to distinct loci or even on different

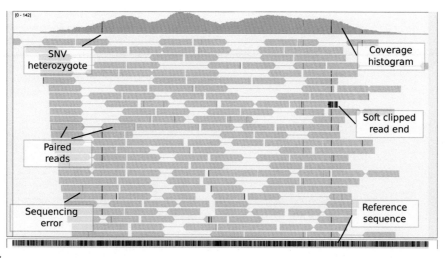

FIGURE 7

Integrative Genomics Viewer window displaying a selected region of a BAM file. Gray horizontal rectangles represent reads matching the reference sequence, colored dashes represent SNVs.

chromosomes; sometimes the direction of the read pair is inconsistent. This kind of mapping can be evidence of structural variants present in the sample. When an SNV or a small indel is reported in the region where incoherent mapping occurred, it has to be carefully investigated, as there is a great probability of getting false positive variants, which can arise from chimeric reads that are occasionally generated during sample preparation.

PROCESSING AND INTERPRETING MAPPING

Owing to the simplicity of the mapping algorithms, mapping results should be processed further to improve subsequent variant calling. This step usually includes [7]:

1. insertions and deletions realignment,
2. duplicates marking, and
3. base quality recalibration.

INSERTIONS AND DELETIONS REALIGNMENT

Mappers usually map reads independently, trying to minimize the number of mismatches with the reference sequence. This may result in incorrect mapping of reads around an indel in the genome or in highly duplicated regions. Local realignment is a step in which reads present in the region of an indel are realigned to minimize the number of variants explaining the observed reads. Ordinarily, it is accomplished by moving indels to the most-left-possible position. The local indel realignment procedure is illustrated in Figure 8. While the most advanced variant calling algorithms should be able to work properly regardless of indel realignment step, it is still advisable to do a variant realignment as it eases manual investigation of the mapping. Local indel realignment can be performed using the IndelRealignment algorithm from the Genome Analysis Toolkit framework (GATK) [20].

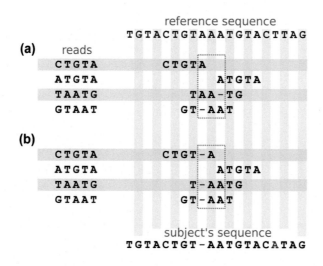

FIGURE 8

Read base mappings a) before and b) after local indel realignment procedure.

Commonly used library preparation kits include a step of cleaving genomic DNA at random positions. The probability that two resulting DNA particles would have exactly the same start and end is low—usually much lower than the probability of inducing a sequencing error in the particle. Therefore, any variant present in duplicated reads, but not present in other reads, is probably a false positive variant. Duplicated reads marking includes information about the uniqueness of the read in the BAM file, making this information available to the variant calling algorithms. As a result, the variant caller handles duplicated reads correctly. An alternative approach would be to remove the duplicated reads.

Finally, it should be noted that there are several factors that can increase the probability of duplications: a) nonrandom cleavage of DNA by target enrichment kits, b) short insert sizes, c) small variance in insert size distribution, d) high coverage, and e) PCR amplification during sample preparation. In such cases, duplicates have to be handled carefully, potentially on a case-by-case basis. Duplications marking and removal can be done using the MarkDuplicates tool from the Picard framework (http://picard.sourceforge.net) or the rmdup command from Samtools [14].

BASE QUALITY RECALIBRATION

The sequencer reports the error probability for each base read. However, these qualities are only an estimation of the probability, not necessarily an illustration of the real probability of the error in the sequenced data. Base quality score recalibration (BQSR) is a procedure for correcting the reported probabilities based on the sequencing results. Error probability is counted by analyzing the covariation among several features of a base, including base position in read, reported quality, and preceding nucleotide. Correct error probabilities support variant calling algorithms to improve both the sensitivity and the specificity. The application useful for proceeding with BQSR is the GATK BaseRecalibrator tool [20]. In general, adequate coverage at the position of interest obviates the need for such corrections, as these can create an additional computational load.

VARIANT CALLING

Variant calling is the key step in the bioinformatic analysis of the NGS data. It aims to detect variants in the aligned reads absent from the reference sequence. Two main factors cause variant calling to be difficult. First, each base is read with some probability of an error. Considering the amount of bases, the number of errors created by the sequencers is relatively high: if all bases were read with the quality of 30 phred, then the estimated frequency of an invalid base would equal 1/1000. This problem is usually controlled by applying a deep sequencing to obtain a coverage of at least 20 reads at each base in the region of interest. If a sequencing error is randomly distributed, a variant caller can recognize and remove invalid bases. Base quality recalibration and duplicates marking steps mentioned before are particularly helpful for limiting this type of error. Incorrect mapping of the reads induces the second type of error, for example, read from paralogous regions of the genome. In such cases, the variant caller would call a variant that is not really present in the sample. The local realignment step addresses this issue to some extent, where adjustment of the mapping in the regions of low complexity is possible. Furthermore, some variant callers ignore reads having low mapping quality, which unfortunately makes some regions of the genome unavailable for the variant calling [21].

MAJOR APPROACHES TO THE VARIANT CALLING

There are two major approaches to variant calling [22]:

1. single-nucleotide-based variant calling, in which a variant caller tries to determine the most probable nucleotide in the sample for each nucleotide in the reference sequence, and
2. haplotype-based, in which a variant caller tries to identify the most probable haplotype for a dynamically defined region of the reference sequence.

While the first solution usually works well for SNVs and very short indels and has very fast implementations, the second can recognize much more complicated variants, such as multiple-nucleotide variants or longer indels, but at a price of much higher computational needs. Haplotype-based algorithms can also recognize and report variant phasing: they can detect variants that are placed on the same chromosome and inherited together. Algorithms for both approaches usually consist of three consecutive steps: (a) detection of possible variants that are present in reads; (b) measure of the probability of getting the observed reads, given different sets of variants detected in the previous step; and (c) selection of the most probable variants set. The only difference between the two approaches is in the method of performing step (a). In the case of single-nucleotide-based approaches, the algorithm traverses the reference sequence and for each reference position it analyzes bases present in reads. Any differences from the reference are passed to step (b) as possible variants. The haplotype-based approach is a generalization of the single-nucleotide approach: instead of focusing on a single base, in this approach a dynamically defined region of the genome is analyzed. First, analysis of mapping positions of read pairs is used for detecting a region containing a potential haplotype. Second, reads mapped to this region are aligned de novo, without a reference sequence, usually using de Bruijn graphs. This reference-free de novo alignment allows discovery of haplotypes unbiased by the content of the reference sequence. Detected haplotypes are then passed to step (b). The probability of the nucleotides/haplotypes given the observed reads is determined using Bayes's rule.

The ploidy of the sequenced organism is important information for the variant caller. Most of the genome in a healthy subject is diploid, with the exception of the mitochondrial genome or sex chromosomes in males. In searches for variants in these regions, the ploidy issue should be carefully considered, for example, by separated variant calling in these regions with proper variant caller settings.

Many of the variant callers (for instance, GATK [23], Freebayes [22]) allow the user to call variants in multiple samples at once and may include prior knowledge about variants present in the population analyzed. It is highly recommended to use these options, as they significantly improve the ability to detect correct variants in low coverage areas. However, because of the high computational requirements of the haplotype-based approach, multisample calling may not be possible for a big population of samples, whereas single-nucleotide-based approaches usually have much higher limits.

VARIANT CALLING FORMAT

A standard format for storing a variant caller result is a Variant Call Format (VCF), illustrated in Figure 9. The VCF file is a human-readable text file that can contain information about variants in multiple samples as well as user-defined annotations [24]. The header line of the file starts with the # sign and contains the definition of the annotations used in the body, but can also contain custom information such as the version of the reference sequence or name of the program used to create the file. The last line of the header

```
##fileformat=VCFv4.1
##FILTER=<ID=LowCoverage,Description="DP < 5">
##FORMAT=<ID=DP,Number=1,Type=Integer,Description="Approximate read depth">
##FORMAT=<ID=GT,Number=1,Type=String,Description="Genotype">
##INFO=<ID=Freq,Number=1,Type=Float,Description="Frequency in population">
   #CHROM  POS   ID     REF    ALT    QUAL     FILTER        INFO        FORMAT   SampleName
a) chr1    22    rs200  A      C      342.24   PASS          Freq=0.06   GT:DP    0/1:32
b) chr1    28    .      A      ATA    247.83   PASS          .           GT:DP    1/1:15
c) chr3    545   .      CCG    C,CCT  102.33   PASS          .           GT:DP    1/2:45
d) chr3    547   .      G      T      33.23    LowCoverage   .           GT:DP    0/1:4
```

Header

Body

FIGURE 9

A Variant Call Format file consists of two parts: a header containing optional annotations and sample name and a body containing a list and description of the detected variants. Variant examples: a) heterozygote SNV A > C; b) homozygote insertion of T; c) two variants, deletion of CG and SNV G > T; d) heterozygote SNV G > T.

contains names of the columns of the body part with identifiers of the samples. The body part information is formatted in tab-separated columns that encompass information about:

- CHROM and POS: chromosome position of the variant.
- ID: variant identification—usually an dbSNP ID.
- REF and ALT: the reference sequence and observed sequence, respectively, at the point of variant; the ALT column may contain multiple comma-separated alleles.
- QUAL: the variant quality measure—the phred-scaled probability of ALT alleles being called wrong.
- FILTER: PASS if the site has passed quality filters, semicolon list of filters that failed; filters are defined in the header and may inform about, for example, observed strand bias or low coverage.
- INFO: additional information about the variant; annotations are defined in the header.
- FORMAT: describes the format of genotype description used in the following columns; the fields of the FORMAT column are described in the header.
- The subsequent columns contain information about the genotypes in samples, according to the format described in the FORMAT column.

As the single variant can be identified in a VCF file in many ways, the automatic analysis of VCF files is difficult. This is illustrated in Figure 9 in lines (c) and (d): both of them describe the same variant, an SNV on chromosome 3 at position 547 that changes G to T. However, this variant can be described as 547G > T as well as 545CCG > CCT. It is very important to remember this feature of VCF files in the case of automatic processing of VCF files: even such basic tasks as checking the existence of a particular variant in a sample may be problematic.

Although VCF is a human-readable text file that can be opened in any text editor, the manual inspection of it and search for a variant responsible for an observed phenotype are certainly not convenient. A better solution is to export data from a VCF file to a spreadsheet or a relational database that supports a tabular view and provides basic utilities such as filtering and sorting.

Several tools to perform variant calling exist—the most commonly used single-nucleotide-based implementations are those in the Samtools and the GenotypeCaller from the GATK framework. Examples of widely used haplotype-based callers are the GATK HaplotypeCaller or the Freebayes. As variant calling is based on heuristics, it should not be surprising that variant calling algorithms, even using the same approach, may give results that differ a lot in the case of variants with low coverage or those that exhibit complex manifestations [25].

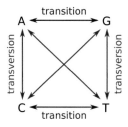

FIGURE 10

Transversions (Tv) and transitions (Ti). Note that there are twice as many transversions as transitions. Empirically Ti/Tv is ~2.0.

Information about the quality of the variant as well as the content of the FILTER column can be used for further filtering of low-quality variants. There are several ways to estimate a number of false positive calls in the VCF file. For instance, measuring the transition/transversion ratio (Ti/Tv ratio) is an effective method. The idea behind this measure is the observation that the frequencies of different kinds of base substitutions in the DNA are not equal: the probability of transition from purine to purine or pyrimidine to pyrimidine is higher than that of transversion from purine to pyrimidine or pyrimidine to purine. During variant calling, the probability of making a random mistake in variant calling is equal for any kind of mistake. Usually, the observed Ti/Tv ratio for the human exome is 2.8, while the Ti/Tv ratio for the whole genome is around 2.0–2.1. In the case of a uniform sequencing error, the Ti/Tv ratio is equal to 0.5 (Figure 10). Bias observed in the Ti/Tv ratio may indicate the existence of a randomly distributed variant calling error and can be used for controlling the cutoff point for false positive variants [23]. Such filtration is highly recommended if variants detected are later used for determining the frequency of variants in a population or other large-scale analysis. However, in the case of clinical applications—finding a variant responsible for an individual phenotype—it is usually recommended to keep more false positive variants that can be confirmed later than risk hiding a true positive variant from further analysis steps.

VARIANT ANNOTATION

Each genome carries thousands of differences from the reference sequence, most of them irrelevant to the phenotype. Manual identification of a couple of variants among thousands of potential candidates would be impossible without further annotation of the data, which helps to narrow the search area and filter out less probable variants. The correct annotation of variants remains quite a challenge. Even assigning basic information, including the exact location of a variant with respect to features such as exon, intron, untranslated region, splice site, promoter, splicing enhancer, or suppressor is not clear-cut owing to alternative splicing. Determining the effect of the mutation as deleterious, neutral, or beneficial also presents difficulties. Ultimately, the aim of the analysis is to quantify the functional/biological effects of the mutations. These steps are not straightforward and are commonly addressed using software such as Annovar [26] or snpEff [27] as well as public or commercial databases and services.

Table 1 lists standard annotations that are used for annotating variants.

It should be noted that different annotation tools give different results, owing to differences in algorithms as well as the databases. For example, Annovar software annotates single variants with a unique function, despite the possible different roles due to alternative splicing. In the case of multiple roles,

Table 1 Standard Variant Annotations

Gene name	The gene located in the place of the variant or closest to the variant
Is coding	Whether the variant is placed in a coding region
Amino acid change	Whether the variant changes a sequence of amino acids
Variant frequency	It is very beneficial to know the frequency of the variant in the analyzed population. There are publicly available databases containing information about variant frequencies, in some selected populations, for instance, 1000 Genomes Project [28], ESP Project [29] (Exome Variant Server URL: http://evs.gs.washington.edu/EVS/). In the case of analyzing multiple samples it is advisable to build one's own database of variant frequencies and use this frequency as another annotation. It may help in finding artifacts introduced by the sequencer and chosen bioinformatic work flow.
SNP ID	The dbSNP database [30] contains information about many SNVs that were observed already. SNP ID may be helpful for finding additional information about the variant on the Internet or in publications. However, it should be emphasized that having an ID assigned does not imply that the variant is not rare or is not pathogenic.
Linkage to known phenotype	Some of the variants are already linked to known phenotypes. There are databases that store such information, such as ClinVar [31], Human Gene Mutation Database [32], or Catalog of Published Genome-Wide Association Studies [33]. Although data in these databases are very valuable, they should be treated with caution as some of the linkage reports may not be confirmed in future studies.
Conservation score	Regions or even single nucleotides playing an important biological function are usually highly conserved. Therefore it is advisable to annotate the variants with a conservation score.
Severity of amino acid change	Many tools score the variant by algorithmically guessing whether the observed amino acid change is benign or deleterious, for instance, the Grantham score, PolyPhen [34], or Sift [35]. Unfortunately, the information gain by using algorithmic measures of amino acid influence is very low: some phenotype-causing variants may be annotated as benign, as well as variants irrelevant to the observed phenotype being annotated as damaging.

Annovar always chooses the most severe one [26]. The snpEff tool represents an alternative approach, in which a single variant is annotated with the information about each transcript it is linked to, which results in a high amount of annotations for each variant [27].

Using annotations is crucial for finding the variants linked with the sample phenotype. Figure 11 presents an example of the number of variants obtained from whole-exome sequencing after the application of standard annotation-based filters.

To prioritize variants an expert who analyzes the results may use the a priori knowledge about the genes that are linked to the observed phenotype as well as the inheritance model of the phenotype. Of great utility will be tools that assign scores for the observed variants, giving a higher score to those that are more probably linked with the observed phenotype. One of these is Exomiser, which algorithmically judges the importance of the variant based on the variant frequency, pathogenicity, quality, inheritance pattern, phenotype data from human and model organisms, and proximity in the interactome to phenotypically similar genes [36]. Another way to reduce the number of variants is analysis of trios (two parents + child). This approach is particularly effective in the

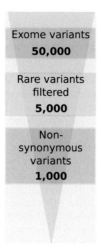

FIGURE 11

Typical numbers of variants in a human sample after the application of standard filtering steps. While it is impossible to manually inspect tens of thousands of variants, standard filters lower the search space to approximately 1000 variants, which can be further limited by using prior knowledge about the genes linked to the phenotype or the known inheritance model.

case of autosomal recessive phenotypes or finding de novo mutations. The PhaseByTransmission algorithm from the GATK framework is an example of a bioinformatic tool specially designed for this kind of analysis.

SOFTWARE AND HARDWARE ISSUES

Last, we would like to discuss the issues concerning software and hardware used for bioinformatic analysis of NGS data. The amount of data generated by the sequencers grows faster than the computational power of the computers. This moves the focus from the sequencers to the bioinformatic analysis of the data, as they become the bottleneck in the analysis. NGS data processing needs not only specialized algorithms but also hardware to perform the computations and store the data. There are two main hardware-related problems that need to be addressed in NGS data analysis: (a) the time cost of the data analysis and (b) security and protection against hardware failure storage.

COMPUTATIONAL ARCHITECTURE

Most of the commonly used tools in NGS data analysis were designed to run on a single computer with fast access to the input files as well as additional data such as the reference sequence. Current implementations optimally use the available computer memory and are able to use multiple CPU cores if available. This single computer approach works well when the size of the analyzed files is on the order of gigabytes. Such simple computational architecture is enough for performing analysis of targeted sequencing: in this case the whole pipeline from FASTQ to annotated list of variants usually takes

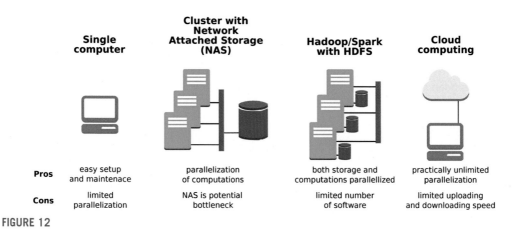

	Single computer	Cluster with Network Attached Storage (NAS)	Hadoop/Spark with HDFS	Cloud computing
Pros	easy setup and maintenace	parallelization of computations	both storage and computations parallellized	practically unlimited parallelization
Cons	limited parallelization	NAS is potential bottleneck	limited number of software	limited uploading and downloading speed

FIGURE 12

Various architectures of computing systems for NGS data processing.

several hours on a modern personal computer. However, as the cost of sequencing goes down, the amount of data produced by sequencers grows rapidly. Multiple samples can be deep sequenced in a single sequencing experiment. It is not unusual, in WGS analysis, to analyze input files with hundreds of millions of reads per sample. The analysis of such input files can take days on a single computer, which may be acceptable for scientific purposes or population analysis, but is unacceptable for clinical applications. Big data processing demands more advanced computational architecture. Four common structures of computational back-end are shown in Figure 12.

Using a single computer is the simplest solution. Although today's computers have enough computational power to analyze targeted sequencing data fast, this solution offers limited scalability. The most popular practice for processing big data is to use a cluster of multiple computers with data stored on an external network-attached storage. This enables scaling the computations, as they can be distributed over several computational nodes. It works especially well for analyzing multiple samples at once. Improving the speed of single-sample processing using the cluster approach is also possible: some of the analysis steps, including read filtering and mapping, can be parallelized by splitting the input into several parts and performing analysis on multiple cluster nodes independently. Such an approach improves the speed of analysis at the small added cost of additional steps for splitting and joining the results. Splitting the input to many computational nodes and joining the results require a fast network for intercomputer communication, which can become a bottleneck as the number of the computational nodes grows.

An alternative approach to analysis of big data sets, borrowed from fields that need high-speed analysis of petabytes of data, such as search engines, removes the access-to-data bottleneck by using a distributed computing environment with distributed file systems: each computational node is also a storage node, containing a part of the stored data. An example of such a computational environment is Hadoop with HDFS storage [37]. Distributed computing systems need specialized data structures for storing data. Therefore a new file format, the ADAM format mentioned previously under Mapping, was proposed as a new standard for storing BAM and VCF data in distributed file systems. The ADAM format is new and, although very promising, most of the commonly used software do not support it as of this writing.

Cloud computing is an interesting alternative to building server rooms and acquiring computers. It also allows easy scaling of computational and storage capacity. However, setting up the computational environment can be challenging. The limited speed of uploading the data to the cloud is another bottleneck. In the case of outsourcing computations and data storage of human sequences, privacy issues take on added importance.

STORAGE ARCHITECTURE

Security against hardware failure is another challenge. The standard approach is to store the data on multiple disks with some redundancy, such as disk mirroring (called RAID1) or more advanced systems in which data are spread on multiple disks of which one (e.g., RAID5) or even two (RAID6) can fail without data loss. Such solutions work very well for NGS data storage. However, owing to the huge size of the NGS data the probability of silent data corruption grows. Silent data corruption is a random undetected hardware error that can occur for multiple reasons, including hardware failure, cosmic radiation, etc. While the probability of the silent data corruption of a single bit is very low, the amount of significant bits being written in the NGS data processing makes it a potential source of data loss. As the data corruption is undetected until the data are read, systems with low redundancy such as RAID1 or RAID5 do not give enough protection. The solution to silent data corruption is to store the data with higher redundancy, preferably in several distinct locations, or to use more specialized backup systems [38].

Secure and safe data storage is expensive; therefore the important question that needs to be addressed is what exactly should be stored. There are several possibilities depending on the demands. For researchers it is advisable to store the raw reads to have the ability to reproduce all the steps in the near or more distant future. Although FASTQ files are usually processed in a compressed form (compressed by gzip), a much better compression ratio can be achieved using tools designed for FASTQ compression, such as fastqz [39] or dsrc [40]. Tools specialized in FASTQ compression can also perform a lossy compression achieving even better compression ratios; for example, lowering the precision of the quality scores can lead to better compression.

For clinical use of NGS, it may be more desirable to store already processed and aligned reads. As mentioned previously, although the BAM format is the most commonly used, there are other file formats specially designed for storing aligned reads, for example, CRAM [41], which performs a reference sequence-based compression. Unfortunately, the CRAM format is not yet supported by most of the commonly used tools. However, it can be easily converted back to the BAM format.

The relatively high cost of data storage also calls into question whether, considering the falling sequencing prices, it is desirable to perform long-term storage of sequencing results at all—it may be cheaper to sequence DNA again when it is needed than to store and protect the data for several years. It might make sense to store just the raw unprocessed sequencing data, which is only a few gigabytes of data, instead of the alignments and other metadata, which can be re-created on demand. The battle between the cost of reanalysis and the storage of analyzed data will determine the path taken and this might be a constantly changing battleground.

We have presented the crucial issues associated with bioinformatic analysis of NGS data starting from sequence files generated by the sequencers to the mapping of the reads, processing of the mapping results, and variant calling, with particular emphasis on their annotation. We have also addressed hardware issues connected with NGS data analysis and storage. We have listed tools that can be used for each task, with analyses of their merits and shortcomings.

There are new instruments based on single-molecule sequencing (PacBio, Oxford Nanopore). Some of these (e.g., PacBio) can generate large image files but the final data are long reads (tens of kilobases) from around a few hundreds of thousands of reads per run. Since the results have more errors, these data sets might need more processing power, for alignments and assembly, but they are not projected to tax storage. Currently they are being used to assist assembly projects, which primarily use deep sequencing data from platforms such as Illumina, and to study large-scale structural variations. There are other techniques (e.g., based on single-molecule imaging techniques such as from OpGen http://opgen.com/) that might supplant such single-molecule sequencing/mapping techniques in the future.

We have tried to present universal issues associated with bioinformatic analysis of NGS data that might remain relevant in the face of technological progress in NGS, computer technologies, and bioinformatic analysis. Despite this, the pace of development of technologies, especially various single-molecule, nanopore-based technologies as well as optical mapping techniques, and computational platforms and tools might lead to a rethinking in the future of genomic analyses in the clinic.

REFERENCES

[1] Daniels MG, Bowman RV, Yang IA, Govindan R, Fong KM. An emerging place for lung cancer genomics in 2013. J Thorac Dis 2013;5(Suppl. 5):S491–7. http://dx.doi.org/10.3978/j.issn.2072-1439.2013.10.06.

[2] Kamalakaran S, Varadan V, Janevski A, Banerjee N, Tuck D, McCombie WR, et al. Translating next generation sequencing to practice: opportunities and necessary steps. Mol Oncol 2013;7:743–55. http://dx.doi.org/10.1016/j.molonc.2013.04.008.

[3] Pant S, Weiner R, Marton MJ. Navigating the rapids: the development of regulated next-generation sequencing-based clinical trial assays and companion diagnostics. Front Oncol 2014;4:78. http://dx.doi.org/10.3389/fonc.2014.00078.

[4] Hert DG, Fredlake CP, Barron AE. Advantages and limitations of next-generation sequencing technologies: a comparison of electrophoresis and non-electrophoresis methods. Electrophoresis 2008;29:4618–26. http://dx.doi.org/10.1002/elps.200800456.

[5] Grada A, Weinbrecht K. Next-generation sequencing: methodology and application. J Invest Dermatol 2013;133:e11. http://dx.doi.org/10.1038/jid.2013.248.

[6] McCarthy MI, Abecasis GR, Cardon LR, Goldstein DB, Little J, Ioannidis JPA, et al. Genome-wide association studies for complex traits: consensus, uncertainty and challenges. Nat Rev Genet 2008;9:356–69. http://dx.doi.org/10.1038/nrg2344.

[7] Van der Auwera GA, Carneiro MO, Hartl C, Poplin R, del Angel G, Levy-Moonshine A, et al. From FastQ data to high-confidence variant calls: the genome analysis toolkit best practices pipeline. In: Current protocols in bioinformatics. John Wiley & Sons, Inc; 2002.

[8] Cock PJA, Fields CJ, Goto N, Heuer ML, Rice PM. The Sanger FASTQ file format for sequences with quality scores, and the Solexa/Illumina FASTQ variants. Nucleic Acids Res 2010;38:1767–71. http://dx.doi.org/10.1093/nar/gkp1137.

[9] Del Fabbro C, Scalabrin S, Morgante M, Giorgi FM. An extensive evaluation of read trimming effects on illumina NGS data analysis. PloS One 2013;8:e85024. http://dx.doi.org/10.1371/journal.pone.0085024.

[10] Schröder J, Bailey J, Conway T, Zobel J. Reference-free validation of short read data. PloS One 2010;5:e12681. http://dx.doi.org/10.1371/journal.pone.0012681.

[11] Lohse M, Bolger AM, Nagel A, Fernie AR, Lunn JE, Stitt M, et al. RobiNA: a user-friendly, integrated software solution for RNA-Seq-based transcriptomics. Nucleic Acids Res 2012;40:W622–7. http://dx.doi.org/10.1093/nar/gks540.

[12] Martin M. Cutadapt removes adapter sequences from high-throughput sequencing reads. EMBnet.J 2011; 17:10–2. http://dx.doi.org/10.14806/ej.17.1.200.

[13] Schroder J, Hsu A, Boyle SE, Macintyre G, Cmero M, Tothill RW, et al. Socrates: identification of genomic rearrangements in tumour genomes by re-aligning soft clipped reads. Bioinformatics 2014;30:1064–72. http://dx.doi.org/10.1093/bioinformatics/btt767.

[14] Li H, Handsaker B, Wysoker A, Fennell T, Ruan J, Homer N, et al. The sequence alignment/map format and SAMtools. Bioinforma 2009;25:2078–9. http://dx.doi.org/10.1093/bioinformatics/btp352.

[15] Fonseca NA, Rung J, Brazma A, Marioni JC. Tools for mapping high-throughput sequencing data. Bioinforma 2012;28:3169–77. http://dx.doi.org/10.1093/bioinformatics/bts605.

[16] Hsi-Yang Fritz M, Leinonen R, Cochrane G, Birney E. Efficient storage of high throughput DNA sequencing data using reference-based compression. Genome Res 2011;21:734–40. http://dx.doi.org/10.1101/gr.114819.110.

[17] Lee W-P, Stromberg MP, Ward A, Stewart C, Garrison EP, Marth GT. MOSAIK: a hash-based algorithm for accurate next-generation sequencing short-read mapping. PloS One 2014;9:e90581. http://dx.doi.org/10.1371/journal.pone.0090581.

[18] Subramanian S, Pierro VD, Shah H, Jayaprakash AD, Weisberger I, Shim J, et al. MiST: a new approach to variant detection in deep sequencing datasets. Nucleic Acids Res 2013. http://dx.doi.org/10.1093/nar/gkt551.

[19] Robinson JT, Thorvaldsdóttir H, Winckler W, Guttman M, Lander ES, Getz G, et al. Integrative genomics viewer. Nat Biotechnol 2011;29:24–6. http://dx.doi.org/10.1038/nbt.1754.

[20] McKenna A, Hanna M, Banks E, Sivachenko A, Cibulskis K, Kernytsky A, et al. The genome analysis toolkit: a MapReduce framework for analyzing next-generation DNA sequencing data. Genome Res 2010;20:1297–303. http://dx.doi.org/10.1101/gr.107524.110.

[21] Derrien T, Estellé J, Marco Sola S, Knowles DG, Raineri E, Guigó R, et al. Fast computation and applications of genome mappability. PloS One 2012;7:e30377. http://dx.doi.org/10.1371/journal.pone.0030377.

[22] Garrison E, Marth G. Haplotype-based variant detection from short-read sequencing. 2012. ArXiv12073907 Q-Bio.

[23] DePristo MA, Banks E, Poplin R, Garimella KV, Maguire JR, Hartl C, et al. A framework for variation discovery and genotyping using next-generation DNA sequencing data. Nat Genet 2011;43:491–8. http://dx.doi.org/10.1038/ng.806.

[24] Danecek P, Auton A, Abecasis G, Albers CA, Banks E, DePristo MA, 1000 Genomes Project Analysis Group, et al. The variant call format and VCFtools. Bioinforma 2011;27:2156–8. http://dx.doi.org/10.1093/bioinformatics/btr330.

[25] O'Rawe J, Jiang T, Sun G, Wu Y, Wang W, Hu J, et al. Low concordance of multiple variant-calling pipelines: practical implications for exome and genome sequencing. Genome Med 2013;5:28. http://dx.doi.org/10.1186/gm432.

[26] Wang K, Li M, Hakonarson H. ANNOVAR: functional annotation of genetic variants from high-throughput sequencing data. Nucleic Acids Res 2010;38:e164. http://dx.doi.org/10.1093/nar/gkq603.

[27] Cingolani P, Platts A, Wang LL, Coon M, Nguyen T, Wang L, et al. A program for annotating and predicting the effects of single nucleotide polymorphisms, SnpEff: SNPs in the genome of *Drosophila melanogaster* strain w1118; iso-2; iso-3. Fly (Austin) 2012;6:80–92. http://dx.doi.org/10.4161/fly.19695.

[28] 1000 Genomes Project Consortium, Abecasis GR, Auton A, Brooks LD, DePristo MA, Durbin RM, Handsaker RE, et al. An integrated map of genetic variation from 1,092 human genomes. Nature 2012;491:56–65. http://dx.doi.org/10.1038/nature11632.

[29] Lehtokari V-L, Kiiski K, Sandaradura SA, Laporte J, Repo P, Frey JA, et al. Mutation update: the spectra of nebulin variants and associated myopathies. Hum Mutat 2014. http://dx.doi.org/10.1002/humu.22693.

[30] Sherry ST, Ward MH, Kholodov M, Baker J, Phan L, Smigielski EM, et al. dbSNP: the NCBI database of genetic variation. Nucleic Acids Res 2001;29:308–11.

[31] Landrum MJ, Lee JM, Riley GR, Jang W, Rubinstein WS, Church DM, et al. ClinVar: public archive of relationships among sequence variation and human phenotype. Nucleic Acids Res 2014;42:D980–5. http://dx.doi.org/10.1093/nar/gkt1113.

[32] Stenson PD, Ball EV, Mort M, Phillips AD, Shiel JA, Thomas NST, et al. Human gene mutation database (HGMD): 2003 update. Hum Mutat 2003;21:577–81. http://dx.doi.org/10.1002/humu.10212.

[33] Welter D, MacArthur J, Morales J, Burdett T, Hall P, Junkins H, et al. The NHGRI GWAS catalog, a curated resource of SNP-trait associations. Nucleic Acids Res 2014;42:D1001–6. http://dx.doi.org/10.1093/nar/gkt1229.

[34] Adzhubei IA, Schmidt S, Peshkin L, Ramensky VE, Gerasimova A, Bork P, et al. A method and server for predicting damaging missense mutations. Nat Methods 2010;7:248–9. http://dx.doi.org/10.1038/nmeth0410-248.

[35] Kumar P, Henikoff S, Ng PC. Predicting the effects of coding non-synonymous variants on protein function using the SIFT algorithm. Nat Protoc 2009;4:1073–81. http://dx.doi.org/10.1038/nprot.2009.86.

[36] Robinson PN, Köhler S, Oellrich A, Sanger Mouse Genetics Project, Wang K, Mungall CJ, Lewis SE, et al. Improved exome prioritization of disease genes through cross-species phenotype comparison. Genome Res 2014;24:340–8. http://dx.doi.org/10.1101/gr.160325.113.

[37] Taylor RC. An overview of the Hadoop/MapReduce/HBase framework and its current applications in bioinformatics. BMC Bioinforma 2010;11(Suppl. 12):S1. http://dx.doi.org/10.1186/1471-2105-11-S12-S1.

[38] Rosenthal DSH. Keeping bits safe: how hard can it be? Commun ACM 2010;53:47–55. http://dx.doi.org/10.1145/1839676.1839692.

[39] Bonfield JK, Mahoney MV. Compression of FASTQ and SAM format sequencing data. PloS One 2013;8:e59190. http://dx.doi.org/10.1371/journal.pone.0059190.

[40] Roguski L, Deorowicz S. DSRC 2–Industry-oriented compression of FASTQ files. Bioinforma 2014;30:2213–5. http://dx.doi.org/10.1093/bioinformatics/btu208.

[41] Cochrane G, Cook CE, Birney E. The future of DNA sequence archiving. GigaScience 2012;1:2. http://dx.doi.org/10.1186/2047-217X-1-2.

ANALYSIS OF STRUCTURAL CHROMOSOME VARIANTS BY NEXT GENERATION SEQUENCING METHODS

Dariusz Plewczynski[1,2,3], Sławomir Gruca[4,5], Przemysław Szałaj[3,6], Krystian Gulik[4], Silviene Fabiana de Oliveira[2,7], Ankit Malhotra[2]

[1]*Centre of New Technologies, University of Warsaw, Warsaw, Poland;* [2]*The Jackson Laboratory for Genomic Medicine, Farmington, CT, USA;* [3]*Centre for Bioinformatics and Data Analysis, Medical University of Bialystok, Bialystok, Poland;* [4]*Department of Immunology, Medical University of Warsaw, Warsaw, Poland;* [5]*Bioinformatics Group, University of Leeds, Leeds, West Yorkshire, UK;* [6]*I-BioStat, Hasselt University, Hasselt, Belgium;* [7]*Laboratório de Genética, Departamento de Genética e Morfologia, Instituto de Ciências Biológicas, Universidade de Brasília, Brasília, Brazil*

CHAPTER OUTLINE

INTRODUCTION

In this chapter, we will describe the impact of observed copy number variations (CNVs) and other structural variations (SVs) in cancer and diseases of genetic origin.

Typically, microarray-based comparative genomic hybridization techniques along with various staining techniques (G banding, fluorescence in situ hybridization (FISH), spectral karyotyping, etc.) and other molecular approaches have been used in the clinic for the detailed cytogenetic view of a cell. Yet, with the costs of next generation sequencing (NGS) studies dropping rapidly the question arises: Can these techniques be replaced by NGS technology-enabled analysis of the whole exome sequencing (WES) or the whole genome sequencing (WGS)?

We will discuss the basic approach taken by several groups first to identify structural variants, then to quantify their impact, and finally to provide clinical diagnosis and suggest possible treatments. We will explore the computational methods, summarizing the data modalities as acquired from various copy number calling methods. The probabilistic and statistical methods are able to link mutations observed in patients (compared to the control group, i.e., nonpathogenic individuals) with phenotypic effects.

The other aspect of the quest for human genomic pathogenicity is related to the normal variability of the human population in the ethnical context (ethnical origin). With the advent of NGS, a high volume of normal human genomes has been sequenced and stored within the 1000 Genomes Project initiative [1]. It was confirmed that genomic SVs are abundant in humans, yet the architecture of most SVs still remains unknown [2].

STRUCTURAL VARIANTS IN THE HUMAN GENOME

Although humans share most of their genomic information, there is a significant degree of diversity between individuals owing to the evolutionary process. Such diversity is represented both by SVs, such as CNVs [2,3], and by single-nucleotide polymorphism (SNPs). Structural variant sizes span several scales: microscale (from a few nucleotides up to 40 kb), mesoscale (40 kb–2 Mb), and macroscale (>2 Mb). An important distinction is between balanced and unbalanced SVs. Balanced SVs include translocations (TRA) and inversions (INV)—changes that do not involve a loss of genetic information but may create chimeric or broken genes. Unbalanced SVs lead to a net decrease/increase of DNA through deletion (DEL)/ duplication (DUP)/ insertion (INS) or higher order amplification. The most common form of SV are CNVs; these are unbalanced changes defined as genomic segments larger than 1 kb whose number varies between individuals [4].

CNVs differ significantly between ethnic populations and can have a dramatic impact on disease susceptibility. The first estimates of the normal human genome variability were done at the population level using tiling oligonucleotide microarrays, comprising 42 million probes. Conrad et al. observed that structural variations in DNA greater than 1 kb account for most bases that vary among human genomes [5]. Among 450 multiethnic individuals they identified 11,700 CNVs with sizes greater than 443 bp, including 30 loci that are candidates for influencing disease susceptibility. For complex traits, the heritability missed by genome-wide association studies and not accounted for by common CNVs can be explained by rare CNVs [5].

STRUCTURAL VARIATION AND HUMAN DISEASE

SVs can cause diverse phenotypes (e.g., differences in organ sizes [6,7], drug metabolic efficiency [8]) and also may be involved in the development of diseases such as autism [9], psoriasis [10], schizophrenia [11,12], obesity [13,14], and Crohn's disease [15]. Not much is known about SVs despite their expected significant impact, for example, on the development of personalized medicines and novel treatments of fatal rare diseases. There is a large group of chromosomal microdeletion-related syndromes, including both Mendelian diseases and many complex traits such as autism and schizophrenia and others.

An example of the potential impact of SVs on susceptibility to a Mendelian disorder was described after modifying the standard NGS data analysis pipeline to determine more accurately the contribution of CNVs to nonsyndromic hearing loss (NSHL) for a cohort of 686 patients [16]. The procedure of exon sequencing for all targeted genes known to be related to NSHL was done by genomic enrichment and detection of SNPs, insertions/deletions (indels), and CNVs using the median read-depth ratios and a sliding-window approach. The results showed that 15.2% of patients had one CNV within a known deafness gene. In the 39% of individuals with a genetic cause of hearing loss, a link to CNV was observed for approximately 18% of the cases. The authors concluded that CNVs are an important genetic reason for NSHL development, and their detection indeed requires the modification of standardized NGS data analysis protocols.

The most profound example of genomic variation-related disease is cancer. The accurate detection of larger somatic CNVs is quickly becoming the routine practice in cancer genome analysis, oncotarget identification, and clinical applications. As it has been stressed [17], the endeavor of clinical genetic analysis of cancer malignancies requires sequencing of a large number of genomes in parallel to identify a wide variety of mutations observed in tumor cells. Indeed, the NGS technology ideally fits into this goal. By decreasing the costs while increasing the throughput of sequencing, NGS provides the means for bringing personalized medicine into clinical practice [18]. For example, medical doctors can detect clinically relevant genomic alterations that could be of diagnostic, prognostic, or therapeutic significance, as reviewed by Chang et al. [18]. We discuss SVs and cancer in more detail later in this chapter.

ANALYSIS OF STRUCTURAL VARIATION BY LEGACY TECHNOLOGIES

In the past few decades, our understanding of the human genome has increased exponentially, together with the rapid growth of sequence databases and bioinformatics tools. At the beginning, large chromatin polymorphisms were identified by light microscopy and SNPs were identified and analysed using Restriction Fragment of Length Polymorphism in the very beginning and after by using polymerase chain reaction (PCR)-based DNA sequencing [19]. In the first decade of the twenty-first century, advances in microarray techniques, both comparative genomic hybridization and SNP genotyping microarrays, allowed us to analyze the human genome with increasing resolution [19]. The legacy technologies were applied for CNV calling, such as karyotyping, PCR, FISH, and microarrays (in particular in the format of aCGH). Yet, all these techniques have their limitations that necessitate the development of new experimental methods for SV calling. The legacy approaches (with the exception of high-resolution microarrays) permit only a very coarse view of SVs in the genome. On the other hand, despite high throughput and sensitive detection of CNVs, microarrays (including aCGH) are fundamentally limited as they cannot discover balanced SVs such as translocations or inversions. The strengths and weaknesses of the microarray techniques have been reviewed [5,20].

In the 2011 review by Hochstenbach et al. [21], authors evaluated the genome-wide array-based identification of CNVs in clinical diagnostics in the context of intellectual disability (DI) and other brain-related disorders. A causative genomic gain or loss was detected in 14–18% of cases; some *de novo* CNVs were found that were not present in healthy subjects. Those structural variants, because of their size, have a major impact on the phenotype by altering multiple genes. Such significant results provide a solid basis for further development of CNV-based biomarkers that could be used for genetic tests for such diseases.

Similar results were also reported for patients with other brain-related diseases, such as autism (5–10% in nonsyndromic, 10–20% in syndromic patients) and schizophrenia (approximately 5%) [21]. Other studies confirmed the correlation between CNVs and idiopathic generalized epilepsy, attention-deficit hyperactivity disorder, major depressive disorder, and Tourette syndrome. Patients tend to have larger CNV numbers and sizes compared to healthy controls.

In clinical setups it is believed that the combination of NGS and microarray-based techniques provides the optimal solution for calling SVs in patients. First, low-cost and high-throughput screening using the aCGH microarrays can be performed, followed by NGS experiments (either WGS or WES type, depending on the goal of the study), which can provide higher-resolution data to confirm and fine-map the microarray calls and discover new SVs not observed in aCGH studies. The importance of such personal genomic data is proven to be crucial for many diseases. For example, exome sequencing in DI and autism patients revealed *de novo* mutations in protein-coding genes in 60% and 20% of cases, respectively [21]. Therefore, it is likely that arrays will be complemented by NGS methods in the present clinical applications, yet further on we expect that NGS will be used as a single diagnostic tool in patients with complex genetic diseases, such as brain-related disorders, revealing both CNVs and mutations in a single test.

STRUCTURAL VARIATION AND NEXT GENERATION SEQUENCING

The basic idea behind NGS is to generate large amounts of DNA sequence data through massive parallelization. The various first, second, and third generation sequencing platforms have enabled the generation of gigabases of sequence at varying levels of read lengths. Thanks to advances in computational methods, currently all types of SVs, including large CNVs, can be effectively detected from either WGS or targeted sequencing data generated from DNA libraries [22]. NGS enables the genome-wide detection of novel SVs with up to base-pair resolution in large populations. There are three basic approaches to SV detection (Figure 1): (1) read-depth (RD), (2) split-read (SR), and (3) paired-end mapping (RP) [2,3,23]. In RD sample sequences are aligned to the reference genome and the depth of coverage at a particular locus is used to quantitate the amount of genomic material (Figure 1(b)), while SR splits and aligns sequences to the reference for gap detection (i.e., deletion; Figure 1(c)). In RP, paired-end DNA sequences are aligned to the reference genome so as to compare the distances between the two ends of the sequences with the expected size from the reference (Figure 1(d)).

Below, we present a review of the available methods and applications/tools to estimate genetic variations using NGS data.

METHODS FOR ESTIMATION OF COPY NUMBER VARIATION FROM NGS DATA

Using NGS data, CNVs are typically identified by RD-based methods. A generic pipeline (as shown in Figure 2(b)) is composed of several steps. It starts with quality control of the sequence fastq files to remove sequencing and library preparation artifacts. This is followed by alignment of the fastq files to a reference genome.

The next step involves counting the reads for predefined windows across the genome. The windows are selected depending on the amount of sequencing coverage, desired resolution, and mapability/repeat content of various parts of the genome. The read counts are then normalized to correct for GC bias and account for different levels of coverage across multiple samples. The normalized RDs are

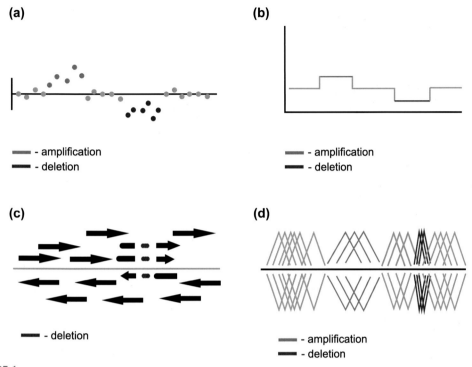

FIGURE 1

A schematic view of SV analysis methods using (a) aCGH, (b) read depth, (c) split read, and (d) paired-end. This illustrates gains of genomic material in this locus.

(a)–(d) were modified from Iskow et al. [3].

converted to log ratios by comparing to a reference control genome. The log ratios across the genome are then segmented by segmentation algorithms and converted into genomic segments that have similar log ratio values across the region.

There are several different algorithms that have been previously used for segmentation of log ratios obtained by previous steps into copy number estimates. Most of these are based on the hypothesis that most of the genome is diploid in nature. There are instances, especially in very aggressive and highly aneuploid tumors, for which this assumption might not be valid, and such cases need to be treated differently. The simplest approach to segmentation is one that estimates the mean and standard deviation of log ratios across the genome/chromosome and uses these numbers to divide the genome into various copy number bins. The Hidden Markov Model-based methods take this idea further by making the various copy number classes into nodes in a state machine and "learning" the weights and probabilities of transitions between states. However, such an approach requires training data sets. Circular binary segmentation was developed to overcome such a requirement. The basic idea is to recursively divide the genome into segments of data points such that all data points are similar within each segment. Rank-based segmentation further improves on this by utilizing the ranks of the various data points instead of using actual normalized log ratio values [24].

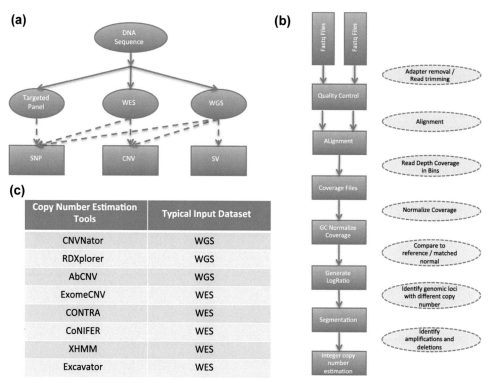

FIGURE 2

Copy number estimation from next generation sequencing experiments. (a) The various kinds of sequencing data sets being generated at the clinic and their applicability for genetic variant identification. (b) Generic copy number detection pipeline. (c) A representative set of tools used to detect CNVs from WES- and WGS-based data sets.

The last step involves taking the segmentation results and assigning integer copy number estimates to the various segments. This can be done in two ways. We can either "learn" the various copy number values using training data or we can try to fit Gaussian mixture models to the distribution of segment log ratios. This generic analysis approach works best for data from WGS experiments, and several approaches have been published over the past years; they are presented in the section on software tools.

STRUCTURAL VARIATION, NGS, CANCER, AND THE CLINIC

NGS technologies are routinely applied in contemporary molecular research and have steadily been penetrating into clinical practice [25]. Detection of clinically relevant genomic alterations is particularly critical for cancer management [17,18,26]. For example, in the case of non-small-cell lung cancer it is possible to stratify patients and guide therapy based on the existence of SVs in the *ALK, MET, RET,* and *ROS* genomic regions [17]. That is particularly relevant in the light of the growing (albeit still limited) repertoire of drugs targeting specific gene products.

Today, targeted sequencing, such as WES, offers significant advantages in the clinic over WGS in terms of sequencing costs, allowing for much higher sequencing depth coverage in regions of interest (exomes or others). A clinically applicable targeting technique for implementing variant detection has been described [27]. Unfortunately, targeting worsens the quality of the signal coming out of the sequenced sample, strongly affecting CNV analysis, which is sensitive to nonuniform coverage [28] and needs modifications of the generic analysis pipelines. These issues arise from the facts that (1) the amount of sequencing across different exons is not homogeneous, (2) the sequencing regions are not uniformly contiguous across the genome, and (3) the human genome does not have a uniform distribution of genes—there are regions that are extremely gene rich or poor. Coverage levels vary across the genome because of varying capture efficiencies in NGS experiments. The noise of deep-targeted sequencing, common for cancer studies, stems from the usually applied PCR procedure step, and attention must be paid to dealing with artifacts created by the procedure appropriately [29]. Targeted approaches are cost saving and thus well suitable for clinical settings; hence methodologies for extracting SV information from targeted areas are of special importance. The peculiarities of amplicon-based NGS (targeting) techniques within the context of available sequencers and kits for constructing libraries and related clinical oncology issues, such as amount of genetic material needed and time required for the preparation, were thoroughly discussed by Chang and Li [18]; the authors regard targeting as the best fit for the applied oncology. The issue of coverage uniformity is also relevant for WGS of single cells, which is a novel approach for cancer research. Single-cell sequencing relates to the subject of liquid biopsies discussed later in this section. Methods for achieving a reasonable uniformity of amplification of a whole genome, required for CNV detection, have already been proposed [30]. Within the context of discussing amplification procedures for NGS and SV detection, it is worth noting that there are significant differences between sequencing platforms and a careful consideration is advised when selecting between them [31,32]. The sequencing machine and amplification method determine read lengths and paired-end sequencing capabilities, which have an impact on downstream data processing [25]. Multiple methods of SV detection rely on information provided by paired-end sequencing. Despite the conceptual similarity of the mate-pair and paired-end sequencing approaches [23], each influences the output of the sequencing step differently, therefore augmenting the final result of the SV analysis [33].

Cancer is the most profound example of a disease involving SVs. It has been shown that practically every tumor carries multiple forms of structural variation, whether they be altered copy numbers of cancer-driving genes or chromosomal rearrangements; although the number of SV events is usually much lower than the count of point mutations in tumor-involved genes [34,35], their role in cancer initiation and progression must not be underestimated, as the events often introduce major changes. It has been reported that in some cases CNV analysis can discriminate between simultaneous tumors in a patient while there is no signal from SNPs or short indels [36].

Chronic myeloid leukemia (CML) is a hematological cancer induced by a structural variation—*BCL–ABL* gene fusion [37]. CML is also a poster example of a successful application of a targeted therapy that works for the majority of the patients [38]. Generic imatinib—a tyrosine kinase inhibitor abolishing the problems caused by the structural variation in CML—is available, and it is hoped to make the therapy even more accessible. What is intriguing is that a fraction of imatinib-treated patients not only go into remission but may be cured and withdrawn from the drug therapy [39,40]. The use of NGS-based gene break-point position estimation was proposed for the design of patient-customized quantitative PCR assays for monitoring the therapy and signaling the timing of a safe drug withdrawal [41]. The clinical success achieved in managing CML might eventually be replicated for more cancers. Analysis of 7000 tumor samples from The Cancer Genome Atlas (TCGA) (NGS RNA-seq data) not only revealed that gene fusions are common but also identified several novel kinase gene fusions, involving *MET*, *PIK3CA*, and *PKN1*, among others [42].

Hematological malignancies have been thoroughly studied thanks to NGS technologies [43], owning this, in part, to the ease of obtaining the material for analysis. However, it has been shown for solid tumors as well that a blood collection procedure alone provides enough material to extract circulating cancer DNA in amount sufficient to monitor the therapy [44–47]. In other words, clinically applicable SV analysis might be turned into a minimally invasive blood test. Such bodily fluid collections, aimed at acquiring genetic material released by growing tumors, have been termed "liquid biopsies," and the cancer-derived genetic material itself—ctDNA (circulating tumor DNA). Liquid biopsies combined with NGS-based SV ctDNA analysis have the potential of becoming a practice in monitoring cancer therapies in the clinic [48]. It is encouraging that the WGS coverage needed to obtain a clinically relevant result may be as low as 0.1 [49,50], what translates into an acceptable cost of the procedure. Moreover, if a diagnosis could be based on circulating genetic material only, without the need for puncturing a tumor, the often hazardous biopsy procedure would become obsolete.

In addition to DNA, tumors release microRNA molecules, both free and packed in microparticles [51] and exosomes [52]. MicroRNA profiling of the bodily fluids (including saliva and urine) can be performed on NGS data and is a promising approach to cancer detection [53].

Apart from oncotarget identification, NGS technologies provide the means for improving clinical management of many cancer patients undergoing already approved therapies. There have been identified genomic signatures, including CNVs, that are not related to cancer, but carry signals concerning drug metabolism capabilities, thus providing a vital piece of information enabling fine-tuning of the dosing of chemotherapeutics for each patient. For example, *CYP2D6* variants (polymorphisms and deletions) are not rare in the U.S. population and strongly influence the metabolism of a widely used breast cancer drug—tamoxifen [54]. Moreover, modern targeted cancer therapies often carry extreme price tags and performing NGS tests for drug metabolism assessment would make those therapies more effective not only clinically (optimal tumor cell targeting with simultaneous minimization of drug side effects), but also economically [35].

The costs of NGS are decreasing rapidly; thus one might envision that the near feature will bring to the clinic comprehensive yet inexpensive NGS-based tests for cancer detection and profiling. Nevertheless, the widespread acceptance of such diagnostic modalities requires scientific consensus, emergence of analysis standards, and sorting out of reimbursement issues [17] might turn out to be rather slow processes, lagging behind scientific discoveries [54]. Last but not least, clinicians must become confident in the application of the new techniques in their practice. Eventually, high-throughput technologies shall reach a status of standard diagnostic tools, like medical imaging.

NGS-BASED STRUCTURAL VARIATION DETECTION SOFTWARE

The approaches employed in NGS-based SV analysis (as described earlier in the chapter) make use of RD, SR, and RP information [23]. Each one has its own capabilities and limitations and the potential to complement the others. RP methods are preferred to detect transposonal origin CNVs and structural events not changing copy number on average, while RD approaches perform well for segmental duplication [28]. Hybrid methods have also been proposed and implemented to mitigate disadvantages of the basic approaches. Table 1 provides a brief overview of selected tools for SV analysis and includes information on pipelines integrating multiple tools/methods for a comprehensive detection: iSVP [59] and SHEAR [60], with the latter supporting genome reconstruction applicable for studies of cancer. Please note that many of the tools do not handle retrotransposition events well, yet structural variation in the form of somatic retrotransposon insertions has been shown to be a significant aspect of tumor biology [61]. By exploiting a richer information set, they usually localize genomic breakages and estimate copy number more precisely, thus providing a more comprehensive detection of structural events. LUMPY [55] is one example of a hybrid RP/SR/RD SV detection tool as well as a CNV detection method, referred in Table 2 as "Nord-et-al." [56], combining RD with partial mapping information. The latter method locates genomic break points with accuracy similar to that of RD-based methods for WGS, despite the presence of the targeting noise.

Gene fusions are a type of SV and specialized tools are needed to detect such variants in sequencing data. The aforementioned kinase gene translocation discoveries based on TCGA data sets were enabled by the application of the STAR aligner [57], see Table 1, able to cope with chimeric alignments characteristic for gene fusions. The tool is also recommended for RNA-seq data alignment for its speed and accuracy. However, there exist several software pieces designed specifically for more complete SV detection, utilizing SR and/or RP information to identify, with the aid of clustering, several variant signatures, reflecting insertions, deletions, inversions, and novel tandem duplications as described by Medvedev et al. [23]. A recent study by Abel et al. [58] provides an excellent comparison of several gene rearrangement tools (including in-depth comparison with FISH) within the context of clinically relevant *ALK* and *KMT2A* rearrangements. They conclude that NGS technologies combined with proper bioinformatics software create a reliable toolset for clinical use, capable of detecting gene translocations with the quality of the FISH method on top of providing additional vital information such as the cancer-relevant gene mutations. They report, for the investigated *KMT2A* rearrangement, that the mean coverage needed to reach and exceed the FISH method's sensitivity (90%) is 330×; they caution, however, that the coverage needed in clinical settings may significantly differ and should be evaluated carefully, as tumor sample cellularity, DNA availability, and targeting approach strongly influence the sensitivity and specificity of the detection.

CNVs often significantly contribute to cell pathology. In some instances of cancers, such as papillary thyroid carcinoma, an exclusivity of genomic changes appears to exist—between protein coding mutations and copy number alterations [77]. Thus, it may be assumed that in some instances, CNVs are major contributors to cancer development and progression and establishing tumor drug resistance. Therefore CNV-aware (SV-aware, to be more precise) analysis is mandatory for a comprehensive cancer profiling. Yet, regular test procedures for variation detection (such as FISH and aCGH) are unlikely to locate CNVs—because of inadequate resolution. The situation is not alleviated by application of the regular NGS data analysis pipelines (for their description please refer to the chapter "Basic Bioinformatic Analyses of NGS Data"), as they are oblivious to most structural variants. Even commercial NGS software packages might not deal well with CNVs—it has been reported, in a sequencing-based CML tyrosine kinase inhibitor resistance study, that two proprietary software tools failed to cope with a 540-bp-long deletion [78]. Therefore, for an NGS software package to be complete, it must be SV analysis capable.

Along with the spread of NGS platforms, a plethora of bioinformatics tools for processing and analyzing the sequencing data has blossomed. In 2014 Pabinger et al. evaluated over 30 software tools for

Table 1 Overview of Selected Software Tools for NGS-Based Structural Rearrangement Detection. Categories of SVs: deletion (DEL), insertion (INS), inversion (INV), translocation (TRA) and duplication (DUP).

Software Tool	Tool Information & Benchmark References	Download URL	Software License	Release Year/Last Updated/Version	SVs detection capabilities (Performance)	Remarks (Assuming 50bp Read Length and 200bp Insert Size)
Break-Dancer [62]	[23,58,59,62,63]	http://breakdancer.sourceforge.net	GNU GPL v3	2009/2013/1.1.2	DEL(■■■□) INS(■■■□) INV(■■□□) TRA(■■□□)	• deletions >50bp, • insertions 60–80bp, • ~100bp breakpoint accuracy • coverage 20x/40x recommended (deletions/insertions) • fast • low memory requirements
Pindel [64]	[23,55,59,64–68]	https://github.com/genome/pindel	GNU GPL v3	2009/2013/0.2.4w	DEL(■■■■) INS(■■□□)	• deletions <10kb • insertions <25bp • high break-point resolution • coverage >20x recommended • slow • high memory requirements
MoDIL [69]	[23,62,65,69]	http://compbio.cs.toronto.edu/modil	NA	2009/2012/2(beta)	DEL(■■□□) INS(■■□□)	• deletions >25bp • insertions 25–75bp
Variation-Hunter [63,70]	[23,62,63,65,70]	http://sourceforge.net/projects/variationhunter	GNU GPL v2	2010/2012/0.4	DEL(■■□□) INS(■■□□) INV(■■□□) DUP(■■□□)	• deletions <500kb • insertions <100bp • inversions <10Mb • slow, high memory requirements • suitable for transposonal duplication discovery
GASVPro [71,72]	[55]	https://code.google.com/p/gasv	GNU GPL v3	2009 (GASV)/2012/1.2	DEL INS DUP INV TRA	• exploits RP and RD information
PEMer [73]	[23,73]	http://sv.gersteinlab.org/pemer	Creative commons (attribution-noncommercial)	2009	DEL(■■■■) INS(■■□□) DUP INV(■■■□) TRA	• for SVs >200bp

Software	Ref.	URL	License	Year	SV types	Notes
SOAPindel [66]	[66]	http://soap.genomics.org.cn/soapindel.html	NA	2011/2014/2.1	DEL(■■■■) INS(■■■■)	• for events <50bp • very slow
LUMPY [55]	[55]	https://github.com/arq5x/lumpy-sv	MIT license	NA/2014/NA	DEL INS DUP INV TRA	• exploits RP, SR, and RD information
Ulysses [33]	[33]	https://github.com/gillet/ulysses	GNU GPL v3	2014/NA/1.0	DEL INS DUP INV TRA	• comprehensive SV detection • detection of low-frequency variants
SVfinder [74]	[74]	https://github.com/cauyrd/SVfinder	NA	2014/NA/NA	DEL INS DUP INV TRA	• integrates genomic information on repeats to increase specificity
GINDEL [68]	[68]	http://sourceforge.net/projects/gindel	NA	2014/NA/0.8	DEL(■■■■) INS(■■■■)	• training set required • appropriate for low coverage (<5×) samples • for events >25 bp • fast
CREST [75]	[58]	http://www.stjuderesearch.org/site/lab/zhang	GNU GPL v2/v3 but requires non-free software components	2011/NA/1.0	DEL INS DUP INV TRA	
GATK Haplotype Caller [76]	[59,66,76]	https://www.broadinstitute.org/gatk	MIT license for noncommercial use	2010/2014/3.3	DEL(■■■■) INS(■■■■) other	• for events <25 bp • slow, very high memory requirements
iSVP [59]	[59]	NA	NA	NA	DEL(■■■■)	• pipeline proposal only, implementation not available • combines several tools (GATK Haplotype Caller, Pindel, BreakDancer) to cope with deletions of any size • validated for deletions

Continued

Table 1 Overview of Selected Software Tools for NGS-Based Structural Rearrangement Detection. Categories of SVs: deletion (DEL), insertion (INS), inversion (INV), translocation (TRA) and duplication (DUP).—cont'd

Software Tool	Tool Information & Benchmark References	Download URL	Software License	Release Year/Last Updated/Version	SVs detection capabilitie (Performance)	Remarks (Assuming 50 bp Read Length and 200 bp Insert Size)
SVMerge [67]	[67]	http://svmerge.sourceforge.net	GNU GPL v3	2010/2012/1.2	DEL INS DUP INV TRA	• comprehensive SV detection for events >50 bp • local de novo assembly to refine break points and increase detection specificity • combines several SV detection tools (Pindel, BreakDancer, RDXplorer and two in-house developed tools for dealing with transposable elements)
SHEAR [60]	[60]	http://vk.cs.umn.edu/SHEAR	NA	2013/2014/0.2.12	DEL INS DUP INV TRA	• uses CREST inside • able to handle tumor heterogeneous data • suitable for duplication discovery
STAR [57]	[57]	https://github.com/alexdobin/STAR	GNU GPL v3	2013/2014/2.4	TRA	• designed for RNA-seq alignment—not strictly an SV detection tool

WGS and WES data analysis selected after surveying over 200 [79]. They discussed the subjects of read quality assessment, alignment, variant identification, annotation and visualization, general functionality and individual features. This review considered emulating the genomic variability important from a clinical standpoint, employing analyses of rare disease WES data for SV identification and two cancer data sets for identifying somatic mutations and SVs; an artificial data set was also used for the purpose of evaluating CNV detection. A review by Liu et al. [26] describes methodologies of NGS-based CNV detection in particular. Other reviews providing comprehensive comparisons and benchmarks for software tools designed for CNV calling for WGS [80] and WES [81,82] allowed us to provide an overview of bioinformatics tools, presented in Table 2. The provided remarks, comments, and, especially, the performance indexes are based on sources referenced in the second column. The deletions, in the form of large and small hemizygous deletions and losses of heterozygosity, are common in cancer genomes [83–85], thus special attention shall be given regarding their detection–therefore the performance metrics are presented separately for deletions and insertions. As there are numerous examples of sometimes well-performing pieces of software, created for a specific project and then orphaned, in order to indicate the status of the tools, we included information regarding the dates of their publication and updates.

We are stressing that performance indexes provided in Tables 1 and 2 are arbitrary, as a consolidation of sparse and incoherent results—multiple comparisons carried out with different test settings on separate data sets processed by nonstandardized hardware cannot be deterministic. The tables serve as indicators of general capabilities of the tools.

Some of the tools for CNV-oriented cancer analysis have the capability of providing results without using the reference data representing healthy tissue of a patient. A word of caution regarding interpretations of calls provided by the tools based on contemporary public CNV databases is warranted, as the information on SVs in those databases might be inaccurate [86].

The present landscape of bioinformatics tools is rather shattered and chaotic. Numerous software pieces and databases are being created and some of them eventually abandoned. Cross-platform issues, documentation, and installation requirements—all complicate the case. Furthermore, there exist robust methodologies for which there are no available implementations. Researchers may decide against releasing their tools for a variety of reasons—possibly because of significant costs associated with polishing and documenting the work; maintenance costs might play a role as well. Considering NGS software for SV analysis, there is a need for an emergence and acceptance of a set of robust methods and tools. This would benefit the clinic as well, where stability of the solutions is appreciated. A crucial aspect of the software world is licensing. It is encouraging to witness growing responsibility of scientists, which is expressed by licensing the developed software in ways compatible with the definition of free software [110,111]. This is of the utmost importance for accelerating the advancement of human health science. Publishing a software tool as a free software gives it a chance to be picked up by the Internet community and, consequently, further developed and maintained.

FUTURE DIRECTIONS

An increasingly appreciated aspect of SVs is their relationship with three-dimensional chromatin structure, especially in cancer cells. Somatic copy number alterations (SCNAs) show striking correlation with the locus distances (i.e., the higher the chromatin contact probability, the more SCNA

Table 2 Overview of Selected Software Tools Created for NGS-Based CNV Detection.

Software Tool	Tool Information & Benchmark References	Download URL	Designed for	Software License	Release Year/Last Updated/Version
CNV-seq [87]	[80,87–89]	http://tiger.dbs.nus.edu.sg/CNV-seq	WGS	GNU GPL v2/v3 (R package) and unknown–open to download (Perl scripts)	2009/2014/0.2–8
CNVnator [28]	[28,80,90–92]	http://sv.gersteinlab.org/cnvnator	WGS	Creative commons (attribution-noncommercial)	2011/2014/v0.3
FREEC [93]	[80,93,94]	http://bioinfo.curie.fr/projects/freec	WGS & WES (Control-FREEC)	GNU GPL v2/v3	2011/2013/7.0 (Control-FREEC)
RDXplorer [95]	[80,88,95]	http://rdxplorer.sourceforge.net/	WGS	Free for academic and noncommercial use	2009/2011/3.2
readDepth [89]	[80,89]	https://code.google.com/p/readdepth	WGS	Apache License 2.0	2011/2012/0.9.8.4
SegSeq [96]	[80,91,96]	http://www.broadinstitute.org/cgi-bin/cancer/publications/pub_paper.cgi?mode=view&paper_id=182	WGS	© 2008 the Broad Institute/Massachusetts Institute of Technology	2008/2009/1.0.1
GENSENG [90]	[90]	http://sourceforge.net/projects/genseng	WGS	Creative commons (attribution-noncommercial)	2012/NA/0.1
mHMM R package [91]	[91]	https://www.stt.msu.edu/users/hengwang/mHMM.html	WGS	GNU GPL v2/v3	2014/NA/1.0
RAIG [97]		http://compbio.cs.brown.edu/projects/raig	WGS	© 2014 Brown University, Providence, RI	2014/2014/1.01
ADTEx [94]		http://sourceforge.net/projects/adtex	WES	GNU GPL v3	2013/2014/1.0.4
ExomeCNV [98]	[81,94,98–103]	http://cran.r-project.org/src/contrib/Archive/ExomeCNV	WES	GNU LGPL v2.1	2011/2012/1.4
EXCAVATOR [103]		http://sourceforge.net/projects/excavatortool	WES	Free for noncommercial use	2013/2014/2.2
CONTRA [99]	[81,99,102]	http://sourceforge.net/projects/contra-cnv	WES	GNU GPL v3	2011/2013/2.0.4
ExomeDepth [101]	[81,101,104]	Comprehensive R Archive Network	WES	GNU GPL v3	2012/2014/1.0.7
CoNIFER [105]	[81,92,104]	http://conifer.sourceforge.net	WES	GNU GPL v3	2011/2012/0.2.2
cn.MOPS [106]	[104]	http://www.bioinf.jku.at/software/cnmops	WGS/WES	GNU LGPL v2	2012/2014/1.12

Software Tool		URL		License	Version
exomeCopy [107]	[81,100,101,104]	http://www.bioconductor.org/packages/release/bioc/html/exomeCopy.html	WES	GNU GPL v2/v3	2011/2013/1.12
cnvOffSeq [92]	[92]	http://sourceforge.net/projects/cnvoffseq	WES	GNU GPL v3	2014/NA/0.1.1
exon Del [100]	[100]	https://github.com/slzhao/ExonDel	WES	NA (said to be "freely available for public use")	2013/2013/1.2
Nord-et-al. [56]	[56]	Available on request	WES	NA	2011/NA/NA

Software Tool	Pros	Cons	Recommended for	Remarks
CNV-seq	• Balanced overall performance		• General CNV analysis	
CNVnator	• Excellent general performance • High break-point resolution (10–200bp) • Works well for low-coverage samples • Modest RAM requirements	• Not suitable for retrotransposonal CNVs • Binaries not available • Non-free software	• Excellent for CNV detection for noncommercial use	• Can be complemented with PEM-based software • Filtering can increase deletion specificity
FREEC	• Very low RAM requirements • CPU threading support • Loss of heterozygosity detection (Control-FREEC)		• When computational resources are limited	• Replaced by Control-FREEC [108]
RDXplorer	• Modest CPU and RAM requirements	• Non-free software	• Analyzing large volumes of data	• Implementing event-wise testing (EWT) [95] • AbCNV provides another EWT implementation [109]
readDepth	• GC bias correction for bisulfide sequencing • CPU threading support	• Low coverage hampers variant detection	• Deep-sequenced samples • When having enough computational power	
SegSeq		• Increasing coverage does not improve performance • High RAM requirements • Needs proprietary programming environment • Non-free software		
GENSENG		• Non-free software		

Continued

Table 2 Overview of Selected Software Tools Created for NGS-Based CNV Detection.—cont'd

Software Tool	Pros	Cons	Recommended for	Remarks
mHMM R package RAIG				
ADTEx	• High specificity	• Non-free software	• Analyzing somatic copy number aberrations in cancer genomes • Searching for CNVs longer than 200 bp • Detection of chromosome arm-level events in cancer	• DWT applied to decrease targeting noise • Extension of CoNVEX method [102]
ExomeCNV		• Low specificity • Requires knowledge of normal DNA contamination of tumor sample	• Cancer sample analysis	• Removed from CRAN repository
EXCAVATOR	• Discriminates types of deletions	• Not designed for rare variant analysis • Non-free software	• Cancer sample analysis	
CONTRA				• Does not assume Poisson distribution for coverage
ExomeDepth	• Easy to use R package			• Does not assume Gaussian distribution of read count ratio
CoNIFER cn.MOPS	• High deletion specificity • Relatively high insertion specificity	• Low sensitivity		
exomeCopy cnvOffSeq	• Excellent performance for deletions >5 kb • Moderate hardware requirements	• Low specificity	• Searching for CNVs longer than 5000 bp	• SVD-based RD normalization
exon Del	• High exonal deletion sensitivity	• Licensing not clear	• Homozygous exon deletion detection	
Nord-et-al. [56]	• High sensitivity and specificity • Does not require sequencing over break points	• Not publicly available	• CNV analysis of targeted high-coverage sequencing data • When break-point locations are needed to be localized	• Hybrid approach, augmenting RD methodology with partial-mapping information

Software Tool	Deletions				Insertions				Processing Speed
	Sensitivity	Specificity	Position Estimation	Count Estimation	Sensitivity	Specificity	Position Estimation	Count Estimation	
CNV-seq	▪▪▫▫▫	▪▪▫▫▫	▪▪▫▫▫	▪▪▫▫▫	▪▪▫▫▫	▪▪▫▫▫	▪▪▫▫▫	▪▪▫▫▫	▪▪▫▫▫
CNVnator	▪▪▫▫▫	▪▪▫▫▫	▪▪▫▫▫	▪▪▫▫▫	▪▪▫▫▫	▪▪▫▫▫	▪▪▫▫▫	▪▪▫▫▫	▪▪▫▫▫
FREEC	▪▪▫▫▫	▪▪▫▫▫	▪▪▫▫▫	▪▪▫▫▫	▪▪▫▫▫	▪▪▫▫▫	▪▪▫▫▫	▪▪▫▫▫	▪▪▫▫▫
RDXplorer	▪▪▫▫▫	▪▪▫▫▫	▪▪▫▫▫	▪▪▫▫▫	▪▪▫▫▫	▪▪▫▫▫	▪▪▫▫▫	▪▪▫▫▫	▪▪▫▫▫
readDepth	▪▪▫▫▫	▪▪▫▫▫	▪▪▫▫▫	▪▪▫▫▫	▪▪▫▫▫	▪▪▫▫▫	▪▪▫▫▫	▪▪▫▫▫	
SegSeq	▪▪▫▫▫	▪▪▫▫▫	▪▪▫▫▫	▪▪▫▫▫	▪▪▫▫▫	▪▪▫▫▫			▪▪▫▫▫
GENSENG	▪▪▫▫▫	▪▪▫▫▫			▪▪▫▫▫	▪▪▫▫▫			
mHMM R package	▪▪▫▫▫	▪▪▫▫▫			▪▪▫▫▫	▪▪▫▫▫			
RAIG									
ADTEx	▪▪▫▫▫	▪▪▫▫▫			▪▪▫▫▫	▪▪▫▫▫			
ExomeCNV									
EXCAVATOR	▪▪▫▫▫	▪▪▫▫▫			▪▪▫▫▫	▪▪▫▫▫			
CONTRA	▪▪▫▫▫	▪▪▫▫▫			▪▪▫▫▫	▪▪▫▫▫			
Exo-meDepth	▪▪▫▫▫	▪▪▫▫▫			▪▪▫▫▫	▪▪▫▫▫			
CoNIFER	▪▪▫▫▫	▪▪▫▫▫			▪▪▫▫▫	▪▪▫▫▫			
cn.MOPS	▪▪▫▫▫	▪▪▫▫▫			▪▪▫▫▫				
exomeCopy	▪▪▫▫▫	▪▪▫▫▫							
cnvOffSeq	▪▪▫▫▫	▪▪▫▫▫					▪▪▫▫▫		
exon Del	▪▪▫▫▫								
Nord-et-al. [56]	▪▪▫▫▫		▪▪▫▫▫		▪▪▫▫▫				

events) [112]. Why SCNAs favor the proximate loci is not known; however, it is thought that ligations between two near break points are more likely to occur and thus alter the genome structure more frequently than distant loci [113]. These findings strongly support the direct link between tumor development and chromatin structure and SVs and prompt us to develop both the experimental and the theoretical techniques explaining the larger variation in genome structure in tumor cells and the impact of SVs on it [113,114].

The future of SV studies might benefit from combining sequencing technologies with other high-throughput methods. For example, the combination of NGS and contact information of spatial proximity of genomic regions, that is, three-dimensional distances between genomic loci, allows augmenting the genomic nucleotide sequence with two-dimensional (2D; contacts between enhancers, promoters, and genes) and 3D (three-dimensional structure of the nucleus) information. Further advances in the HiC methodology lie not only in improving the resolution of contact identification for population implementation of the HiC method [115], but also on introducing the single-cell HiC [116]. This novel approach allows for estimation of cell-to-cell variability of 2D contact maps within a single cell line (the same tissue). The structural variability of the whole genome within the population of cells can be analyzed using 2D experimental data (HiC) or 3D chromatin structure, when experimental methods are combined with computational biophysical polymer-based algorithms [116].

Since hot spots of multiple genomic interactions or higher order conformations are likely to have long loops or repetitive sequences, therefore accurate long read-length sequencing is vital for 3D chromatin modeling (which might not be feasible using the conventional short read-length NGS platforms). The PacBio long-range sequencing technology provides a long mean reading length of ~8500 bp [117], permitting an effective *de novo* assembly, even within the repetitive regions. We hypothesize that combination of the HiC approach with genotyping from both PacBio (long reads) and Illumina (high throughput, low sequencing cost per base pair [32]) sequencers shall significantly enhance the modeling accuracy and identification of novel conformational changes between individual cancer cells that were not possible using conventional approaches. Combining PacBio sequencing with other high-throughput platforms has already been suggested for *de novo* assembly of bacterial genomes [118].

Along with the increasing affordability of computing power and the improvements in software platforms, the methods for SV detection that are not currently feasible to use, because of prohibitively high computational requirements, may become usable in the near future. The emergence of open standards for simplifying parallelization of computer programs will only speed up this process. *OpenCL* stands as an excellent example here, as it enables true cross-platform portability—in terms of a programming language, operating system, and computing hardware. This is of great importance as we have been experiencing commoditization of general processing units, which essentially are massively parallel processors.

The decreasing costs of sequencing, discovery of new biomarkers and targeted therapeutics shall eventually make personalized medicine on a mass scale a reality. Today's clinical cancer diagnostics, predominantly based on histology, is progressively being augmented by molecular diagnostic tools. The outlook for the further development and commoditization of personalized cancer treatment is very promising, but to accomplish this promise, close collaborations between researchers and clinicians are essential, to develop databases and tools for genomic variation analysis, linking the genomic changes with their associated phenotypes.

ACKNOWLEDGMENTS

This work was supported by the Polish National Science Center (Grants UMO-2013/09/B/NZ2/00121 and 2014/15/B/ST6/05082), COST BM1405 EU action, and the EU's Seventh Framework Programme for Research (FP7-REGPOT-2012-CT2012-316254-BASTION). SFO was partially supported by the Conselho Nacional de Desenvolvimento Cientifico e Tecnologico (CNPq). PS and DP were supported by funds from the National Leading Research Center in Bialystok and the European Union under the European Social Fund.

REFERENCES

[1] The 1000 Genomes Project Consortium, et al. A map of human genome variation from population-scale sequencing. Nature 2010;467(7319):1061–73.
[2] Mills RE, et al. Mapping copy number variation by population-scale genome sequencing. Nature 2011;470(7332):59–65.
[3] Iskow RC, Gokcumen O, Lee C. Exploring the role of copy number variants in human adaptation. Trends Genet 2012;28(6):245–57.
[4] Raphael BJ. Chapter 6: structural variation and medical genomics. PLoS Comput Biol 2012;8(12):e1002821.
[5] Conrad DF, et al. Origins and functional impact of copy number variation in the human genome. Nature 2010;464(7289):704–12.
[6] Cooper GM, et al. A copy number variation morbidity map of developmental delay. Nat Genet 2011;43(9):838–46.
[7] Cooper GM, et al. Corrigendum: a copy number variation morbidity map of developmental delay. Nat Genet 2014;46(9):1040.
[8] Johansson I, et al. Inherited amplification of an active gene in the cytochrome P450 CYP2D locus as a cause of ultrarapid metabolism of debrisoquine. Proc Natl Acad Sci USA 1993;90(24):11825–9.
[9] Sebat J, et al. Strong association of de novo copy number mutations with autism. Science 2007;316(5823):445–9.
[10] de Cid R, et al. Deletion of the late cornified envelope LCE3B and LCE3C genes as a susceptibility factor for psoriasis. Nat Genet 2009;41(2):211–5.
[11] International Schizophrenia, C. Rare chromosomal deletions and duplications increase risk of schizophrenia. Nature 2008;455(7210):237–41.
[12] Stefansson H, et al. Large recurrent microdeletions associated with schizophrenia. Nature 2008;455(7210):232–6.
[13] Bochukova EG, et al. Large, rare chromosomal deletions associated with severe early-onset obesity. Nature 2010;463(7281):666–70.
[14] Jacquemont S, et al. Mirror extreme BMI phenotypes associated with gene dosage at the chromosome 16p11.2 locus. Nature 2011;478(7367):97–102.
[15] McCarroll SA, et al. Deletion polymorphism upstream of IRGM associated with altered IRGM expression and Crohn's disease. Nat Genet 2008;40(9):1107–12.
[16] Shearer AE, et al. Copy number variants are a common cause of non-syndromic hearing loss. Genome Med 2014;6(5):37.
[17] Pfeifer JD. Clinical next generation sequencing in cancer. Cancer Genet 2013;206(12):409–12.
[18] Chang F, Li MM. Clinical application of amplicon-based next-generation sequencing in cancer. Cancer Genet 2013;206(12):413–9.
[19] Stankiewicz P, Lupski JR. Structural variation in the human genome and its role in disease. Annu Rev Med 2010;61:437–55.

[20] Zhao M, et al. Computational tools for copy number variation (CNV) detection using next-generation sequencing data: features and perspectives. BMC Bioinformatics 2013;14(Suppl. 11):S1.

[21] Hochstenbach R, et al. Genome arrays for the detection of copy number variations in idiopathic mental retardation, idiopathic generalized epilepsy and neuropsychiatric disorders: lessons for diagnostic workflow and research. Cytogenet Genome Res 2011;135(3–4):174–202.

[22] Abel HJ, Duncavage EJ. Detection of structural DNA variation from next generation sequencing data: a review of informatic approaches. Cancer Genet 2013;206(12):432–40.

[23] Medvedev P, Stanciu M, Brudno M. Computational methods for discovering structural variation with next-generation sequencing. Nat Methods 2009;6(Suppl. 11):S13–20.

[24] Olshen AB, et al. Circular binary segmentation for the analysis of array-based DNA copy number data. Biostatistics 2004;5(4):557–72.

[25] Rizzo JM, Buck MJ. Key principles and clinical applications of "next-generation" DNA sequencing. Cancer Prev Res (Phila) 2012;5(7):887–900.

[26] Liu B, et al. Computational methods for detecting copy number variations in cancer genome using next generation sequencing: principles and challenges. Oncotarget 2013;4(11):1868–81.

[27] de Vree PJ, et al. Targeted sequencing by proximity ligation for comprehensive variant detection and local haplotyping. Nat Biotechnol 2014;32(10):1019–25.

[28] Abyzov A, et al. CNVnator: an approach to discover, genotype, and characterize typical and atypical CNVs from family and population genome sequencing. Genome Res 2011;21(6):974–84.

[29] Smith EN, et al. Biased estimates of clonal evolution and subclonal heterogeneity can arise from PCR duplicates in deep sequencing experiments. Genome Biol 2014;15(8):420.

[30] Zong C, et al. Genome-wide detection of single-nucleotide and copy-number variations of a single human cell. Science 2012;338(6114):1622–6.

[31] Glenn TC. Field guide to next-generation DNA sequencers. Mol Ecol Resour 2011;11(5):759–69.

[32] Liu L, et al. Comparison of next-generation sequencing systems. J Biomed Biotechnol 2012;2012:251364.

[33] Gillet-Markowska A, et al. Ulysses: accurate detection of low-frequency structural variations in large insert-size sequencing libraries. Bioinformatics 2014.

[34] Sima J, Gilbert DM. Complex correlations: replication timing and mutational landscapes during cancer and genome evolution. Curr Opin Genet Dev 2014;25:93–100.

[35] Vogelstein B, et al. Cancer genome landscapes. Science 2013;339(6127):1546–58.

[36] Sehn JK, Abel HJ, Duncavage EJ. Copy number variants in clinical next-generation sequencing data can define the relationship between simultaneous tumors in an individual patient. Exp Mol Pathol 2014;97(1):69–73.

[37] Hehlmann R, et al. Chronic myeloid leukaemia. Lancet 2007;370(9584):342–50.

[38] Santos FP, et al. Evolution of therapies for chronic myelogenous leukemia. Cancer J 2011;17(6):465–76.

[39] Branford S, et al. Early molecular response and female sex strongly predict stable undetectable BCR-ABL1, the criteria for imatinib discontinuation in patients with CML. Blood 2013;121(19):3818–24.

[40] Ross DM, et al. Safety and efficacy of imatinib cessation for CML patients with stable undetectable minimal residual disease: results from the TWISTER study. Blood 2013;122(4):515–22.

[41] Ellery P, Forbes M, Gerrard G, Aitman T, Kasperaviciute D, Milojkovic D, et al. NGS-assisted DNA-based digital qPCR facilitates stratification of CML patients in long-term molecular remission based on the presence of detectable BCR-ABL1 DNA. Blood 2013;122:4006.

[42] Stransky N, et al. The landscape of kinase fusions in cancer. Nat Commun 2014;5:4846.

[43] Braggio E, et al. Lessons from next-generation sequencing analysis in hematological malignancies. Blood Cancer J 2013;3:e127.

[44] Heitzer E, et al. Complex tumor genomes inferred from single circulating tumor cells by array-CGH and next-generation sequencing. Cancer Res 2013;73(10):2965–75.

[45] Ni X, et al. Reproducible copy number variation patterns among single circulating tumor cells of lung cancer patients. Proc Natl Acad Sci USA 2013;110(52):21083–8.

[46] Dawson SJ, et al. Analysis of circulating tumor DNA to monitor metastatic breast cancer. N Engl J Med 2013;368(13):1199–209.

[47] Dago AE, et al. Rapid phenotypic and genomic change in response to therapeutic pressure in prostate cancer inferred by high content analysis of single circulating tumor cells. PLoS One 2014;9(8):e101777.

[48] Heitzer E, Auer M, Ulz P, Geigl JB, Speicher MR. Circulating tumor cells and DNA as liquid biopsies. Genome Med 2013;5(73).

[49] Heitzer E, et al. Tumor-associated copy number changes in the circulation of patients with prostate cancer identified through whole-genome sequencing. Genome Med 2013;5(4):30.

[50] Navin N, et al. Tumour evolution inferred by single-cell sequencing. Nature 2011;472(7341):90–4.

[51] Barteneva NS, et al. Circulating microparticles: square the circle. BMC Cell Biol 2013;14:23.

[52] Zen K, Zhang CY. Circulating microRNAs: a novel class of biomarkers to diagnose and monitor human cancers. Med Res Rev 2012;32(2):326–48.

[53] Jarry J, et al. The validity of circulating microRNAs in oncology: five years of challenges and contradictions. Mol Oncol 2014;8(4):819–29.

[54] McLeod HL. Cancer pharmacogenomics: early promise, but concerted effort needed. Science 2013;339(6127):1563–6.

[55] Layer RM, et al. LUMPY: a probabilistic framework for structural variant discovery. Genome Biol 2014;15(6):R84.

[56] Nord AS, et al. Accurate and exact CNV identification from targeted high-throughput sequence data. BMC Genomics 2011;12:184.

[57] Dobin A, et al. STAR: ultrafast universal RNA-seq aligner. Bioinformatics 2013;29(1):15–21.

[58] Abel HJ, et al. Detection of gene rearrangements in targeted clinical next-generation sequencing. J Mol Diagn 2014;16(4):405–17.

[59] Mimori T, et al. iSVP: an integrated structural variant calling pipeline from high-throughput sequencing data. BMC Syst Biol 2013;7(Suppl. 6):S8.

[60] Landman SR, et al. SHEAR: sample heterogeneity estimation and assembly by reference. BMC Genomics 2014;15:84.

[61] Helman E, et al. Somatic retrotransposition in human cancer revealed by whole-genome and exome sequencing. Genome Res 2014;24(7):1053–63.

[62] Chen K, et al. BreakDancer: an algorithm for high-resolution mapping of genomic structural variation. Nat Methods 2009;6(9):677–81.

[63] Hormozdiari F, et al. Next-generation VariationHunter: combinatorial algorithms for transposon insertion discovery. Bioinformatics 2010;26(12):i350–7.

[64] Ye K, et al. Pindel: a pattern growth approach to detect break points of large deletions and medium sized insertions from paired-end short reads. Bioinformatics 2009;25(21):2865–71.

[65] Hayes M, Pyon YS, Li J. A model-based clustering method for genomic structural variant prediction and genotyping using paired-end sequencing data. PLoS One 2012;7(12):e52881.

[66] Li S, et al. SOAPindel: efficient identification of indels from short paired reads. Genome Res 2013;23(1):195–200.

[67] Wong K, et al. Enhanced structural variant and breakpoint detection using SVMerge by integration of multiple detection methods and local assembly. Genome Biol 2010;11(12):R128.

[68] Chu C, Zhang J, Wu Y. GINDEL: accurate genotype calling of insertions and deletions from low coverage population sequence reads. PLoS One 2014;9(11):e113324.

[69] Lee S, et al. MoDIL: detecting small indels from clone-end sequencing with mixtures of distributions. Nat Methods 2009;6(7):473–4.

[70] Hormozdiari F, et al. Combinatorial algorithms for structural variation detection in high-throughput sequenced genomes. Genome Res 2009;19(7):1270–8.

[71] Sindi S, et al. A geometric approach for classification and comparison of structural variants. Bioinformatics 2009;25(12):i222–30.

[72] Sindi SS, et al. An integrative probabilistic model for identification of structural variation in sequencing data. Genome Biol 2012;13(3):R22.

[73] Korbel JO, et al. PEMer: a computational framework with simulation-based error models for inferring genomic structural variants from massive paired-end sequencing data. Genome Biol 2009;10(2):R23.

[74] Yang R, et al. Integrated analysis of whole-genome paired-end and mate-pair sequencing data for identifying genomic structural variations in multiple myeloma. Cancer Inf 2014;13(Suppl. 2):49–53.

[75] Wang J, et al. CREST maps somatic structural variation in cancer genomes with base-pair resolution. Nat Methods 2011;8(8):652–4.

[76] McKenna A, et al. The Genome Analysis Toolkit: a MapReduce framework for analyzing next-generation DNA sequencing data. Genome Res 2010;20(9):1297–303.

[77] Agrawal N, Akbani R, Aksoy BA, Ally A, Arachchi H, Asa SL, et al. Integrated genomic characterization of papillary thyroid carcinoma. Cell 2014;159:676–90.

[78] Kastner R, et al. Rapid identification of compound mutations in patients with Philadelphia-positive leukaemias by long-range next generation sequencing. Eur J Cancer 2014;50(4):793–800.

[79] Pabinger S, et al. A survey of tools for variant analysis of next-generation genome sequencing data. Brief Bioinform 2014;15(2):256–78.

[80] Duan J, et al. Comparative studies of copy number variation detection methods for next-generation sequencing technologies. PLoS One 2013;8(3):e59128.

[81] Samarakoon PS, et al. Identification of copy number variants from exome sequence data. BMC Genomics 2014;15:661.

[82] Tan R, et al. An evaluation of copy number variation detection tools from whole-exome sequencing data. Hum Mutat 2014;35(7):899–907.

[83] Bignell GR, et al. Signatures of mutation and selection in the cancer genome. Nature 2010;463(7283):893–8.

[84] Solimini NL, Xu Q, Mermel CH, Liang AC, Schlabach MR, Luo J, et al. Recurrent hemizygous deletions in cancers may optimize proliferative potential. Science 2012;337:104–9.

[85] Mauro JA, et al. Copy number loss or silencing of apoptosis-effector genes in cancer. Gene 2014.

[86] Bastida-Lertxundi N, et al. Errors in the interpretation of copy number variations due to the use of public databases as a reference. Cancer Genet 2014;207(4):164–7.

[87] Xie C, Tammi MT. CNV-seq, a new method to detect copy number variation using high-throughput sequencing. BMC Bioinformatics 2009;10:80.

[88] Magi A, et al. Read count approach for DNA copy number variants detection. Bioinformatics 2012;28(4):470–8.

[89] Miller CA, et al. ReadDepth: a parallel R package for detecting copy number alterations from short sequencing reads. PLoS One 2011;6(1):e16327.

[90] Szatkiewicz JP, et al. Improving detection of copy-number variation by simultaneous bias correction and read-depth segmentation. Nucleic Acids Res 2013;41(3):1519–32.

[91] Wang H, Nettleton D, Ying K. Copy number variation detection using next generation sequencing ead counts. BMC Bioinformatics 2014;15:109.

[92] Bellos E, Coin LJ. cnvOffSeq: detecting intergenic copy number variation using off-target exome sequencing data. Bioinformatics 2014;30(17):i639–45.

[93] Boeva V, et al. Control-free calling of copy number alterations in deep-sequencing data using GC-content normalization. Bioinformatics 2011;27(2):268–9.

[94] Amarasinghe KC, et al. Inferring copy number and genotype in tumour exome data. BMC Genomics 2014;15:732.

[95] Yoon S, et al. Sensitive and accurate detection of copy number variants using read depth of coverage. Genome Res 2009;19(9):1586–92.

[96] Chiang DY, et al. High-resolution mapping of copy-number alterations with massively parallel sequencing. Nat Methods 2009;6(1):99–103.

[97] Wu HT, Hajirasouliha I, Raphael BJ. Detecting independent and recurrent copy number aberrations using interval graphs. Bioinformatics 2014;30(12):i195–203.

[98] Sathirapongsasuti JF, et al. Exome sequencing-based copy-number variation and loss of heterozygosity detection: ExomeCNV. Bioinformatics 2011;27(19):2648–54.

[99] Li J, et al. CONTRA: copy number analysis for targeted resequencing. Bioinformatics 2012;28(10):1307–13.

[100] Guo Y, et al. Detection of internal exon deletion with exon Del. BMC Bioinformatics 2014;15:332.

[101] Plagnol V, et al. A robust model for read count data in exome sequencing experiments and implications for copy number variant calling. Bioinformatics 2012;28(21):2747–54.

[102] Amarasinghe KC, Li J, Halgamuge SK. CoNVEX: copy number variation estimation in exome sequencing data using HMM. BMC Bioinformatics 2013;14(Suppl. 2):S2.

[103] Magi A, et al. EXCAVATOR: detecting copy number variants from whole-exome sequencing data. Genome Biol 2013;14(10):R120.

[104] Guo Y, et al. Comparative study of exome copy number variation estimation tools using array comparative genomic hybridization as control. Biomed Res Int 2013;2013:915636.

[105] Krumm N, et al. Copy number variation detection and genotyping from exome sequence data. Genome Res 2012;22(8):1525–32.

[106] Klambauer G, et al. cn.MOPS: mixture of Poissons for discovering copy number variations in next-generation sequencing data with a low false discovery rate. Nucleic Acids Res 2012;40(9):e69.

[107] Love MI, et al. Modeling read counts for CNV detection in exome sequencing data. Stat Appl Genet Mol Biol 2011;10(1).

[108] Boeva V, et al. Control-FREEC: a tool for assessing copy number and allelic content using next-generation sequencing data. Bioinformatics 2012;28(3):423–5.

[109] Malhotra A, et al. Chromosomal structural variations during progression of a prostate epithelial cell line to a malignant metastatic state inactivate the NF2, NIPSNAP1, UGT2B17, and LPIN2 genes. Cancer Biol Ther 2013;14(9):840–52.

[110] Foundation, F.S. Free software definition. 2014. Available from: http://www.gnu.org/philosophy/free-sw.html.

[111] Wikipedia. Free software license. 2014. Available from: http://en.wikipedia.org/wiki/Free_software_license.

[112] Beroukhim R, et al. The landscape of somatic copy-number alteration across human cancers. Nature 2010;463(7283):899–905.

[113] Fudenberg G, et al. High order chromatin architecture shapes the landscape of chromosomal alterations in cancer. Nat Biotech 2011;29(12):1109–13.

[114] Malhotra A, et al. Breakpoint profiling of 64 cancer genomes reveals numerous complex rearrangements spawned by homology-independent mechanisms. Genome Res 2013;23(5):762–76.

[115] Dixon JR, et al. Topological domains in mammalian genomes identified by analysis of chromatin interactions. Nature 2012;485(7398):376–80.

[116] Nagano T, et al. Single-cell Hi-C reveals cell-to-cell variability in chromosome structure. Nature 2013;502(7469):59–64.

[117] Koren S, et al. Hybrid error correction and de novo assembly of single-molecule sequencing reads. Nat Biotechnol 2012;30(7):693–700.

[118] Miyamoto M, et al. Performance comparison of second- and third-generation sequencers using a bacterial genome with two chromosomes. BMC Genomics 2014;15:699.

NEXT GENERATION SEQUENCING IN ONCOLOGY

4

Eliza Glodkowska-Mrowka[1], Tomasz Stoklosa[2]

[1]Department of Laboratory Diagnostics and Clinical Immunology of Developmental Age, Medical University of Warsaw, Warsaw, Poland; [2]Department of Immunology, Medical University of Warsaw, Warsaw, Poland

CHAPTER OUTLINE

Why are you torturing us with all those mutations?

*A piece of a reviewers' remark on the manuscript published finally in **Nature** in 2008 describing for the first time the whole genome of an acute myeloid leukemia patient according to Professor Tim Ley, pioneer in next generation sequencing and author of this landmark study.*

The effective treatment of cancer represents one of the most important challenges for human society. To develop new therapies we have to better understand the pathogenesis of cancer, but this useful term actually describes an extremely complex and heterogeneous group of diseases. Cancer is characterized by genetic and epigenetic changes that are acquired by a single cell (or a group of cells) and that lead to clonal expansion and uncontrolled growth. The description of an atypical minute chromosome, later called the Philadelphia chromosome, in chronic myeloid leukemia by P. Nowell and D. Hungerford in 1960 is considered to be the first report showing a causative link between a genetic abnormality and human cancer [1]. Therefore, the genetic characterization of cancer became one of the major goals in the long-lasting war on cancer—to paraphrase the famous U.S. National Cancer Act signed by President Richard Nixon in 1971. Unfortunately, for many decades it was not feasible to decipher the cancer genome with the available technology and the discovery of the Philadelphia chromosome's role in chronic myeloid leukemia pathogenesis was rather an exception than a rule in our knowledge about the genetic basis of cancer. The twenty-first century brought unprecedented progress in the field of cancer genetics. The first human cancer genome that was sequenced in 2008 with a next generation sequencing (NGS) approach was for a cytogenetically normal acute myeloid leukemia. It was accomplished by T. Ley's group from Washington University in St. Louis, Missouri, USA [2]. High initial costs of NGS (approximately US $1 million per genome) were responsible for initial skepticism about the potential use of this technology in cancer diagnostics. It was rather obvious that at least for the next few years the technique would be used only in cancer research and would not be available in routine clinical settings. Thanks to extremely rapid advances in technology and a significant drop in the cost of NGS, it is more and more often considered as a powerful tool for cancer diagnosis and monitoring in clinical settings.

The introduction of so-called targeted therapies and the idea of personalized medicine have definitively changed our views on cancer diagnostics and treatment. Although there are opinions that chemotherapy and radiotherapy should also be called "targeted," as both actually target cancer cells, most experts agree that the era of personalized oncology began in the mid-1990s with the introduction of anticancer monoclonal antibodies. In 1997 and 1998 two monoclonal antibodies, rituximab (anti-CD20) and trastuzumab (anti-HER2), were approved by regulatory agencies in United States and Europe to treat non-Hodgkin lymphoma and breast cancer patients, respectively. This was accompanied by the introduction of companion diagnostic tests for the identification of patients eligible for the treatment [3]. Since then a number of targeted therapies have been registered and many new molecules are being tested in clinical trials. All of them require detailed testing to check patients' eligibility for the treatment (companion diagnostics). Therefore molecular/genetic testing plays a key role in diagnosis and classification of cancer patients as well as in personalized medicine. Although the genome-wide analysis of cancer samples encounters many difficulties, NGS can provide a more accurate picture of cancer as a somatic genetic disease than any other method used to date.

Comprehensive profiling of a large number of tumor specimens using high-throughput technologies became the goal of several projects such as The Cancer Genome Atlas (TCGA), International Cancer Genome Consortium (ICGC), and Pediatric Cancer Genome Project (PCGP). These projects aim at large-scale genomic analysis of multiple adult and pediatric cancer samples to identify somatic mutations that drive cancer. These cancer sequencing efforts not only yield an unparalleled insight into the altered signaling pathways in cancer but also help identify new therapeutic targets as well as new biomarkers useful in cancer diagnosis, monitoring, and prognosis [4–7]. High-throughput genetic testing allowed us to better understand tumor biology and created new opportunities to use the new techniques in clinics. This chapter summarizes the current state of knowledge regarding the use of NGS in cancer research as well as the rapidly developing use of this technique in clinical oncology.

NGS IN CANCER RESEARCH

The identification of new targets for personalized therapy requires comprehensive understanding of mechanisms underlying carcinogenesis and disease progression. High-throughput sequencing techniques made possible extensive studies on cancer development, evolution, and formation of metastases as well as response to the treatment. Cancer genomes may contain a broad spectrum of aberrations, from point mutations, through insertions/deletions, up to genomic rearrangements and viral-genome insertions. All these mutations can be detected by deep sequencing of DNA from cancer cells. Although more and more genomic data on tumor biology area being published, we are just beginning to understand how to use these data to understand the mechanisms of oncogenesis and employ the knowledge in clinical settings.

NGS techniques are widely used for research purposes to uncover driver mutations in cancer and study genomic instability, tumor evolution, and cancer epigenetics as well as to discover new therapeutic targets. Moreover, novel techniques have made it possible to study cancer with single-cell resolution to avoid the admixture of normal cell genomes and thoroughly study cancer evolution.

As a result of extensive research efforts in oncogenomics, a vast number of publications cataloging genomic disruptions in cancer and their possible effects on cancer cell biology have been published. As cancer is a complex and heterogeneous somatic genetic disease, it is impossible to extensively describe the full spectrum of cancer-associated mutations in a concise form (e.g., publication or book chapter). The only possibility to gather all the available knowledge in one place is to organize the data in the form of a computerized database. There are several comprehensive databases, established in consequence of the avalanche of data from high-throughput genomic analyses, summarizing our knowledge on cancer-associated somatic mutations derived from both publications available in the scientific literature and large-scale experimental screens such as TCGA and PCGP.

The largest databases include COSMIC (Catalogue of Somatic Mutations in Cancer), held by the Sanger Institute, and a database provided by the ICGC. These databases serve as an unparalleled source of information on known mutations and might be invaluable help in the analysis of NGS data in clinical settings. Worldwide efforts to catalog mutations in multiple cancer types are under way and this is likely to lead to new discoveries that will be translated to new diagnostic, prognostic, and therapeutic targets.

In addition to databases focused on indexing all described mutations in cancer, there are also databases gathering information about genetic testing, for example, The National Institutes of Health Genetic Testing Registry (GTR; available online at http://www.ncbi.nlm.nih.gov/gtr/). The GTR database maintains comprehensive information about testing offered worldwide for disorders with a genetic basis, including genetic tests employed in the diagnosis and treatment of cancer. The database provides details of each test (e.g., its purpose, target populations, and methods; what it measures; analytical validity; clinical validity; clinical utility; ordering information) and laboratory (e.g., location, contact information, certifications, and licenses) [8].

NGS IN CLINICAL SETTINGS

Even though as of this writing there is no U.S. Food and Drug Administration (FDA)-approved NGS-based test to diagnose cancer, several laboratories, both public and private, around the world offer NGS-based cancer testing for patients. Reported applications of NGS in clinical practice include familial genetic testing for cancer susceptibility such as the *BRCA1* gene, which should allow us to overcome the limitations of current *BRCA1* screening (missing rare/new mutations). Another possible application is pharmacogenomic tests. This is illustrated by determining the mutational status of selected genes to recruit patients for molecularly targeted treatment (e.g., *KRAS* or *BRAF* mutations for anti-epidermal growth factor receptor (EGFR) or anti-BRAF^{V600E} therapies, respectively), determining susceptibility to drug toxicity, assessment of prognosis, and prediction of resistance, as well as the study of mosaicism. Another potential use for NGS in oncology concerns patients who have exhausted all possible options to identify putative genes that could serve as druggable targets. Below we try to summarize all these approaches.

FAMILIAL GENETIC TESTING

Hereditary cancer syndromes account for approximately 5–10% of all cancer diagnoses. Currently, mutations in several genes that increase the risk for developing several types of cancer have been described; however, we have not yet identified the genetic causes of all types of familial predisposition for cancer. In the classical approach based on Sanger (capillary) sequencing an individual with a family history of cancer is tested only for the genes that are known to be linked with hereditary cancer syndrome, starting with the most likely gene. If no mutation is found, additional genes may be taken into consideration. The advantage of NGS over the classical approach is obvious, as the use of massive parallel sequencing allows the evaluation of multiple genes simultaneously, including genes not linked to a particular type of cancer before.

Usually the testing offered for patients at high risk of hereditary cancer syndrome is based on gene panels rather than whole-genome or whole-exome sequencing (WGS or WES). The first germ-line NGS panels introduced into the diagnostics of cardiomyopathies proved that the method is cost-effective and not only allows the detection of many known mutations but also comprises novel genes involved in the pathogenesis of hereditary syndromes. Several manufacturers have released commercial panels for hereditary cancer syndromes. However, the offered tests include not only highly penetrant genes that increase the risk of cancer by 18–20 times, but also low or moderately penetrant genes. The detection of mutations in those moderately penetrant genes might be challenging for clinicians to interpret, as there are no established clinical management guidelines. Moreover, it is also unclear how to determine who should be offered NGS panel testing and who will pay for the test (reimbursement issues).

According to National Comprehensive Cancer Network, NGS-based testing should be offered as a second-tier test for individuals at high risk of hereditary cancer syndrome. Therefore, these tests should be offered only to the selected group of patients and analyzed by experienced clinicians who are able to interpret the data according to the latest knowledge. As of this writing, there are no clear recommendations on the choice of patients for NGS testing for hereditary mutations predisposing to cancer. The first attempt to establish clinical guidelines for identifying high-risk patients who should be offered NGS testing was made by Caitlin Mauer et al. [9]. They proposed clinical criteria for ordering NGS hereditary cancer panels based on the patient's history and negative results of standard tests for highly penetrant and well-known genetic mutations.

MOLECULAR DISEASE CLASSIFICATION

The classical approach to cancer classification was based on the tissue of origin. Therefore, we distinguished cancers of various organs, such as lung cancer, breast cancer, colorectal cancer, etc. Since 2005 we have learned a lot about cancer-driving mutations and cancer genomes. This knowledge has led us to the conclusion that two patients with virtually the same disease, for example, colorectal cancer, may be totally different in terms of driving mutations and drug sensitivity. Hence, there is a need to establish new molecular disease classifications that might be helpful in clinical decision-making and patient care.

It needs to be emphasized that identification of new mutations is only the first step. NGS-based tests must prove useful in improving patient outcomes. However, the clinical significance of novel mutations is initially difficult to determine, even if they occur in well-described pathways involved in carcinogenesis.

PHARMACOGENOMICS

Since the introduction of first targeted therapies, the algorithms for tumor diagnosis necessitate not only standard morphological and/or immunophenotypic assessment of tumor samples but also detailed genotyping and mutation profiling. The treatment of various solid tumors, for example, thyroid carcinoma, lung cancer, melanoma, and colorectal cancer, requires the knowledge of the mutational status in genes important for drug sensitivity/resistance. When our knowledge expands and more and more targeted therapies become available, genomic studies for cancer diagnosis will become necessary for all types of tumors.

Detailed information on the use of NGS in pharmacogenomics has been given in a separate chapter (see Chapter 11, "Next Generation Sequencing in Pharmacogenomics").

TESTING IN "LAST-RESORT" PATIENTS

If a patient has not responded to conventional therapies approved for the treatment of a particular type of cancer, there may be a rationale for performing NGS testing to identify potential druggable targets and potentially find an off-label drug or appropriate clinical trial. Although the benefits from such testing seem to be obvious, interpretation of the results of NGS testing is still fraught with many problems.

Although off-label use of targeted drugs might seem to be the best practice for the patient who did not respond to a few lines of standard treatment, their use has many well-known risks including the risk of serious side effects and inefficacy. As the biology of the majority of cancers is still not well understood, we are not able to confidently transfer the use of a specific drug from one clinical setting to another in the absence of clinical trial results. For example, the presence of a *KRAS* mutation in metastatic colon cancer predicts the inefficacy of anti-EGFR drugs. Although logically *KRAS* mutational status should predict the response to anti-EGFR therapy in early colon cancer, the results of clinical trials showed otherwise [10]. Even though we cannot be sure if the treatment will be able to help those patients, in last-resort situations the use of off-label drugs seems justified. However, to maximize the utility of such an approach, the patients treated with off-label drugs based on NGS results should participate in registries documenting the results of such treatment (e.g., the registry NCT01851213 established by Frampton et al. [11]).

CIRCULATING CELL-FREE TUMOR DNA AND CIRCULATING TUMOR CELLS

Genotyping of tumor tissue for somatic mutations using Sanger sequencing and now also NGS has become routine practice in cancer treatment. Although sequencing the DNA from a tumor sample is the gold standard in clinical and experimental oncology, the method brings many problems in terms of acquisition of the sample and analysis of the material. As an invasive procedure, tumor biopsy is not only expensive for the health care system and inconvenient for the patient, but may also generate clinical complications. The adverse events rate in tumor biopsies is relatively high and ranges from 1.6% for abdominal samples to even 17.1% for thoracic samples [12]. The other problems include small tissue amounts, relatively low amounts of tumor cells in each biopsy sample, and sample preservation in formalin-fixed paraffin-embedded blocks, inducing DNA cross-linking and diminishing its quality for molecular analyses. In addition, cancer heterogeneity, including both intratumoral (different genetic profiles within the same tumor) and intrametastatic heterogeneity (different genetic profiles between primary and secondary tumors), poses a major limitation of molecular analysis of tumor biopsies.

The alternatives to tissue biopsy for genetic testing include genotyping of circulating cell-free DNA (cfDNA) or circulating cells (in the case of oncology—circulating tumor cells—CTCs). Such analyses have been used in the investigation of fetal DNA to uncover the presence of germ-line mutations in early pregnancy without the need to perform invasive procedures. More and more studies show the potential utility of these methods in other fields of medicine, such as nephrology [13], cardiology [14], and oncology [15].

Fragmented circulating cfDNA is found in cell-free blood plasma in health and disease, but the amount of cfDNA increases dramatically with cellular injury or necrosis as cfDNA is shed into the blood by apoptotic and necrotic cells. Tumor volume increases parallel to cellular turnover rate and increasing number of apoptotic and necrotic cells. Therefore, the load of circulating tumor DNA (ctDNA), which constitutes a significant part of cfDNA in cancer patient blood, is correlated with tumor staging and prognosis and might be useful to assess tumor dynamics [15]. Distinguishing ctDNA from cfDNA and from normal tissues is possible owing to the presence of mutations in tumor-derived DNA; however, detection of ctDNA might be challenging because of insufficient representation of ctDNA in total cfDNA. The ctDNA content in total cfDNA varies greatly, from 0.01% up to 90% and depends on the stage of the disease [15]. Therefore, in the majority of cases standard genotyping techniques such as Sanger sequencing are not sensitive enough to detect ctDNA, especially in patients with a low tumor burden.

The advent of NGS techniques made it possible to use ctDNA genotyping in clinical settings. Theoretically, ctDNA genotyping provides the same genetic information as a tissue biopsy and may be useful in personalized medicine [16]. Blood sampling is minimally invasive and is a source of fresh DNA of good quality for molecular analyses. Considering that ctDNA contains the same genetic defects as tumor tissue, ctDNA sampling is often compared to standard tissue biopsy and called liquid biopsy.

There are many advantages to liquid biopsy over standard tissue biopsy. The blood sample can be easily drawn multiple times during the course of the disease/therapy, allowing dynamic monitoring of genetic changes in the tumor instead of static information obtained from single-tissue biopsy [17]. Moreover, ctDNA in a plasma sample consists of DNA fragments collected from all tumors in a patient (primary and metastatic tumors), so the result of genotyping represents the molecular heterogeneity of the disease.

Both CTCs and ctDNA might be used as a source of DNA for genotyping in clinical settings. However, there are several important differences between CTCs and ctDNA as a sample for clinical diagnostics. First, ctDNA analysis allows one to detect only DNA alterations (point mutations, insertions/deletions, amplifications, rearrangements), not changes in cancer morphology or protein expression possible to assess in CTCs. Moreover, wild-type cancers are undetectable in ctDNA genotyping as their genome is synonymous with healthy tissues and it is not possible to distinguish between normal tissue-derived and cancer-derived DNA. On the other hand, ctDNA is easier from a technical point of view, as it does not require special handling such as separation and purification of viable cancer cells before DNA isolation. Therefore, ctDNA has a better potential to be widely used as a method for liquid biopsy sampling compared to CTC-based methods [18]. It is noteworthy that CTC-derived DNA and ctDNA cannot be treated as the same material for biomarker assessment, as ctDNA is often present in the sera of patients without detectable CTCs [19].

Liquid biopsies might be used in various applications, including companion diagnostics for qualification for personalized therapy [20–22], detection of occult or minimal residual disease, monitoring of resistance and tumor heterogeneity [23,24], early prediction of response to the therapy, and assessing and monitoring of tumor burden [25,26]. The last use is supported by the results of several clinical trials performed in patients with ovarian [24,27], breast [24,28], lung [24], and colorectal [28] cancer. It was confirmed that rapid increases in ctDNA levels correspond to disease progression, while declines in ctDNA levels are correlated with successful pharmacological or surgical treatment.

The main disadvantage of liquid biopsies is their relatively low sensitivity. In patients with advanced stages of the disease, sensitivity of ctDNA genotyping is almost 100%. However, the sensitivity of ctDNA measurements in early stage tumors decreases significantly and it is not possible to detect ctDNA if its content drops below 0.01%. On the other hand, the short half-life of ctDNA (approximately 2 h) makes it a perfect biomarker for tumor burden changes monitored in hours rather than weeks or months after clinical intervention. Moreover, as mutations in ctDNA are specific for the individual's tumor, there are fewer false-positive results observed for other circulating biomarkers or image diagnostics [18].

Low amounts of DNA, both ctDNA and DNA isolated from CTCs, in the sample pose the major technical challenge in the use of liquid biopsies in clinical settings. It was suggested that sequential isolation, ex vivo culture, and further characterization of CTCs might be used to facilitate the monitoring of cancer evolution in clinical settings. Published results of a proof-of-concept study performed in six patients with breast cancer confirmed that this approach might be helpful in identifying the best therapies for individual cancer patients over the course of their disease, but further optimization is required [29].

TREATMENT MONITORING

Potential future applications of NGS in clinical oncology involve the development of sensitive assays to detect early relapse of the disease or measurement of disease burden [15]. Moreover, measurement of CTCs or ctDNA with deep-sequencing methods may be used to monitor the risk of relapse. The presence of CTCs and ctDNA in the peripheral blood of cancer patients is a hallmark of metastatic dissemination. Although circulating tumor cells are extremely rare and estimated to account for one cell in a billion cells that are circulating in the blood, data have shown that their molecular characterization is feasible and can be potentially used in clinical settings to provide real-time information on biomarker status [19]. By sequencing the tumor genome of a patient, clinicians are able to design patient-specific probes that might be further employed to monitor the progress of a patient's treatment and provide early detection of any signs of relapse or progression.

In addition, the assessment of ctDNA content in cancer patient blood samples has important clinical applications, enabling a noninvasive approach to assess the mutation status at various stages of the treatment, for example, at the time of tumor recurrence. The NGS-based approach offers a relatively safe and noninvasive method of cancer monitoring in comparison to standard methods such as tumor biopsy or biopsy of metastatic lesions. ctDNA analysis might become a good alternative to repetitive tissue sampling necessary for systematic identification of therapeutic targets and drug resistance by serial testing of blood during the course of treatment.

BIOMARKERS IN CLINICAL DECISION-MAKING

More and more NGS data point toward the discovery of new cancer biomarkers. These biomarkers can potentially be applied as predictive, prognostic, and pharmacogenomic biomarkers helping in clinical decision-making regarding the type of therapy, dosage, and patient eligibility for the treatment. However, the number of putative biomarkers confirmed to be useful in clinical trials is relatively low.

It is important to emphasize that not all biomarkers and mutations are clinically relevant and can be implemented in the clinics. For predictive markers that are used in companion diagnostics, that is, the markers used for patient stratification for certain targeted drug or other intervention, the usefulness of a particular biomarker has been sufficiently proven in clinical trials. For example, the IPASS trial has demonstrated that patients with EGFR mutation had longer progression-free survival if taking gefitinib in addition to chemotherapy [30]. These results make us sure that EGFR can be adopted as a marker of treatment suitability for gefitinib.

The enormous amount of cancer genomes published each year have led to the suggestion that many new biomarkers can be important in cancer diagnosis and treatment. However, frequently novel technologies develop new biomarkers that show great promise but fail in clinical trials and are insufficient for clinical decision-making. Currently, the testing of a broad range of markers is suitable only for clinical trials and not for clinical practice.

CURRENT OBSTACLES PREVENTING NGS FROM WIDE APPLICATION IN CLINICAL SETTINGS

As we discussed above, it is quite obvious now that for the majority of tumors single mutations in individual genes are not sufficient for the development of a malignant phenotype. Oncogenic transformation requires many events leading to the accumulation of mutations in oncogenes, tumor suppressors, and genes involved in cell cycle control, signaling, and metabolism. Therefore, considering the broad heterogeneity of cancer, clinical genetic analysis of malignancies requires sequencing of a large number of genes. This task could be best addressed with a massively parallel sequencing approach. How is it possible that, despite thousands of cancer genomes sequenced for research purposes, we still have limited access to NGS in the clinics?

Although high-throughput sequencing methods provide a more accurate picture of cancer than any other method used so far, detailed analysis of NGS data on cancer samples is difficult and time-consuming. NGS techniques have already proven their worth in cancer research; however, their application in clinical and routine laboratory practice is still in its early stages.

There are two groups of problems limiting the wide use of NGS methods in clinical settings. The first problem is reimbursement. Insurance providers are rigorously focused on only those clinical tests that provide information directly influencing the patient's care. In the absence of evidence-based data confirming that a particular variant or test is linked with a treatment improving patient outcome, most probably the test will not be reimbursed. Moreover, even though there is more and more evidence that the NGS approach may improve clinical outcomes [31–33], deep economic analyses of NGS testing in oncology have not been performed yet.

The second problem is related to regulatory issues. Personalized medicine requires the knowledge of individual genetic characteristics to tailor targeted therapy. Knowing that cancer is a genetically complex and heterogeneous disease, it seems obvious that personalized treatment requires the knowledge of the whole cancer exome and transcriptome. However, regulatory agencies register new targeted therapies together with companion diagnostic tests based on the rule "one drug—one gene." According to the Genetic Testing Registry database, as of September 2014 approximately 50 laboratories around the world offered more than 3000 laboratory-developed clinical tests utilizing NGS methodology. Nevertheless, there is still no FDA-approved NGS-based diagnostic test to be used in oncology. The situation is even more complicated as only a single NGS instrument received CE/IVD (Certified for In Vitro Diagnostic) marking to be used in clinical diagnostics [34].

TARGETED VERSUS GENOME-WIDE APPROACH

NGS enthusiasts believe that the potential of the technique is so broad that genome-wide analyses will be soon introduced into routine clinical practice. However, bearing in mind the problems with analysis of large data sets and our lack of knowledge on the clinical relevance of the majority of genetic abnormalities, it is rather unlikely that genome-wide analyses will be introduced into clinical practice in the near future. Nevertheless, this does not mean that NGS techniques are not useful in clinical settings. On the contrary, the targeted approach, limited to a few hundred genes of known clinical significance, seems to be rapidly developing and more and more tests are becoming available.

Someone may ask, why are we still considering sequencing of relatively small panels of genes instead of WGS or WES? The cost of sequencing per base pair and the cost of NGS platforms will continue to drop, while bioinformatics tools and library preparation will become automated and easier to use. There is no doubt that even now the cost of exome sequencing of a tumor is only slightly higher than the cost of sequencing of a panel of several hundred genes. So why should we bother using a targeted instead of a genome-wide approach? It is mainly because of problems with data analysis and reporting of detected variants. Clinical reporting requires careful interpretation of the discovered variants. At the level of a single clinical laboratory, it is almost impossible to ensure that the data are evidence-based and up to date, so it is easier and safer to use a targeted approach.

The main drawback of the disease-targeted testing approach, however, is that it requires not only validation of a new targeted panel for each disease but also constant updates of existing panels with newly identified genes. As it generates incremental costs for laboratories, whole or clinically relevant exome sequencing seems more cost-effective. To avoid the problems with interpretation of data, the laboratory might perform WES and analyze only the genes that are known to be relevant for the patient's phenotype. In such case it is not only much easier to develop a new targeted panel or update the previous one, but it is also possible to reanalyze the data obtained earlier including newly described variants [8]. To support this approach, a few commercial panels covering clinically relevant genes are available, for example, TruSight Exome by Illumina (examples of such panels are described in Chapter 11, "Next Generation Sequencing in Pharmacogenomics").

FUTURE OF ONCOGENOMICS

The massively parallel sequencing approach is currently revolutionizing the diagnostics of cancer and is directly involved in therapeutic advances with targeted therapy in various cancers. Even though NGS technologies continue to evolve and new possibilities are forthcoming, the future of oncogenomics will be full of challenges. In the era of NGS, the question if there is any place left for traditional sequence analysis, such as Sanger sequencing or PCR-based genotyping methods, will be answered soon.

PROS AND CONS OF TRADITIONAL VERSUS NGS SEQUENCING APPROACH

The low sequencing capacity of Sanger sequencing limits the usage of this method to single-gene or hot-spot analysis. Significant costs and difficulties in the application of this technique to multiple gene analysis are major drawbacks for its use in oncogenomics. On the other hand, although the cost of NGS per base is extremely low in comparison to Sanger sequencing, the high baseline cost of running a single NGS test makes this approach profitable only if the test involves a large amount of sequences. Because of that, Sanger sequencing is still more cost-effective in the case of tests involving fewer than 5–10 genes, and there are barely any NGS-based tests involving fewer than 10 genes. However, it seems natural that with decreasing cost and hassle of the analysis the majority of Sanger-based tests will be replaced with NGS-based analyses [8].

CHALLENGES IN CLINICAL ONCOGENOMICS

One of the major problems with the application of NGS in cancer diagnostics is the availability of samples or small sample size. The biology of cancer itself makes the analysis complicated mainly because of the limited amount of DNA, problems with paired tumor–normal tissue comparisons as well as intratumor clonal heterogeneity, and also poor quality of formalin-fixed paraffin-embedded (FFPE) samples. Some of these obstacles can be overcome by NGS as it can be performed on an array of tumor samples, including FFPE samples, and is applicable to small specimens, fine-needle aspiration samples, circulating tumor cells, and circulating free DNA in plasma.

Other issues that need to be resolved include regulatory status (as we mentioned before, there is no FDA-approved NGS-based test for cancer testing) and the still relatively high costs of analysis. Last but not the least are ethical issues (discussed in Chapter 17, "Next Generation Sequencing - Ethical and Social Issues"), which include not only the appropriate information for patients regarding the complexity of genome-wide testing but also a policy for whether and how to inform patients about incidental findings.

REFERENCES

[1] Nowell PC, Hungerford DA. Minute chromosome in human chronic granulocytic leukemia. Science 1960;132(3438):1497.
[2] Ley TJ, et al. DNA sequencing of a cytogenetically normal acute myeloid leukaemia genome. Nature 2008;456(7218):66–72.
[3] Slamon DJ, et al. Use of chemotherapy plus a monoclonal antibody against HER2 for metastatic breast cancer that overexpresses HER2. N Engl J Med 2001;344(11):783–92.

[4] Cancer Genome Atlas, N. Comprehensive molecular portraits of human breast tumours. Nature 2012;490(7418):61–70.

[5] Cancer Genome Atlas Research, N. Comprehensive molecular profiling of lung adenocarcinoma. Nature 2014;511(7511):543–50.

[6] Cancer Genome Atlas Research, N. Comprehensive genomic characterization of squamous cell lung cancers. Nature 2012;489(7417):519–25.

[7] Cancer Genome Atlas Research, N. Comprehensive molecular characterization of clear cell renal cell carcinoma. Nature 2013;499(7456):43–9.

[8] Rehm HL. Disease-targeted sequencing: a cornerstone in the clinic. Nat Rev Genet 2013;14(4):295–300.

[9] Mauer CB, et al. The integration of next-generation sequencing panels in the clinical cancer genetics practice: an institutional experience. Genet Med 2014;16(5):407–12.

[10] Alberts SR, et al. Effect of oxaliplatin, fluorouracil, and leucovorin with or without cetuximab on survival among patients with resected stage III colon cancer: a randomized trial. JAMA 2012;307(13):1383–93.

[11] Frampton GM, et al. Development and validation of a clinical cancer genomic profiling test based on massively parallel DNA sequencing. Nat Biotechnol 2013;31(11):1023–31.

[12] Overman MJ, et al. Use of research biopsies in clinical trials: are risks and benefits adequately discussed? J Clin Oncol 2013;31(1):17–22.

[13] Tovbin D, et al. Circulating cell-free DNA in hemodialysis patients predicts mortality. Nephrol Dial Transplant 2012;27(10):3929–35.

[14] Jing RR, et al. A sensitive method to quantify human cell-free circulating DNA in blood: relevance to myocardial infarction screening. Clin Biochem 2011;44(13):1074–9.

[15] Diehl F, et al. Circulating mutant DNA to assess tumor dynamics. Nat Med 2008;14(9):985–90.

[16] Rothe F, et al. Plasma circulating tumor DNA as an alternative to metastatic biopsies for mutational analysis in breast cancer. Ann Oncol 2014;25(10):1959–65.

[17] Heidary M, et al. The dynamic range of circulating tumor DNA in metastatic breast cancer. Breast Cancer Res 2014;16(4):421.

[18] Luke JJ, et al. Realizing the potential of plasma genotyping in an age of genotype-directed therapies. J Natl Cancer Inst 2014;106(8).

[19] Bettegowda C, et al. Detection of circulating tumor DNA in early- and late-stage human malignancies. Sci Transl Med 2014;6(224):224ra24.

[20] Nannini M, et al. Liquid biopsy in gastrointestinal stromal tumors: a novel approach. J Transl Med 2014;12(1):210.

[21] Douillard JY, et al. Gefitinib treatment in EGFR mutated Caucasian NSCLC: circulating-free tumor DNA as a surrogate for determination of EGFR status. J Thorac Oncol 2014;9(9):1345–53.

[22] Couraud S, et al. Noninvasive diagnosis of actionable mutations by deep sequencing of circulating free DNA in lung cancer from never-smokers: a proof-of-concept study from BioCAST/IFCT-1002. Clin Cancer Res 2014;20(17):4613–24.

[23] Yoo C, et al. Analysis of serum protein biomarkers, circulating tumor DNA, and dovitinib activity in patients with tyrosine kinase inhibitor-refractory gastrointestinal stromal tumors. Ann Oncol 2014;22(11):2272–7.

[24] Murtaza M, et al. Non-invasive analysis of acquired resistance to cancer therapy by sequencing of plasma DNA. Nature 2013;497(7447):108–12.

[25] Klevebring D, et al. Evaluation of exome sequencing to estimate tumor burden in plasma. PLoS One 2014;9(8):e104417.

[26] Spellman PT, Gray JW. Detecting cancer by monitoring circulating tumor DNA. Nat Med 2014;20(5):474–5.

[27] Forshew T, et al. Noninvasive identification and monitoring of cancer mutations by targeted deep sequencing of plasma DNA. Sci Transl Med 2012;4(136):136ra68.

[28] Leary RJ, et al. Detection of chromosomal alterations in the circulation of cancer patients with whole-genome sequencing. Sci Transl Med 2012;4(162):162ra154.

[29] Yu M, et al. Cancer therapy. Ex vivo culture of circulating breast tumor cells for individualized testing of drug susceptibility. Science 2014;345(6193):216–20.

[30] Fukuoka M, et al. Biomarker analyses and final overall survival results from a phase III, randomized, open-label, first-line study of gefitinib versus carboplatin/paclitaxel in clinically selected patients with advanced non-small-cell lung cancer in Asia (IPASS). J Clin Oncol 2011;29(21):2866–74.

[31] Hagemann IS, et al. Stabilization of disease after targeted therapy in a thymic carcinoma with KIT mutation detected by clinical next-generation sequencing. J Thorac Oncol 2014;9(2):E12–6.

[32] Tsimberidou AM, et al. Personalized medicine in a phase I clinical trials program: the MD Anderson Cancer Center initiative. Clin Cancer Res 2012;18(22):6373–83.

[33] Kothari N, et al. Comparison of KRAS mutation analysis of colorectal cancer samples by standard testing and next-generation sequencing. J Clin Pathol 2014;67(9):764–7.

[34] Pant S, Weiner R, Marton MJ. Navigating the rapids: the development of regulated next-generation sequencing-based clinical trial assays and companion diagnostics. Front Oncol 2014;4:78.

NEXT GENERATION SEQUENCING IN HEMATOLOGICAL DISORDERS

5

Iwona Malinowska[1], Eliza Glodkowska-Mrowka[2]

[1]Department of Pediatrics, Hematology and Oncology, Medical University of Warsaw, Poland; [2]Department of Laboratory Diagnostics and Clinical Immunology of Developmental Age, Medical University of Warsaw, Warsaw, Poland

CHAPTER OUTLINE

INTRODUCTION

The past several years have shown a revolution in sequencing technologies, which has led to a better understanding of genetics and genome biology [1]. Next generation sequencing (NGS) allows for a broad analysis of a genome by whole-genome sequencing (WGS), exome sequencing, transcriptome sequencing, and epigenomics. These techniques have given better insight into tumor biology by allowing the identification of genetic variants driving cancer progression. There is great potential for NGS in clinical application, including its use for the efficient detection of either inherited or somatic mutations in cancer genes. There is a hope that this knowledge will shift to clinical utility and improve diagnostics, prognosis, monitoring of minimal residual disease, and the identification of new targets for therapy.

Several large-scale consortia, including the International Cancer Genome Consortium [2], The Cancer Genome Atlas project [3], the Pediatric Cancer Genome Project [4], and the Children's Oncology Group–National Cancer Institute Therapeutically Applicable Research to Generate Effective Treatments (TARGET) project [5] have cataloged thousands of cancer genomes of various neoplastic diseases in children and adults. These discoveries have unmasked novel driver mutations, leading to a better under-standing of the pathogenesis of acute myeloid leukemia (AML), chronic lymphocytic lymphoma (CLL), acute lymphoblastic leukemia (ALL), and non-Hodgkin lymphoma (NHL) and providing a firm footing for further research.

NGS uses several different methodologies for the analysis of genomics, transcriptomics, and epigenomics [6]. These different omics techniques are necessary to define the full landscape of genetic alterations. The predominant clinical use of NGS in patients with hematological malignancies is sequencing cancerous genomic DNA and that of normal tissue. WGS allows for the detection of sin-gle-nucleotide variants, indels (insertion or deletions), complex structural rearrangements, and copy number changes. Sequencing of the whole genome provides a huge amount of data. Some of this genetic information can be interpreted, but a significant amount is still novel or of unknown clinical significance.

Transcriptomics is the study of transcriptome, the complete set of RNA transcripts that are produced by the genome of a specific cell. Comparison of transcriptomes allows the identification of genes that are expressed differentially in distinct cell populations or in response to different treatments.

Epigenomics is a method of analysis of DNA methylation, mapping of transcription factors, modified histones, and epigenetic regulators. The role of epigenetic aberrations in the pathogenesis of neoplastic diseases has been revealed by recurrent gene mutations that highlight epigenetic pathways, as well as by the clinical success of therapies that work through epigenetic mechanisms [6].

A significant amount of further study needs to be undertaken before these discoveries can fully integrate into clinical medicine and help clinicians with their management of patients. The large-scale sequencing of cancer genomes will provide opportunities for patient stratification and personalized approaches to treatment that are based on individual mutational profiles.

CHILDHOOD AND ADULT ACUTE LYMPHOBLASTIC LEUKEMIAS

The ALLs are a group of malignant hematological disorders, in which the blasts originate from precursors of the lymphoid lineage. ALL is the most prevalent malignancy in children. About 80–85% of childhood ALL cases originate from B-cell precursor cells (BCP-ALL), whereas 10–15% of cases originate from T-cell precursors (T-ALL) [7]. ALL remains the leading cause of cancer-related mortality in childhood. Overall survival rates for children with ALL have reached 85–90% owing to the use of multiagent chemotherapy protocols. However, the relapse rate is approximately 20% and is associated with high mortality.

In adults, this disease is less common than in children. However, because there are more adults than children, the number of cases seen in adults is comparable to that seen in children, even though the incidence is significantly lower. ALL is slightly more common in males than in females.

Current risk stratification tools have been in place for several years and are based on clinical features including age at diagnosis, leukocyte count, initial response to treatment, and presence of recurrent structural chromosomal alterations, including aneuploidy and translocations [7,8].

ALL is characterized by recurrent genetic abnormalities, including balanced translocations and aneuploidies. Based on the World Health Organization (WHO) classification, BCP-ALL is categorized into ALL with hyperdiploidy (>50 chromosomes), ALL with hypodiploidy (<44 chromosomes), and ALL with translocation t(9;22) (q34;q11.2) encoding *BCR–ABL1*, t(12;21) (p13;q22) encoding *TEL–AML1*, t(1;19) (q23;p13.3) encoding *E2A–PBX1*, t(5;14) (q31;q32) encoding *IL3–IGH*, and rearrangement of *MLL* at 11q23, with a diverse range of partner genes [9,10]. T-ALL is a malignant disease of thymic cells accounting for 25% of all childhood ALL and 15% of adult ALL. T-ALL originates from a developing thymocyte; therefore alterations in genes responsible for normal thymocyte development and lineage determination are suspected to contribute to T-ALL pathomechanisms. Common alterations include rearrangement of the T-cell receptor gene loci to transcription factor genes including *TLX1*, *TLX3*, *LYL1*, *TAL1*, and *MLL* [11].

Since only about three-quarters of children and a lower proportion of adults with B-ALL and half of T-ALL patients carry gross chromosomal alterations in conventional cytogenetic analysis, attention has been drawn to smaller, not cytogenetically detectable aberrations. Microarray and candidate gene sequencing has identified a number of recurrent gene mutations. Many of the affected genes encode proteins engaged in lymphoid development (PAX5, IKZF1, EBF1, LMO2), tumor suppression (CDKN2A/CDKN2B, PTEN, RB1, TP53), regulation of apoptosis (BTG1), signaling (BTLA, CD200, TOX), and transcriptional regulation and coactivation (TBL1XR1, ETV6, ERG), as well as regulators of chromatin structure and epigenetic regulators (CTCF, CREBBP) [12,13].

Sanger sequencing studies of B-lineage ALL have identified recurring mutations in genes controlling lymphoid development (*PAX5, IKZF1*), Ras signaling (*NRAS, KRAS, NF1*), cytokine receptor signaling (*IL7R, JAK20*), and tumor suppression (*TP53*) [13].

In T-ALL, activating mutations of *NOTCH1, FBXW7, PTEN*, and *WT1* have been found [14,15].

NGS studies have been performed in pediatric and adult patients with ALL [4,5]. The data obtained from these studies have shown that alterations in multiple cellular pathways, including cytokine receptor and Ras signaling, tumor suppression, lymphoid development, and epigenetic regulation, are all hallmarks of multiple ALL subtypes [16]. In addition to the importance of identifying somatic mutations and pathways, NGS was also shown to be a powerful tool to differentiate related diseases [17a]. The following paragraphs describe these findings.

An NGS platform for targeted personalized therapy of leukemia has been launched to set up a network of physicians and scientists devoted to the implementation of the most advanced DNA/RNA NGS technologies in hematology for both preclinical research and diagnostics. The project focuses on the depiction of a comprehensive catalog of diagnostic and prognostic markers to guide the personalized approach to defining cohorts of leukemia patients and on highlighting interindividual variability that may play a role in the differential outcome of diverse therapeutic strategies. Furthermore, the aim of the platform is to provide new knowledge of the etiology of hematological diseases and origins of interpersonal variability in response to therapy. The long-term goal is to optimize tools to guide targeted treatments of leukemia patients [17b].

T-CELL ACUTE LYMPHOBLASTIC LEUKEMIA

Early T-cell precursor (ETP) ALL has been recognized as a subtype of T-cell ALL. It is associated with a high level of poor outcomes. Immunophenotypic features of ETP ALL include expression of T-lineage markers (cytoplasmic CD3) but absence of other T-ALL markers such as CD1 and CD8, weak or lack of expression of CD5, and aberrant expression of myeloid and stem cell markers (CD13, CD33, CD34, CD117). ETP ALL lacks a known unifying genetic alteration. Zhang and co-workers performed WGS on samples from 12 patients diagnosed with ETP ALL [18]. They found activating mutations in genes regulating cytokine receptors and Ras signaling in 67% of cases, inactivating mutations causing disturbances in hematopoietic development in 58%, and mutations in histone-modifying genes in 48% of cases. Inactivating mutations causing disturbances in hematopoietic development most commonly involved *ETV6*, *GATA3*, *IKZF1*, and *RUNX1*. Activating mutations in genes regulating cytokine receptors and Ras signaling involved *NRAS*, *KRAS*, *FLT3*, *JAK1*, *JAK3*, and *IL7R*. Mutations in *IL7R* cause constitutive activation of the Janus family kinase (JAK)–STAT signaling pathway. The authors concluded that the spectrum of mutations is similar to that in myeloid leukemias, and comparison of the transcriptional profile of ETP ALL with those of normal human hematopoietic progenitors showed significant similarity to hematopoietic stem and early myeloid progenitors. Thus, ETP ALL probably represents part of a spectrum of immature, stem cell-like leukemias [18]. The potential new therapeutic approaches, including epigenetic drugs and agents targeting JAK–STAT signaling, are currently under way [17a].

A high frequency of somatic alteration is observed in epigenetic modifiers, including *EED*, *EZH2*, and *SUZ12* [19], and a novel recurrent somatic mutation in *DNM2* involved in lymphoid development or leukemogenesis was also identified [18].

Whole-exome sequencing (WES) of five paired ETP ALL adult samples identified novel recurrent mutations in *FAT1* (25%), *FAT3* (20%), *DNM2* (35%), and genes associated with epigenetic regulation (*MLL2*, *BMI1*, and *DNMT3A*). Mutations in epigenetic regulators support clinical trials, including epigenetic-oriented therapies, for this high-risk subgroup. Interestingly, more than 60% of adult patients with ETP ALL harbor at least a single genetic lesion in *DNMT3A*, *FLT3*, or *NOTCH1*, which points to the design of targeted therapies [20].

Among the most ground-breaking findings in T-ALL biology was the identification of a NOTCH1–FBXW7 pathway disruption in over a half of all T-ALL cases as well as its possible impact on a patient's outcome [21]. The products of *NOTCH1* and *FBXW7* are involved in the same signaling pathway and are known to play a major role in the molecular pathomechanisms of T-ALL. NOTCH1 is a cell-membrane receptor responsible for signal transduction, thus taking part in the control of hematopoietic cell proliferation, maturation, adhesion, and apoptosis. *NOTCH1* mutations have been reported in about 50% of all T-ALL cases [22,23]. *FBXW7* encodes a component of the ubiquitin ligase complex, which plays a role in the degradation of such proteins as c-MYC, cyclin E, and NOTCH1—molecules crucial for T-ALL pathogenesis [21]. *NOTCH1* mutations, together with those of *FBXW7*, result in elevated activity of the intracellular NOTCH1 subunit and overexpression of its downstream targets, which in turn leads to tumor development. *FBXW7* and *NOTCH1* mutations are also relatively frequently observed in the same patient—60% of patients carrying a *FBXW7* mutation also carry a *NOTCH1* mutation, suggesting an association between these two mutational events, together leading to the development of T-ALL [21,22].

Aberrant NOTCH1 signaling has been identified in hematopoietic and nonhematopoietic neoplasms. Mutations in *NOTCH1* have been identified in 50% of adult T-ALL, CLL, myeloid cell leukemia

(MCL), and Burkitt lymphoma [2]. Results of studies evaluating the association of *NOTCH1* mutation and clinical outcome are conflicting. In T-ALL, *NOTCH1* mutations were associated with an improved response to glucocorticoids. Additional studies are required to assess its prognostic role in other neoplasms. Small-molecule NOTCH inhibitors have been used in combination with inhibitors of the phosphatidylinositol-3-OH kinase (PI3K)–AKT–mTOR pathway in animal T-ALL modeling, with promising results [24].

BCR–ABL1-LIKE ACUTE LYMPHOBLASTIC LEUKEMIA

Studies have identified a subgroup of patients with ALL, particularly large among children, which, despite the absence of a BCR–ABL1 fusion protein, is characterized by a high level of probability of disease progression. This subtype of leukemia was named "BCR–ABL1-like" because the gene expression profile is similar to that of BCR–ABL1-positive leukemia [25–29]. In this subtype of ALL, the dysfunction of genes involved in signal transduction from the receptors for cytokines, such as activating mutations of JAKs and *CRLF2* gene rearrangements (cytokine receptor-like factor 2), is often present. Another common factor in BCR–ABL1 leukemia-positive and BCR–ABL1-like ALL is a particularly frequent presence of deletions and point mutations in the gene *IKZF1*, which result in JAK–STAT signaling being activated [25,26]. Such signaling may be prone to therapy with JAK inhibitors (most notably, ruxolitinib) and is currently being researched further for its therapeutic application.

A wide range of genetic alterations activating cytokine receptor and tyrosine signaling was identified by transcriptome sequencing and WGS of 15 BCR–ABL1-like ALL cases, 12 of which lacked *CRLF2* rearrangement. Most commonly, the rearrangements resulted in chimeric fusion genes deregulating tyrosine kinases (*NUP214–ABL1*, *ETV6–ABL1*, *RANBP2–ABL1*, *RCSD1–ABL1*, *BCR–JAK2*, *PAX5–JAK2*, *STRN3–JAK2*, and *EBF1–PDGFRB*) and cytokine receptors (*IGH–EPOR*). Up to 20% of BCR–ABL1-like cases lacked a chimeric fusion on mRNA-seq analysis. Also, sequence mutations (e.g., activating mutations of *FLT3* and *IL7R*) and structural alterations (e.g., focal deletions of *SH2B3* or *LNK*, which constrains JAK signaling) that activate signaling have been identified in fusion-negative cases. This variety of genetic alterations activates a limited number of signaling pathways, notably ABL1 and PDGFRB (both of which may be inhibited with the tyrosine kinase inhibitors (TKIs) imatinib and dasatinib) and JAK–STAT signaling. These rearrangements have been shown to activate signaling pathways in model cell lines and in primary leukemic cells, and xenografts of BCR–ABL1-like ALL are highly sensitive to TKIs in vivo [30]. Recently, a child was reported with refractory *EBF1–PDGFRB*-positive ALL, which was exquisitely sensitive to imatinib. This emphasizes the potential clinical utility of TKI therapy in BCR–ABL1-like ALL [31]. Ongoing studies involve conducting NGS on childhood and adult ALL to comprehensively define the variety of kinase-activating alterations in BCR–ABL1-like ALL and to develop clinical trials to apply appropriate TKI therapy to patients with BCR–ABL1-like ALL.

HYPODIPLOID ACUTE LYMPHOBLASTIC LEUKEMIA

Up to 3% of ALL cases are characterized by hypodiploidy with fewer than 44 chromosomes and are associated with poor prognosis. Based on the severity of aneuploidy, two subgroups of hypodiploid ALL have been identified: a near-haploid subgroup with 24–31 chromosomes and a low-hypodiploid subgroup with 32–39 chromosomes [32–34]. The nature of additional genetic alterations that cause leukemogenesis and poor outcome in hypodiploid ALL remains yet unknown. Microarray and NGS analysis of a large number of hypodiploid ALL samples (more than 120) has demonstrated that the majority of near-haploid cases contain mutations activating Ras signaling, particularly in *NF1* and the IKAROS family gene *IKZF3* [35]. Conversely, low-hypodiploid cases have near universal mutation of the tumor suppressor *TP53* (p53), with the mutations present in the germ line in approximately 50% of the cases and inactivating mutations of a third IKAROS family member, *IKZF2*. Parallel analysis of primary hypodiploid xenografts (primagrafts) demonstrated activation of Ras–Raf–MEK–ERK and PI3K signaling in the majority of hypodiploid primagrafts that was sensitive to PI3K and PI3K/mTOR inhibitors, but not to MEK inhibitors, which suggests that PI3K inhibition represents a novel therapeutic approach [35]. In addition, the identification of low-hypodiploid ALL as a manifestation of Li–Fraumeni syndrome indicates a need for testing for *TP53* mutational status in all children with low-hypodiploid ALL [36,37]. Genome sequencing studies have also identified other germ-line mutations in familial leukemia, such as a germ-line *PAX5* p.Gly183Ser mutation in two kindreds with autosomal dominant pre-B ALL [17a,38].

RELAPSED ACUTE LYMPHOBLASTIC LEUKEMIA

While several subtypes of ALL are associated with a particularly high risk of treatment failure, relapse occurs across the whole range of ALL subtypes and is associated with a very poor outcome. Moreover, ALL genomes are not static but exhibit acquisition of new chromosomal abnormalities over time. Single-nucleotide polymorphism microarray profiling studies of matched diagnosis–relapse ALL samples show that most ALL cases exhibit changes in the patterns of structural genomic alterations from diagnosis to relapse and that many relapse-acquired lesions, including those targeting genes associated with high-risk ALL (*IKZF1*, *IKZF2*, *CDKN2A*, and *CDKN2B*), are detectable at the diagnosis [39–41]. This suggests that genetically determined tumor heterogeneity is a key predictor of failure of treatment and relapse. These findings have been confirmed by the sequencing of 300 genes in matched diagnosis–relapse samples, which also identified mutations in the transcriptional coactivator and acetyltransferase *CREBBP* (CREB-binding protein) as a relapse-acquired lesion in up to 20% of relapsed ALL samples [42,43]. *CREBBP* mutations can also be found at diagnosis in NHL, particularly diffuse large B-cell lymphoma (DLBCL), and impair histone acetylation. CREBBP in part alleviates the transcriptional response to glucocorticoids, and histone deacetylase inhibitors were effective at destroying steroid-resistant ALL cell lines. Relapse-acquired mutations in *NT5C2* that encode a 5′-nucleotidase enzyme responsible for the inactivation of nucleoside-analog drugs have been identified by two distinct study groups [44,45].

ACUTE MYELOID LEUKEMIA

AML is a clonal disease characterized by excessive proliferation and accumulation of immature trans-formed hematopoietic stem or progenitor cells. AML accounts for 75–80% of acute leukemias in adults and 15–20% in children. The incidence of AML in young adults is 3 or 4 cases per 10,000 per year and increases with age. The incidence of AML is over 30-fold higher in people between 80 and 90 years of age, compared with the rate in younger adults (ages 20–40). AML occurs more frequently in men than in women (with a ratio of 3:2) [11,46,47].

The incidence of AML in children is lower in comparison to adults. The highest incidence of AML in children occurs in infancy and is reported as 1.6 cases per 100,000 per year and then declines and accounts for 0.4 cases per 100,000 per year by age 10 [47].

Although most patients respond to chemotherapy, the overall outcome is poor, with the rate of 5 year survival being below 50% in young adults and below 20% in patients older than 60 years. In young children the outcome of AML is also low, with a long-term survival rate of 60%. AML accounts for about 35% of childhood deaths from leukemia. Mortality is a consequence of resistant progressive disease or treatment-related toxicity.

AML is a genetically heterogeneous disease, characterized by an arrest in differentiation and uncontrolled proliferation of hematopoietic progenitor cells with a variety of somatic cytogenetic and molecular alterations [11,48–50].

Chromosomal abnormalities identified by conventional cytogenetics include balanced translocations, inversions, insertions, monosomies, and trisomies, which are present in approximately 55% of adult cases and 80% of children with AML. These are the strongest prognostic factors for response to treatment and survival in multivariate analysis. The 2008 WHO classification categorized AML based on cytogenetic or molecular abnormalities [11].

Favorable cytogenetic alterations in AML include t(8;21), inv(16)(p13;q22), and t(16;16)(p13;q22) [51,52]. These alterations encode components of the core binding factor transcription complex and give rise to the *RUNX1–RUNX1T1* and *CBFB–MYH11* fusion transcripts, which results in transcriptional repression. The other favorable translocation is t(15;17)(q22;q11–q21), encoding the *PML–RARA* fusion transcript, leading to a differentiation block in acute promyelocytic leukemia.

The mixed-lineage leukemia gene (*MLL*) is located on chromosome 11q23 and is associated with a poor prognosis that is modified by the fusion partner. Other unfavorable risk cytogenetics include a complex karyotype (defined as three or more acquired chromosomal abnormalities in the absence of recurrent genetic changes designated by WHO classification) [11].

A normal karyotype (NK) is commonly found in 45% of AML patients, however. Accordingly, mutational studies have been able to increase the reliability of the prognosis in this subset of patients. Hybridization techniques, such as fluorescence in situ hybridization (FISH) and single-nucleotide polymorphism (SNP) microarrays, have improved the sensitivity of detection. Mutations leading to loss of tumor suppressor genes such as *NF1*, *CDKN2*, and *TP53* were detected. Currently, 85% of patients with NK AML have a mutation, although the prognostic significance of some of these is unclear [49–55].

Somatic mutations in genes involved in regulation of hematopoiesis were found in adults and children with AML [54]. Fms-like tyrosine kinase 3 (FLT3) is a class III kinase receptor. The *FLT3* gene is located on the 13q12 chromosomal band. Mutations on chromosome 13 in the tyrosine kinase *FLT3* receptor generally map to the kinase domain as point mutation D835 or to the juxtamembrane domain as an internal tandem duplication (ITD). Both types of mutations lead to the constitutive activation of the receptor and the transformation of hematopoietic cell lines to cytokine independence.

FLT3-ITD is the most common mutation in adult AML, with the highest frequency in NK AML (30–40%), whereas a D835 mutation is present in 8–14% of AML patients. FLT3-ITD occurs in 15% of childhood AMLs. Patients harboring an FLT3-ITD mutation have poor prognosis compared to that of patients with wild-type cases, whereas the prognostic significance of FLT3 D835 mutations is not clear. The effect of FLT3-ITD depends on the presence of other mutations (for example *NPM1*).

Mutations in *NPM1* (nucleophosmin) occur in 30% of adult and 8% of childhood AMLs. An *NPM1* mutation in the absence of FLT3-ITD is associated with a better outcome in AML patients with a normal karyotype [50].

Currently, there are a number of FLT3 inhibitors at various stages of clinical development, such as PKC412 (midostaurin), CEP-701 (lestaurtinib), or MLN518 (tandutinib). TKIs are promising agents in the treatment of AML patients with an FLT3-ITD mutation, especially when they are combined with chemotherapy [50].

GENETIC CONCEPT OF ACUTE MYELOID LEUKEMIA PATHOGENESIS

Cytogenetic analysis and molecular methods are routinely used in tandem in hematological clinical practice. Identification of chromosomal aberrations and point mutations in the beginning phases of the disease is of crucial importance for the proper diagnosis of AML in accordance with the WHO classification. Before the era of NGS, two functional groups of mutations were identified in leukemic blasts based on cytogenetic and mutational analysis [54]. Class I mutations, including *FLT3*, *KIT*, *RAS*, and *BCR–ABL*, were responsible for the activation of signaling pathways that drive proliferation and result in uncontrolled growth and survival of hemopoietic progenitors.

Class II mutations, such as t(8;21), t(16;16), t(15;17), *CEBPA*, or *RUNX1*, affect transcription factors or components of coactivation complexes of transcription and cause impaired hemopoietic differentiation.

Concomitant occurrence of both classes of mutations is usually observed in AML blasts and leads to malignant transformation. Many studies have confirmed that a single mutation is insufficient for the emergence of acute leukemia. Additional aberrations are needed for a clone to become deregulated and to transform to AML. Conventional cytogenetics, SNP analysis, candidate gene sequencing, and WGS can identify AML clones.

The first description of the results of the sequencing of a complete cancer genome using NGS was published in 2008 by Timothy Ley and coauthors [55], who performed sequencing of a neoplastic and normal-cell genome from a patient with FAB M1 AML (acute myeloid leukemia without maturation). At diagnosis, the cytogenetic analysis revealed a normal karyotype 46, XX. The patient initially achieved complete remission, but relapsed at 11 months. At relapse, a new cytogenetic abnormality was detected, t(10;12) (p12;p13). The authors identified 10 nonsynonymous somatic mutations in the patient's AML genome. Two of the mutations were known previously: the internal tandem duplication of *FLT3 ITD* and a 4-base insertion in exon 12 of the *NPM1* gene, contributing to AML progression. The other eight somatic mutations, present in virtually all tumor cells at presentation and relapse, had not been previously described in the AML genome (*PTPRT*, *CDH24*, *PCLKC*, *SLC15A1*, *KNDC1*, *GPR123*, *EBI2*, *GRINL1B*). The importance of those new somatic mutations in AML pathogenesis is still unknown. The recurrence of a somatic mutation in other AML samples or other cancers is the best test of its relevance in the pathogenesis of neoplasm. Functional studies will be necessary to assess their role in causing AML. The authors emphasized that the parallel sequencing of the patient's normal genome allowed the determination of the mutations that were somatic and indicate germ-line (inherited) variants.

One year later, the same group analyzed an additional 187 AML samples and identified mutations in *IDH1* in 16% of cytogenetically normal AML. A mutation in *IDH1*, encoding isocitrate dehydrogenase 1, which is predicted to affect the arginine residue at position 132 was found earlier in malignant gliomas. It was found that mutation in *IDH1* is a marker of poor prognosis in AML. They also identified a somatic mutation in *DNMT3A*, encoding DNA methyltransferase, in cells from a patient with NK AML, followed by the sequencing of 280 additional patients with *de novo* AML. Twenty-two percent of patients had mutations in the *DNMT3A* gene, which was found to be an independent marker of poor prognosis [56].

These mutations are rare or absent in childhood AML and indicate significant differences between childhood and adult AML.

The Cancer Genome Atlas Research Network reported 200 *de novo* AML genome sequences. AML blasts and matched skin samples from 50 patients and WES for another 150 paired samples of AML blasts and skin were performed [3]. The data show that alterations in multiple genes are hallmarks of multiple AML subtypes. AML genome analysis revealed an average of 13 mutations per individual, with 23 genes considerably mutated. Additionally, 237 genes were mutated in two or more individuals. All gene mutations were categorized into nine groups of aberrations relevant for AML development. These included transcription factor fusions, nucleophosmin *NPM1* mutation, tumor suppressor gene mutations, DNA-methylation-related gene mutation, signaling gene mutation, chromatin-modifying gene mutations, myeloid transcription factor gene mutation, cohesion-complex gene mutation, and spliceosome-complex genes mutation.

Because of significant differences between childhood and adult AML, two distinct projects were launched to characterize the genomic and epigenomic profile of pediatric AML. The Pediatric Cancer Genome Project and TARGET studies of AML have provided new data. Cryptic translocation between the *CBFA2T3* and the *GLIS2* genes in almost 30% of children with FAB M7 AML were identified [4,5]. While initial studies showed that this fusion is associated with adverse outcome, further studies are needed to confirm this finding.

It is particularly notable that studies by The Cancer Genome Atlas and others have determined that 44% of patients with AML exhibit mutations in genes that regulate methylation of genomic DNA. In particular, frequent mutation has been observed in the genes encoding DNA methyltransferase 3A (*DNMT3A*), isocitrate dehydrogenase 1 (*IDH1*), and isocitrate dehydrogenase 2 (*IDH2*), as well as Tet oncogene family member 2. The presence of *DNMT3A*, *IDH1*, or *IDH2* mutations may confer sensitivity to novel therapeutic approaches, including the use of demethylating agents. Understanding the roles of these mutations in AML biology will lead to more rational therapeutic approaches targeting molecularly defined subtypes of the disease [57–63].

The heterogeneous treatment response in AML results from a complex set of factors, including the clinical characteristics of the patients (e.g., age at diagnosis), cytogenetic and molecular features of the leukemic blasts, epigenetic changes or overexpression of a number of genes, and changes in cell drug disposition and tissue accumulation The current data on induction and consolidation therapy of AML suggest that further dose intensification of currently available cytotoxic drugs is unlikely to improve the clinical outcome. It is plausible that the efficacy of AML treatment will be enhanced with the use of molecularly targeted treatment methods for genetic mutations. NGS methods have allowed entire genome analysis and the identification of new candidate genes that may further elucidate the pathogenesis of AML and identify potential therapeutic targets [5].

Although the diagnostic and prognostic implications of genetic aberrations observed in AML are still not fully elucidated, more and more evidence suggests they may become important factors determining optimal therapy. The genetic complexity of AML makes it a perfect candidate for NGS-based diagnostics as it allows for screening of large numbers of genes in multiple patients, in acceptable time frames and at an affordable cost. So far several amplicon-based NGS assays to screen genes involved in AML pathogenesis have been released; however, because of the paucity of studies on reproducibility, robustness, and precision of these tests, it is still unclear if they represent a reliable diagnostic option to be used in clinical settings. For more details on technical issues related to the use of NGS methodology in these settings, see the section "Implementation of NGS-based techniques in clinical practice in hematology."

BRAF MUTATION IN HAIRY CELL LEUKEMIA

BRAF NGS studies of lymphoid malignancies allowed the identification of a *BRAF* mutation in hairy cell leukemia (HCL). In a study described by Tiacci and co-workers, this mutation was present in 100% of the patients with HCL who were examined. This mutation was previously identified in more than half of the papillary carcinomas of the thyroid, malignant melanomas, Langerhans cell histiocytosis, and in some myelomas. Detection of *BRAF* is a diagnostic and prognostic tool. The *BRAFV600E* mutation results in the constitutive activation of its kinase activity. The BRAF inhibitor vemurafenib is one of the kinase inhibitors used in clinical trials in patients with HCL refractory to therapy with purine analogs [64].

CSF3R MUTATION IN CHRONIC NEUTROPHILIC LEUKEMIA

Mutation in *CSF3R* was found in 59% of patients with chronic neutrophilic leukemia (CNL) or atypical chronic myeloid leukemia (CML), both classified as myeloproliferative neoplasms in accordance with the WHO classification. CNL is characterized by an expansion of neutrophils in the bone marrow and peripheral blood. Atypical CML increased the number of neutrophil precursors and granulocytic dysplasia. Until recently the genetic basis of these diseases was unknown. In both, the absence of Philadelphia chromosome with translocation t(9;22) (*BCR–ABL1*) rearrangement and integration of functional and genomic analysis of leukemic cells allowed for identification of the *CSF3R* mutation as causing mutation in CNL and atypical CML. These mutations were identified in two distinct regions of *CSF3R* and lead to downstream kinase signaling through the SRC family or JAK kinases. These downstream kinase pathways make *CSF3R* mutations an attractive marker for TKIs as new therapeutic agents—currently in clinical trial [65a].

NGS IN CHRONIC LYMPHOCYTIC LYMPHOMA

Sutton et al. developed an NGS assay to identify genetic variants associated with CLL, particularly with the aggressive disease course, including new recurrent mutations. These authors used targeted resequencing to assess the mutation status of genes with prognostic potential. Nine genes, *ATM*, *BIRC3*, *MYD88*, *NOTCH1*, *SF3B1*, *TP53*, *KLHL6*, *POT1*, and *XPO1*, were included in the HaloPlex panel (Agilent Technologies, Santa Clara, CA, USA) designed by the SureDesign service (https://earray.chem.agilent.com/sure-design/home.htm). These authors further examined samples from 188 chronic lymphocytic leukemia patients with poor prognostic features. The analysis revealed that 114/180 (63%)

patients carried at least one mutation, with mutations in *ATM*, *BIRC3*, *NOTCH1*, *SF3B1*, and *TP53* accounting for 149/177 (84%) of all mutations. All findings were further validated by Sanger sequencing [65b].

NON-HODGKIN LYMPHOMAS

NHL consists of a diverse group of malignant neoplasms of the lymphoid tissues, variously derived from B-cell progenitors, T-cell progenitors, mature B-cells, mature T-cells, and natural killer cells. The clinical presentation of NHL varies tremendously depending on the type of lymphoma and the areas of involvement. Some NHLs behave indolently with lymphadenopathy waxing and waning over the years. Others are highly aggressive, resulting in death within weeks if left untreated. In typical cases, aggressive lymphomas commonly present acutely or subacutely with a rapidly growing mass, systemic B symptoms (such as fever, night sweats, weight loss), and/or elevated levels of serum lactate dehydrogenase and uric acid. Examples of lymphomas with this aggressive or highly aggressive presentation include diffuse large B-cell lymphoma, Burkitt lymphoma, adult T-cell leukemia–lymphoma, and precursor B- and T-lymphoblastic leukemia/lymphoma.

Indolent lymphomas are often insidious, presenting only with slow-growing lymphadenopathy, hepatomegaly, splenomegaly, or cytopenias. Examples of lymphomas that typically have indolent presentations include follicular lymphoma, chronic lymphocytic leukemia/small lymphocytic lymphoma, and splenic marginal zone lymphoma.

Unlike in adults, in whom low-grade, clinically indolent NHL subtypes predominate, most pediatric NHL cases are of high grade and have aggressive clinical behavior [9]. NHL is the fifth most common diagnosis of cancer in children under the age of 15 years, and it accounts for approximately 7% of childhood cancers in the developed world [9]. In the United States, approximately 800 new cases of pediatric NHL are diagnosed annually, with an incidence of 10–20 cases per million people per year. This incidence appears to be increasing overall, largely thought to reflect a rise in NHL among adolescents. There is a male predominance, and whites are more commonly affected than African-Americans [66,67].

In general, the most common subtypes of NHL in children originate from B-cell progenitors. In the United States and other developed countries, the most common subtypes are Burkitt lymphoma (BL), diffuse large B-cell lymphoma, lymphoblastic T-cell or B-cell lymphoma, and anaplastic large-cell lymphoma. Other subtypes (follicular lymphoma, marginal zone lymphoma) are less common, accounting for approximately 7% of pediatric NHL [67].

BL is an aggressive B-cell malignancy that predominantly affects the pediatric population. Although most children are cured with intensive chemotherapy, up to 20% die of relapsed or refractory disease. The molecular hallmark of BL is the translocation of the *MYC* proto-oncogene to the immunoglobulin heavy or one of the light-chain genes, leading to constitutive MYC activation. Additional molecular alterations that may counteract MYC-induced proapoptotic signals are likely to be relevant in the pathogenesis of BL. RNA sequencing has been performed to investigate the genetic landscape of BL via the use of a cohort that combined pediatric and adult cases. In contrast to adult cases, which typically have a simple karyotype, 60–90% of pediatric tumors have secondary chromosomal abnormalities, the consequences of which are less well characterized [68].

Giulino-Roth and co-workers reported results of targeted sequencing of libraries prepared using a custom SureSelect kit (Agilent) on the HiSeq platform (Illumina), focusing specifically on pediatric

BL. They studied 182 cancer-related genes on 29 formalin-fixed, paraffin-embedded primary pBL samples. Ninety percent of the cases had at least one mutation or genetic alteration, most commonly involving *MYC* and *TP53*. Alterations were also identified in tumor-related genes not previously described in BL, including truncating mutations in *ARID1A*, identified as a tumor suppressor, and the amplification of *MCL1* [69]. Mutations in *ARID1A* were found in 17% of pBL cases. Mutations were distributed throughout the gene and all resulted in protein truncation, consistent with their role as tumor suppressor. As has been proposed for other malignancies, haploinsufficiency of *ARID1A* may be enough for cellular transformation.

Recurrent amplification was found in *MCL1*, a member of the BCL2 family. MCL1 and related proteins inhibit apoptosis by blocking the cell death mediators BAK and BAX. The importance of *MCL1* as an oncogene has been implicated in transgenic mice that develop aggressive B-cell lymphomas. *MCL1* is located on 1q21.2, a genomic region amplified in approximately 25% of pBL cases. Amplification of *MCL1* has been described in a BL subline, but has not been reported in primary BL samples. *MCL1* overexpression may be clinically relevant because it has been linked to chemotherapy resistance, and several inhibitors that may target *MCL1* are in clinical development.

Alterations were also found in other cancer-related genes, including point mutations in *LRP6*; truncating alterations in *LRP1B*, *PTPRD*, *PTEN*, *NOTCH*, and *ATM*; amplifications of *RAF1*, *MDM4*, *MDM2*, *KRAS*, *IKBKE*, and *CDK6*; and a deletion of *CDKN2A*, many of which are targetable by therapies in clinical trials.

Love and his team described the first completely sequenced genome from a Burkitt lymphoma tumor and germ-line DNA, taken from the same patient. They also sequenced the exomes of 59 Burkitt lymphoma tumors and compared them to those from 94 DLBCL tumors. This study identified 70 genes that were recurrently mutated in Burkitt lymphomas, including *ID3*, *GNA13*, *RET*, *PIK3R1*, *ARID1A*, and *SMARCA4*. They described a number of genes for the first time, including *CCT6B*, *SALL3*, *FTCD*, and *PC*. *ID3* mutations occurred in 34% of Burkitt lymphomas and not in DLBCLs. The authors showed that *ID3* mutations promote cell cycle progression and proliferation. Their work showed that gene-coding mutations commonly occur in Burkitt lymphoma and implicated *ID3* as a new tumor suppressor gene [69].

Schmitz and co-workers found that mutations affecting the transcription factor TCF3 (E2A) or its negative regulator ID3 fostered TCF3 dependency in 70% of sBL cases. TCF3 activated the prosurvival PI3K pathway in BL, partly through the augmentation of constitutive B-cell receptor signaling [70].

Gene expression studies identified three subtypes of DLBCL: activated B-cell, germ-line center B-cell, and primary mediastinal B-cell lymphoma. The GC-subtype predominately occurs in children and accounts for 75–83% of all DLBCL cases. By way of comparison, 50% of all DLBCL cases in adults are characterized by a GC-subtype. Pediatric DLBCL responds well to treatment and has an excellent prognosis (with the relative 5-year survival rate being 85%). Clinical outcome is poorer and more heterogeneous in adult patients. Numerous studies used candidate gene mutation analysis, WGS, WES, and transcriptosome analysis to identify recurring aberrations in DLBCL [71–83]. These observations indicate that activated B-cell, germ-line center B-cell, and primary mediastinal B-cell lymphoma originate differently. Mutations in genes controlling B-cell receptor signaling (*CD79B*) and constitutive activity of the NF-κB pathway (mutations in *CARD11* and *MYD88*) are the key signature observed in activated B-cell DLBCL.

Morin and co-workers found that germ-line center B-cell lymphomas are characterized by recurrent mutations in *EZH2*, *BCL2*, *TP53*, *TNFRSF14*, *MEF2B*, *SGK1*, and *GNA13* [74]. The same group analyzed a separate group of patients to identify the additional gene *EP300* and loss of *CREBBP*. More commonly, mutated histone-modifying genes reinforced the importance of these alterations in the pathogenesis of lymphoid neoplasms [75].

INHERITED BONE MARROW FAILURE SYNDROMES

Inherited bone marrow failure syndromes (IBMFSs) are heterogeneous disorders characterized by cytopenia within at least one of the lineages of hematopoietic cells. Some IBMFS, such as Fanconi anemia (FA), dyskeratosis congenita (DC), and Diamond–Blackfan anemia (DBA), are associated with a high risk of developing a malignancy.

Before the NGS technology was introduced, linkage analysis techniques and candidate gene sequencing were used to confirm the genetic background of IBMFS. NGS enabled the discovery of mutations of genes associated with DC (mutations in the *CTC1* and *RTEL1* genes), DBA, and thrombocytopenia-absent radius (TAR) [84].

FANCONI ANEMIA

FA is the most common form of IBMFS. It is a chromosomal instability disorder, which is caused by a germ-line mutation leading to DNA instability and defective repair.

FA is typically diagnosed in the first decade of a person's life. Clinical features include congenital malformations, progressive bone marrow failure, and a high risk of developing squamous cell cancers of the head, neck, and anogenital region, as well as hematopoietic malignancies. Congenital defects are present in 60–70% of patients and include short stature, hypopigmented spots and café-au-lait spots, microcephaly, abnormality of thumbs, and developmental delay.

FA is caused by germ-line mutation in any one of 16 known genes encoding proteins responsible for genomic stability, mainly for repair of DNA interstrand cross-links. These genes belong to complementation groups (Fa-A to Fa-Q) and tend to be inherited in autosomal recessive patterns, except for subtype B, which is commonly inherited as X-linked recessive. Mutations in any of the FA genes result in defective DNA repair and reduced or absent protein function [85].

The *FANCA* gene was discovered by linkage analysis of an FA gene to chromosome 16q24.3. Complementation analysis or protein analysis allowed the discovery of subsequent mutations. These methods are unfortunately costly and time-consuming. The genetic cause of FA is discoverable in 95% of patients.

Several groups have applied NGS technology to improve and accelerate molecular diagnostics of FA patients without known genetic etiology. NGS allowed the discovery of new mutations in the *ERCC4* and *XRCC2* genes, which are critical for DNA repair [86–88].

Knies and his team identified germ-line mutations in four patients with FA diagnosed based on nonmolecular analysis who were not assigned to any complementation group [85].

De Rocco and co-workers sequenced 16 genes involved in FA pathogenesis using the PGM (Life Technologies) technology in 100 samples obtained from patients with FA and identified not only known mutations, but also six cases of novel mutations. They concluded that the NGS procedure should be applied routinely in molecular screening of FA in the nearest future and will allow the identification of mutations in the majority of cases [86].

DYSKERATOSIS CONGENITA

DC is an inherited disorder, characterized by reticulated or mottled skin hyperpigmentation in the facial, neck, shoulder, and trunk regions; nail dystrophy involving both hands and feet; and mucosal leukoplakia.

Patients with DC have a high risk of developing bone marrow failure and cancer. Other characteristic somatic abnormalities include a short stature, pulmonary fibrosis, dental abnormalities, immune deficiencies, esophageal stricture, premature graying or hair loss, osteoporosis, liver cirrhosis, urinary tract anomalies, and hyperhidrosis. Most somatic abnormalities are absent at birth and develop with age. Diagnosis of DC is frequently delayed until adolescence or adulthood.

Approximately 90% of patients develop bone marrow failure. Bone marrow failure, being the most common cause of death, occurs in almost three-quarters of patients with the X-linked variant of DC and less frequently in the other two forms. The median age of bone marrow failure is 10 years of age. In most patients with DC, bone marrow failure typically begins with thrombocytopenia or anemia, followed by pancytopenia.

Malignancies can develop and usually occur in the third and fourth decades of life. In patients with DC, the risk of malignancy is increased 11 times compared to the general population. Typical malignancies include myelodysplastic syndrome, AML, and squamous cell carcinoma primarily of the oropharynx, skin, and gastrointestinal tract.

Known clinical variants of DC include Hoyeraal–Hreidarsson syndrome, which presents in early infancy with a low birth weight, immunodeficiencies, cerebellar hypoplasia, developmental delay, and bone marrow failure. Revesz syndrome in turn is characterized by the typical symptoms characteristic of DC, in addition to exudative retinopathy and Coats plus disease (also known as cerebroretinal microangiopathy with calcification and cysts).

Reports show three patterns of inheritance for DC: X-linked, autosomal dominant, and autosomal recessive. Nine known genes are associated with DC: *DKC1, TERC, TERT, TINF2, WRAP53, NOP10, NHP2, CTC1*, and *RTEL1*. The mutation of such genes, responsible for the functioning and maintenance of telomeres, results in reduction of telomere length in patients with DC.

Telomere length is shortened in patients with bone marrow failure, especially in patients with DC (generally less than the first percentile). Furthermore, telomere length may be associated with disease severity in DC and declines with advancing age. Because maintenance of telomere length is required for the indefinite proliferation of human cells, such defects affect highly regenerative tissues, such as the skin and bone marrow.

It is estimated that the genetic cause of DC is known in about 70% of patients. Prior to the use of NGS, the genetic cause was known in about 50% of patients [84].

The disease may progress in severity and develop at an earlier age over generations, a phenomenon known as disease anticipation. This is due to the transmission of progressively shorter telomeres from generation to generation.

Based on WES, mutations in *RTEL1* have been independently identified by several groups in patients with Hoyeraal–Hreidarsson. The RTEL1 protein regulates the elongation of telomeres and DNA repair [88,89]. WES discoveries revealed a *CTC1* mutation as a causative cause of Coats plus several variants of DC [84,89].

THROMBOCYTOPENIA ABSENT RADIUS SYNDROME

TAR syndrome is an autosomal recessive disorder characterized by severe thrombocytopenia and bilateral absent radii, although the thumbs are always present [92]. Other abnormalities of the upper limbs may be present, including hypoplasia or absence of the ulna or abnormal or absent humerus. Congenital heart disease, usually atrial septal defect or tetralogy of Fallot, occurs in approximately one-third of affected patients.

Severe thrombocytopenia at birth or during the first postnatal week occurs in 59% of patients with TAR. The thrombocytopenia is caused by dysmegakaryocytopoiesis, with differentiation blockade at the stage of an early megakaryocyte precursor. The defect does not involve either thrombopoietin or the thrombopoietin receptor. The bone marrow shows normal erythroid and myeloid precursors with absent or decreased megakaryocytes.

Mortality is significant in the neonatal period and early infancy, primarily owing to intracranial hemorrhage. If the patient survives this period, spontaneous resolution of the thrombocytopenia usually occurs after the first year of age. Treatment is supportive, with platelet transfusions administered when needed [84,91].

The genetic alteration leading to TAR was discovered in 2007 [90]. TAR syndrome has been associated with a deletion of a segment of the 1q21.1 cytoband. The 1q21.1 deletion syndrome phenotype includes TAR and other features such as mental retardation, autism, and microcephaly. It has been shown that the TAR syndrome is caused by the compound inheritance of a low-frequency noncoding SNP and a rare null allele in *RBM8A*, a gene encoding the exon–junction complex subunit member Y14 located in the deleted region [91,92].

DIAMOND–BLACKFAN ANEMIA

DBA is a congenital erythroid aplasia characterized by progressive normochromic, usually macrocytic anemia. Approximately 90% of patients with DBA are diagnosed within their first year of life, with 35% diagnosed within the first month. In general, patients are significantly anemic at the time of presentation and have reticulocytopenia. Bone marrow is of normal cellularity, with markedly decreased or absent erythroid precursors. White blood cell count is generally normal; platelet counts are generally normal but can be increased or decreased. Fifty percent of patients with DBA also have congenital malformations. These findings occur mainly in the head and upper limb area and include craniofacial abnormalities, ophthalmological anomalies, neck anomalies, cardiac anomalies, thumb abnormalities, genitourinary malformations, and pre- and postnatal growth failure. These features are variable and appear to be related to the DBA genotype.

Associated malignancies include acute myelogenous leukemia, myelodysplastic syndrome, and solid tumors including colon cancer, female genital cancers, and osteosarcoma. The cumulative incidence of solid tumor or leukemia is approximately 20% by age 46 years. The risk of solid tumors in nontransplanted patients begins to rise around 30 years of age. The cancer risk is substantially higher than in the healthy population but is lower than that for Fanconi anemia or dyskeratosis congenita. However, the degree of risk for cancer has not been fully established, so it is not yet possible to make recommendations for cancer surveillance [85,93].

DBA is a ribosomopathy, caused by germ-line mutations affecting genes encoding components of the 40S and 60S ribosomal subunits. Germ-line mutations in any of nine known genes (*RPS19*, *RPS17*, *RPS24*, *RPS26*, *RPS10*, *RPS7*, *RPL35A*, *RPL5*, *RPL11*) cause DBA. About 50% of DBA patients have no known mutation. About 45% of cases are familial, usually displaying autosomal dominant inheritance, but with a wide range of severity within a family (reduced penetrance) [94]. Therefore, some cases that appear to be sporadic may prove to be familial if genetic testing is performed.

The gene encoding ribosomal protein 19 (*RPS19*) is mutated in 25% of patients with DBA. Disease-causing mutations in genes encoding the large (*RPL35A*, *RPL5*, *RPL11*) and small (*RPS24*, *RPS17*, *RPS7*, *RPS10*, *RPS26*, *RPS29*) ribosomal subunits have been described [98]. These include deletions and copy-number variations, which are not consistently detected by sequencing. Genotype–phenotype correlations exist, with patients harboring mutations in *RPL5* or *RPL11* presenting more frequently with somatic malformations. For example, in one series 21 of 24 patients with DBA and cleft palate had an *RPL5* mutation, whereas no patient with a cleft palate had an *RPS19* mutation [95]. Investigators have generated induced pluripotent stem cells (iPSCs) from DBA patients carrying mutations in *RPS19* and *RPL5*. These patient-derived iPSCs recapitulate the hematopoietic abnormalities of DBA, are amendable to genetic rescue, and may shed light on the nonhematopoietic abnormalities of DBA.

Studies applying WES identified a mutation at a splice site of *GATA1* in three male patients with DBA. Because two of the patients were siblings and the parents were unaffected, this suggests X-linked or autosomal recessive inheritance (in contrast to the ribosomal mutations described above, which typically have autosomal dominant inheritance). Previously, an identical mutation in *GATA1* was described in an X-linked form of dyserythropoietic anemia and thrombocytopenia. WES also has been used to successfully identify a mutation in a gene encoding a ribosomal protein, *RPS29* [84,96].

COMMERCIAL NGS-BASED ASSAYS FOR CLINICAL USE IN HEMATOLOGY PRACTICE

The introduction of NGS has totally changed our understanding of hematological diseases. However, the technology not only is being used for research purposes, but also is being implemented in clinical practice. Increasing numbers of NGS-based diagnostic assays, covering genes involved in the pathogenesis of hematological diseases, are being developed. As the market for NGS-based diagnostic procedures is rapidly growing, it is impossible to list all the available tests. Therefore, in this section we present selected examples of how the technique can be implemented in clinical diagnostics.

Similar to assays developed for diagnostics of solid tumors, the majority of tests offered for hematologic diseases consist of targeted gene panels based on the analysis of the published data on WGS and WES. For example, the BloodCenter of Wisconsin offers a HemeOnc Panel containing 30 genes associated with several myeloid hematological malignancies, including coding and junctional regions of the following genes: *ABL1*, *ASXL1*, *CBL*, *CEBPA*, *CSF3R*, *DNMT3A*, *ETV6*, *EZH2*, *FLT3*, *GATA2*, *IDH1*, *IDH2*, *JAK2*, *JAK3*, *KIT*, *KRAS*, *MPL*, *NPM1*, *NRAS*, *PHF6*, *PTPN11*, *RUNX1*, *SETBP1*, *SF3B1*, *SH2B3*, *SRSF2*, *TET2*, *TP53*, *U2AF1*, and *WT1*. The majority of the covered genes are related to AML pathogenesis; however, the test can also be useful for other diseases involving myeloid lineage including myeloproliferative neoplasms, myelodysplastic syndromes, chronic myeloid leukemia, severe congenital neutropenia, and chronic neutrophilic leukemia.

Another new assay, Rapid Heme Panel, has been developed at Dana–Farber/Brigham and Women's Cancer Center. The assay is based on an existing molecular testing platform developed for a Dana–Farber/Brigham and Women's Cancer Center genotyping research program (Profile) that sequences 305 genes implicated in various cancer types. The Rapid Heme Panel is one of the largest hematological NGS panels available today, as it has been designed to identify single mutations or DNA alterations in 95 genes that are frequently mutated in blood cancers mainly of myeloid lineage. The assay was developed to help with diagnostic procedures, treatment planning, and identification of patients who may be eligible for clinical trials, including early phase trials of new targeted therapies.

The FoundationOne Heme assay is the most comprehensive panel available so far. It was designed to study the entire coding sequence of 405 genes as well as selected introns of 31 genes involved in rearrangements. In addition to DNA sequencing, the FoundationOne Heme assay utilizes RNA sequencing to interrogate 265 genes somatically altered in human hematologic malignancies, sarcomas, and pediatric cancers.

In addition, panels covering genes typically mutated both in hematological malignancies and in solid tumors are also available. For example, SYMGENE68™ is a targeted NGS panel developed in the CellNetix laboratory covering 68 frequently mutated genes, including *TET2*, *EZH2*, *ASXL1*, *PIK3CA*, *JAK2*, *PTPN11*, *DNMT3A*, *EZH2*, and *PHF6*.

IMPLEMENTATION OF NGS-BASED TECHNIQUES IN CLINICAL PRACTICE IN HEMATOLOGY

It is as clear as day that clinical hematology could strongly benefit from broad implementation of next generation sequencing in clinical practice; however, the data on the technical performance of this methodology in clinical diagnostic settings are still limited. Many mutations detected in these tests are of unknown significance, making it difficult to implement the results in the clinical decision-making process. With a growing number of assays and barely any clinical guidelines to use them, NGS-based techniques cannot become the basis of almost any clinically relevant decision. Moreover, all the tests listed above are available only in the issuing laboratory, making the diagnostic process limited to a single center and increasing the cost and duration of analysis owing to the need to send the sample to the appropriate laboratory. Although there is an urgent need to evaluate the performance of NGS-based assays in clinical settings, only a few studies concerning that issue have been published so far and will be reviewed below.

In 2011, Kohlman et al. published the results of the Interlaboratory Robustness of Next-Generation Sequencing (IRON) study, which assessed the robustness, precision, and reproducibility of *TET2*, *CBL*, and *KRAS* amplicon NGS in chronic myelomonocytic leukemia samples [97]. The study was performed across 10 laboratories in eight countries, including Brazil, the United States, and six European Union countries, using the GS FLX platform and a 454 Life Sciences titanium emulsion PCR setup. The results showed a high concordance of mutation detection in all laboratories and high sensitivity of the NGS method in comparison to Sanger sequencing (mutations detected at frequency of 1–2% vs 20%). Overall, the results of the IRON study confirmed the technical feasibility and utility of amplicon-based deep sequencing in hematological settings and pointed toward the possibility of rapid implementation of this method in clinical diagnostics.

Another attempt to assess the utility of NGS in diagnostics of hematological malignancies has been made. A consortium of 10 European laboratories involved in routine AML molecular diagnostics has performed an interlaboratory comparison (round robin test) of the detection of known AML-related variants. The aim of the Evaluation of NGS in AML Diagnostics (ELAN) study was to assess the sensitivity, reproducibility, and accuracy of a highly multiplexed, single-tube assay simultaneously amplifying a total of 568 amplicons covering 54 entire genes or hot-spot gene regions involved in leukemia (TruSight myeloid sequencing panel; Illumina) in eight samples containing both AML patient material and commercially available test DNA with 10 known mutations. The results of this study showed a high sensitivity (variant frequency of 5%) and impressive quantitative accuracy of the studied assay and, according to the authors, this NGS-based approach appears feasible, even in the clinical setting [98,99].

CONCLUSION

Implementation of NGS technology enabled the discovery of several novel somatic mutations of previously unrecognized genes and shed new light on our understanding of the pathogenesis of hematological diseases. Many of these mutations are of pathogenetic importance in hematological malignancies, which makes them particularly interesting in terms of targeted treatment development as well as diagnosis and disease monitoring. Although the benefits of NGS testing in hematology are almost countless, including early disease diagnosis, risk stratification, and development of personalized therapeutic interventions, just to mention a few, there is still a huge gap between the release of new assays issued by various manufacturers/laboratories and clinical practice guidelines. Without well-designed, multicenter studies confirming the utility of these assays in clinical hematology, NGS-based techniques will still serve as an auxiliary method but not a standard approach in the clinical decision-making process.

REFERENCES

[1] Kobold DC, Steinberg KM, Larson DE, Wilson RK, Mardis ER. The next generation sequencing revolution and its impact on genomics. Cell 2013;155:27–38.
[2] International Cancer Genome Consortium, Hudson TJ, Anderson W, et al. International network of cancer genome project. Nature 2010;464:993–8.
[3] Cancer Genome Atlas Research Network. Genomic and epigenomic landscapes of adult de novo acute myeloid leukemia. N Engl J Med 2013;368:2059–74.
[4] Downing JR, Wilson RK, Zhang, et al. The pediatric cancer genome project. Nat Genet 2012;44:619–22.
[5] http://ocg.cancer.gov/programs/target.
[6] Braggio E, Egan JB, Fonseca R, Stewart AK. Lessons from next-generation sequencing analysis in hematological malignancies. Blood Cancer J 2013;3:e127. http://dx.doi.org/10.1038/bcj.2013.26.
[7] Pui CH, Relling MV, Downing JR. Acute lymphoblastic leukemia. N Engl J Med 2004;350:1535–48.
[8] Inaba H, Greaves M, Mulighan CG. Acute lymphoblastic leukemia. Lancet 2013;381:1943–55.
[9] Russell LJ, Capasso M, Vater I, et al. Deregulated expression of cytokine receptor gene, CRLF2, is involved in lymphoid transformation in B-cell precursor acute lymphoblastic leukemia. Blood 2009;114:2688–98.
[10] Mulligan CG, Collins-Underwood JR, Phillips LA, et al. Rearrangement of CRLF2 in B-progenitor- and Down syndrome-associated acute lymphoblastic leukemia. Nat Genet 2009;41:1243–6.
[11] Swerdlow SH, Campo E, Harris NL, Jaffe ES, Pileri SA, Stein H, et al., editors. WHO classification of tumours of haematopoietic and lymphoid tissues. Lyon: IARC; 2008.
[12] Mulligan CG, Goorha S, Radtke I, et al. Genome-wide analysis of genetic alterations in acute lymphoblastic leukaemia. Nature 2007;446:758–64.

[13] Kuiper RP, Schoenmakers EF, van Reijmersdal SV, et al. High-resolution genomic profiling of childhood ALL reveals novel recurrent genetic lesions affecting pathways involved in lymphocyte differentiation and cell cycle progression. Leukemia 2007;21:1258–66.

[14] Zhang J, Mullighan CG, Harvey RC, et al. Key pathways are frequently mutated in high-risk childhood acute lymphoblastic leukemia: a report from the children's oncology group. Blood 2011;118:3080–7.

[15] Weng AP, Ferrando AA, Lee W, et al. Activating mutations of NOTCH1 in human T cell acute lymphoblastic leukemia. Science 2004;306:269–71.

[16] Gutierez A, Sanda T, Grebliunaite R, et al. High frequency of PTEN, PI3K and AKT abnormalities in T-cell acute lymphoblastic leukemia. Blood 2009;114:647–50.

[17] [a] Mullighan CG. Genome sequencing of lymphoid malignancies. Blood 2013. http://dx.doi.org/10.1182/blood-2013-08-460311.
[b] http://cordis.europa.eu/result/rcn/153287_en.html.

[18] Zhang J, Ding L, Holmfeldt L, et al. The genetic basis of early T-cell precursor acute lymphoblastic leukemia. Nature 2012;481(7380):157–63. http://dx.doi.org/10.1038/nature10725.

[19] Ntziachristos P, Tsirigos A, Van Vlierberghe P, et al. Genetic inactivation of the polycomb repressive complex 2 in T cell acute lymphoblastic leukemia. Nat Med 2012;18:298–301.

[20] Neumann M, Heesch S, Schlee C, et al. Whole-exome sequencing in adult ETP-ALL reveals a high rate of DNMT3A mutations. Blood 2013;121:4749–52.

[21] Kraszewska MD, Dawidowska M, Kosmalska M, et al. BCL11B, FLT3, NOTCH1 and FBXW7 mutation status in T-cell acute lymphoblastic leukemia patients. Blood Cells Mol Dis 2013;50:33–8.

[22] Zuurbier L, Homminga I, Calvert V, et al. NOTCH1 and/or FBXW7 mutations predict for initial good prednisone response but not for improved outcome in pediatric T-cell acute lymphoblastic leukemia patients treated on DCOG or COALL protocols. Leukemia 2010;24:2014–22.

[23] O'Neil J, Look AT. Mechanisms of transcription factor deregulation in lymphoid cell transformation. Oncogene 2007;26:6838–49.

[24] Cullion K, Dreiham KM, Hermance N, et al. Targeting the NOTCH1 and mTOR pathways in a mouse T-ALL model. Blood 2009;113:6172–81.

[25] Mullighan CG, Miller CB, Radtke I, et al. BCR-ABL1 lymphoblastic leukaemia is characterized by the deletion of Ikaros. Nature 2008;453:110–4.

[26] Mullighan CG, Su X, Zhang J, et al. Children's oncology group. Deletion of IKZF1 and prognosis in acute lymphoblastic leukemia. N Engl J Med 2009;360:470–80.

[27] Den Boer ML, van Slegtenhorst M, De Menezes RX, et al. A subtype of childhood acute lymphoblastic leukaemia with poor treatment outcome: a genome-wide classification study. Lancet Oncol 2009;10:125–34.

[28] Loh ML, Zhang J, Harvey RC, et al. Tyrosine kinase sequencing of pediatric acute lymphoblastic leukemia: a report from the Children's oncology group TARGET Project. Blood 2013;121:485–8.

[29] Harvey RC, Mullighan CG, Chen IM, et al. Rearrangement of CRLF2 is associated with mutation of JAK kinases, alteration of IKZF1, Hispanic/Latino ethnicity, and a poor outcome in pediatric B-progenitor acute lymphoblastic leukemia. Blood 2010;115:5312–21.

[30] Roberts KG, Morin RD, Zhang J, et al. Genetic alterations activating kinase and cytokine receptor signaling in high-risk acute lymphoblastic leukemia. Cancer Cell 2012;22:153–66.

[31] Weston BW, Hayden MA, Roberts KG, Bowyer S, Hsu J, Fedoriw G, et al. Tyrosine kinase inhibitor therapy induces remission in a patient with refractory EBF1-PDGFRB-positive acute lymphoblastic leukemia. J Clin Oncol 2013;31:e413–6.

[32] Harrison CJ, Moorman AV, Broadfield ZJ, et al. Childhood and adult leukaemia working parties. Three distinct subgroups of hypodiploidy in acute lymphoblastic leukaemia. Br J Haematol 2004;125:552–9.

[33] Heerema NA, Nachman JB, Sather HN, et al. Hypodiploidy with less than 45 chromosomes confers adverse risk in childhood acute lymphoblastic leukemia: a report from the children's cancer group. Blood 1999;94:4036–45.

[34] Raimondi SC, Zhou Y, Mathew S, et al. Reassessment of the prognostic significance of hypodiploidy in pediatric patients with acute lymphoblastic leukemia. Cancer 2003;98:2715–22.

[35] Holmfeldt L, Wei L, Diaz-Flores E, et al. The genomic landscape of hypodiploid acute lymphoblastic leukemia. Nat Genet 2013;45:242–52.

[36] Villani A, Tabori U, Schiffman J, et al. Biochemical and imaging surveillance in germline TP53 mutation carriers with Li-Fraumeni syndrome: a prospective observational study. Lancet Oncol 2011;12:559–67.

[37] Powell BC, Jiang L, Muzny DM, et al. Identification of TP53 as an acute lymphocytic leukemia susceptibility gene through exome sequencing. Pediatr Blood Cancer 2013;60:E1–3.

[38] Shah S, Schrader KA, Waanders E, et al. A recurrent germline PAX5 mutation confers susceptibility to pre-B cell acute lymphoblastic leukemia. Nat Genet 2013. [published online ahead of print September 8, 2013].

[39] Raimondi SC, Pui CH, Head DR, Rivera GK, Behm FG. Cytogenetically different leukemic clones at relapse of childhood acute lymphoblastic leukemia. Blood 1993;82:576–80.

[40] Mulligan CG, Phillips LA, Su X, Ma J, Miller CB, Shurtleff SA, et al. Genomic analysis of the clonal origins of relapsed acute lymphoblastic leukemia. Science 2008;322:1377–80.

[41] Yang JJ, Bhojwani D, Yang W, et al. Genome-wide copy number profiling reveals molecular evolution from diagnosis to relapse in childhood acute lymphoblastic leukemia. Blood 2008;112:4178–83.

[42] Mulligan CG, Zhang J, Kasper LH, et al. CREBBP mutations in relapsed acute lymphoblastic leukaemia. Nature 2011;471:235–9.

[43] Inthal A, Zeitlhofer P, Zeginigg M, et al. CREBBP HAT domain mutations prevail in relapse cases of high hyperdiploid childhood acute lymphoblastic leukemia. Leukemia 2012;26:1797–803.

[44] Meyer JA, Wang J, Hogan LE, et al. Relapse-specific mutations in NT5C2 in childhood acute lymphoblastic leukemia. Nat Genet 2013;45:290–4.

[45] Tzoneva G, Perez-Garcia A, Carpenter Z, et al. Activating mutations in the NT5C2 nucleotidase gene drive chemotherapy resistance in relapsed ALL. Nat Med 2013;19:368–71.

[46] Grimwade D, Hills RK, Moorman AV, et al. Refinement of cytogenetic classification in acute myeloid leukemia: determination of prognostic significance of rare recurring chromosomal abnormalities among 5876 younger adult patients treated in the United Kingdom Medical Research Council trials. Blood 2010;116:354.

[47] Bhatia S, Neglia JP. Epidemiology of childhood acute myeloid leukemia. J Pediatr Hematol Oncol 1995;17:94–100.

[48] Frohling S, Scholl C, Gilliland DG, Levine RL. Genetics of myeloid malignancies: pathogenetic and clinical implications. J Clin Oncol 2005;23:6285–95.

[49] Schlenk RF, Dohner K, Krauter K, et al. Mutations and treatment outcome in cytogenetically normal acute myeloid leukemia. N Engl J Med 2008;358:1909–18.

[50] Dohner H. Implications of the molecular characterization of acute myeloid leukemia. Hematology 2007:412–7.

[51] Nucifora G, Rowley JD. AML1 and the 8;21 and 3;21 translocations in acute and chronic myeloid leukemia. Blood 1995;86:1.

[52] Nucifora G, Birn DJ, Erickson P, et al. Detection of DNA rearrangements in the AML1 and ETO loci and of an AML1/ETO fusion mRNA in patients with t(8;21) acute myeloid leukemia. Blood 1993;81:883.

[53] Marcucci G, Haferlach T, Dohner H. Molecular genetics of adult acute myeloid leukemia:prognostic and therapeutic implications. J Clin Oncol 2011;29:475–86.

[54] Meyer SC, Levine RL. Translational implications of somatic genomics in acute myeloid leukemia. Lancet Oncol 2014;15:e382–94.

[55] Ley TJ, Mardis ER, Ding L. DNA sequencing of a cytogenetically normal acute myeloid leukaemia genome. Nature 2008;456.

[56] Ley TJ, Ding L, Walter MJ, et al. DNMT3A mutations in acute myeloid leukemia. N Engl J Med 2010;363:2424–33.

[57] Jasielec J, Saloura V, Godley LA. The mechanistic role of DNA methylation in myeloid leukemogenesis. Leukemia 2014. http://dx.doi.org/10.1038/leu.2014.163.

[58] Welch JS, Ley TJ, Link DC, et al. The origin and evolution of mutations in acute myeloid leukemia. Cell 2012;150:264–78. http://dx.doi.org/10.1016/j.cell.2012.06.023.

[59] Abdel-Wahab O, Levine RL. Muttions in epigenetic modifiers in the pathogenesis and therapy of acute myeloid leukemia. Blood 2013;121:3563–72.

[60] Cagnetta A, Adamia S, Acharya C, et al. Role of genotype-based approach in the clinical management of adult acute myeloid leukemia with normal cytogenetics. Leukemia Res 2014;38:649–59.

[61] Ding L, Ley TJ, Larson DE, et al. Clonal evolution in relapsed acute myeloid leukemia revealed by whole genome sequencing. Nature 2012;481:506–10. http://dx.doi.org/10.1038/nature10738.

[62] The Cancer Genome Atlas Research Network. Genomic and epigenomic landscapes of adult de novo acute myeloid leukemia. N Engl J Med 2013;368:2059–74.

[63] Mardis ER, Ding L, Dooling DJ, et al. Recurring mutations found by sequencing and acute myeloid leukemia genome. N Engl J Med September 10, 2009;361(11):1058–66. http://dx.doi.org/10.1056/NEJMoa0903840.

[64] Tiacci E, Trifonov V, Schiavoni G, et al. Braf mutations in hairy cell leukemia. N Engl J Med 2011;364(24):2305–15. http://dx.doi.org/10.1056/NEJMoa1014209.

[65] [a] Maxson JE, Gotlib J, Pollyea DA. Oncogenic CSF3R mutations in chronic neutrophilic leukemia and atypical CML. N Engl J Med 2013;368(19):1781–90. http://dx.doi.org/10.1056/NEJMoa1214514.
[b] Sutton L-A, Ljungström V, Mansouri L, et al. Targeted next-generation sequencing in chronic lymphocytic leukemia: a high-throughput yet tailored approach will facilitate implementation in a clinical setting. Haematologica 2015;100:370–6. http://dx.doi.org/10.3324/haematol.2014.109777.

[66] Morton LM, Wang SS, Devesa SS, Hartge P, Weisenburger DD, Linet MS. Lymphoma incidence patterns by WHO subtype in the United States, 1992–2001. Blood 2006;107:265.

[67] Kaatsch P. Epidemiology of childhood cancer. Cancer Treat Rev 2010;36:277.

[68] Love C, Sun Z, Jima D, et al. The genetic landscape of mutations in Burkitt lymphoma. Nat Genet 2012;44:1321–5. http://dx.doi.org/10.1038/ng.2468. [Epub November 11, 2012].

[69] Giulino-Roth L, Wang K, MacDonald TY, et al. Targeted genomic sequencing of pediatric Burkitt lymphoma identifies recurrent alterations in antiapoptotic and chromatin-remodeling genes. Blood 2012;120:5181–4.

[70] Schmitz R, Young RM, Ceribelli M. Burkitt lymphoma pathogenesis and therapeutic targets from structural and functional genomics. Nature 2012;490:116–20. http://dx.doi.org/10.1038/nature11378. [Epub August 12, 2012].

[71] Campo E, Swerdlow SH, Harris NL, Pileri S, Stein H, Jaffe ES. The 2008 WHO classification of lymphoid neoplasms and beyond: evolving concepts and practical applications. Blood 2011;117:5019–32.

[72] Pasqualucci L, Dominguez-Sola D, Chiarenza A, et al. Inactivating mutations of acetyltransferase genes in B-cell lymphoma. Nature 2011;471:189–95.

[73] Pasqualucci L. The genetic basis of diffuse large B-cell lymphoma. Curr Opin Hematol 2013;20:336–44.

[74] Morin RD, Johnson NA, Severson TM, et al. Somatic mutations altering EZH2 (Tyr641) in follicular and diffuse large B-cell lymphomas of germinal-center origin. Nat Genet 2010;42(2):181–5.

[75] Morin RD, Mendez-Lago M, Mungall AJ, et al. Frequent mutation of histone-modifying genes in non-Hodgkin lymphoma. Nature 2011;476:298–303.

[76] Pasqualucci L, Trifonov V, Fabbri G, et al. Analysis of the coding genome of diffuse large B-cell lymphoma. Nat Genet 2011;43:830–7.

[77] Lohr JG, Stojanov P, Lawrence MS, et al. Discovery and prioritization of somatic mutations in diffuse large B-cell lymphoma (DLBCL) by whole-exome sequencing. Proc Natl Acad Sci USA 2012;109:3879–84.

[78] Zhang J, Grubor V, Love CL, et al. Genetic heterogeneity of diffuse large B-cell lymphoma. Proc Natl Acad Sci USA 2013;110:1398–403.

[79] Lenz G, Wright GW, Emre NC, et al. Molecular subtypes of diffuse large B-cell lymphoma arise by distinct genetic pathways. Proc Natl Acad Sci USA 2008;105:13520–5.

[80] Davis RE, Ngo VN, Lenz G, et al. Chronic active B-cell-receptor signalling in diffuse large B-cell lymphoma. Nature 2010;463:88–92.

[81] Trinh DL, Scott DW, Morin RD, et al. Analysis of FOXO1 mutations in diffuse large B-cell lymphoma. Blood 2013;121:3666–74.

[82] Sneeringer CJ, Scott MP, Kuntz KW, et al. Coordinated activities of wild-type plus mutant EZH2 drive tumor-associated hypertrimethylation of lysine 27 on histone H3 (H3K27) in human B-cell lymphomas. Proc Natl Acad Sci USA 2010;107:20980–5.

[83] Huether R, Dong L, Chen X. The landscape of somatic mutations in epigenetic regulators across 1000 pediatric cancer genomes. Nat Commun April 5, 2014:3630. http://dx.doi.org/10.1038/ncomms4630.

[84] Khincha PP, Savage SA. Genomic characterization of the inherited bone marrow failure syndromes. Semin Hematol 2013;50:333–47.

[85] Knies K, Schuster B, Amezine N, et al. Genotyping of fanconi anemia patients by whole exome sequencing: advantages and challenges. PLoS One 2012;7(12):e52648. http://dx.doi.org/10.1371/journal.pone.0052648.

[86] De Rocco D, Bogetta R, Cappelli E, et al. Molecular analysis of fanconi anemia: the experience of the bone marrow failure study group of the Italian association of pediatric onco-hematology. Hematologica 2014;99:1022–31.

[87] Chandrasekharappa SC, Lach FP, Kimble DC, Kamat A, Teer JK, Donovan FX. Massively parallel sequencing, aCGH, and RNA-Seq technologies provide a comprehensive molecular diagnosis of Fanconi anemia. Blood 2013;121:e138–48.

[88] Ballew BJ, Yeager M, Jacobs K, et al. Germline mutations of regulator of telomere elongation helicase 1, RTEL1, in dysceratosis congenita. Hum Genet 2013;132:473–80.

[89] Walne AJ, Vulliamy T, Kirwan M, Plagnol V, Dokal I. Constitutional mutations in RTEL1 cause severe dysceratosis congenita. Am J Hum Genet 2013;92:448–53.

[90] Klopocki E, Schulze H, Strauss G. Complex inheritance pattern resembling autosomal recessive inheritance involving a microdeletion in thrombocytopenia-absent radius syndrome. Am J Hum Genet 2007;80:232–40.

[91] Albers CA, Paul DS, Schulsze H. Compound inheritance of a low-frequency regulatory SNP and a rare null mutation in exon-junction complex subunit RBM8A causes TAR syndrome. Nat Genet 2012;44:435–9.

[92] Albers CA, Newbury-Ecob R, Ouwehand WH, Gheveart C. New insights into the genetic basis of TAR syndrome. Curr Opin Genet Dev 2013;23:316–23.

[93] Lipton JM, Ellis SR. Diamond-Blackfan anemia: diagnosis, treatment, and molecular pathogenesis. Hematol Oncol Clin N M 2009;23:261–82.

[94] Vlachos A, Rosenberg PS, Atsidaftos E, Alter BP, Lipton JM. Incidence of neoplasia in Diamond Blackfan anemia: a report from the Diamond Blackfan anemia Registry. Blood 2012;119:3815–9.

[95] Gazda HT, Preti M, Sheen MR. Frameshift mutation in p53 regulator RPL26 is associated with multiple physical abnormalities and specific pre-ribosomal RNA processing defects in Diamond–Blackfan anemia. Hum Mutat 2012;33:1037–44.

[96] Sankaran VG, Ghazvinian R, Do R, et al. Exome sequencing identifies GATA1 mutations resulting in Diamond-Blackfan anemia. J Clin Invest 2012;122:2439–43.

[97] Kohlmann A, Klein HU, Weissmann S, et al. The Interlaboratory Robustness of Next-generation sequencing (IRON) study: a deep sequencing investigation of TET2, CBL and KRAS mutations by an international consortium involving 10 laboratories. Leukemia December 2011;25(12):1840–8.

[98] Kohlmann A, Klein HU, Weissmann S, et al. Leukemia December 2011;25(12):1840–8.

[99] Thiede C, Bullinger L, Hernadez-Rivas J, et al. Blood (ASH Annual Meeting Abstracts). 2014. [Abstract number: 2374].

NEXT GENERATION SEQUENCING IN NEUROLOGY AND PSYCHIATRY

Krystyna Szymańska[1,2], Krzysztof Szczałuba[3,4], Anna Kostera-Pruszczyk[5], Tomasz Wolańczyk[2]

[1]*Department of Experimental and Clinical Neuropathology, Mossakowski Medical Research Centre, Polish Academy of Sciences, Warsaw, Poland;* [2]*Department of Child Psychiatry, Medical University of Warsaw, Poland;* [3]*MEDGEN Medical Center, Warsaw, Poland;* [4]*Medical Genetics Unit, Mastermed, Białystok, Poland;* [5]*Department of Neurology, Medical University of Warsaw, Poland*

CHAPTER OUTLINE

INTRODUCTION

Progress in molecular medicine has been accompanied by subsequent changes in the approach to neuropsychiatric diseases. The advent of genomics has made it possible to study these disorders from a biological perspective, to propose new nosology, and to identify genetic biomarkers and novel treatment targets. The breakthrough in sequencing technologies laid the foundation for a fast advance in our understanding of the pathogenic basis of numerous neurological disorders, yielding a limited set of underlying pathways and processes that appear to be implicated in neuronal dysfunction and neuronal loss. According to the recognized molecular pathology neuropsychiatric disorders may be classified as:

1. Defects in the development and maturation of the nervous system
2. Defects in functional proteins of the brain (transporters and receptors of neurotransmitters, ion channels, signal transduction pathways)
3. Defects in metabolic pathways (mitochondrial, lysosomal, peroxisomal disorders, enzymopathies, amino-acidopathies, organic acidurias, inherited defects in metabolism of purines and pyrimidines, creatine, vitamins, copper, and iron)
4. Neurodegenerative disorders
5. Neuromuscular disorders
6. Neuropsychiatric disorders with complex genetic backgrounds.

GENETIC TESTING IN NEUROLOGY AND PSYCHIATRY

Many neurological and psychiatric disorders follow a Mendelian pattern of inheritance but others have polygenic and multifactorial etiologies. Single-gene mutations, single-nucleotide polymorphisms (SNPs), trinucleotide repeats, chromosomal structural variations, or copy number variations (CNVs) constitute the genetic background of those disorders. Genetic testing is indicated for establishing the diagnosis and prognosis, enabling genetic counseling for patients and their families, prenatal testing, and development of novel therapeutic strategies. The use of genetic tests depends on a differential diagnosis, frequency of disease subtypes, test costs, and test availability. The selection of appropriate tests should be guided by the clinical phenotype, inheritance pattern of the suspected condition, and other features including magnetic resonance imaging (MRI) findings and metabolic testing. The comparative genome hybridization array (aCGH) technique helps to identify chromosomal structural variation or CNVs, largely replacing karyotyping. A specific sequence (subtelomeric anomalies or microdeletions) can be detected through fluorescence in situ hybridization (FISH) or multiplex ligation-dependent probe amplification (MLPA). For some triplet-repeat conditions, a gold standard has been the Southern blot. Sanger sequencing is the key platform for single-gene analysis, but with limited capacity.

In contrast, next generation sequencing (NGS) offers the opportunity to simultaneously sequence multiple genes, full exomes, or even whole human genomes and provides a powerful tool in the investigation of complex hereditary neuropsychiatric disorders. NGS helps in pathogenic variant identification and further prioritization of candidate variants [1]. As the classification criteria of different hereditary neurological diseases overlap, the majority of them can be clearly diagnosed only by panel testing of variants. Panels of multiple genes responsible for a certain neurological and psychiatric trait can be applied to target a spectrum of phenotypes such as epilepsy, movement disorders, extrapyramidal syndrome, developmental and peripheral nervous system disorders, etc. Various studies have demonstrated that whole-exome sequencing (WES) and whole-genome sequencing (WGS) can be successfully employed to identify causative mutations in previously known genes associated with disorders of the nervous system [1]. It is expected that WES will have important consequences for everyday clinical practice in neurology.

Defects in the development of the nervous system

Congenital malformations of the central nervous system (CNS) constitute a group of disorders defined as structural abnormalities caused by disturbances in cell organization or function within the brain and/or spinal cord. The current classification of CNS malformations integrates morphological and genetic data [2]. The framework of the classification was established in relation to the subsequent stages of the development of the CNS—disorders of neural tube formation and segmentation, aberrations in cell lineages, disorders of symmetry, malformations secondary to abnormal neuronal and glial proliferation or apoptosis, and malformations due to abnormal neuronal migration and secondary to abnormal postmigrational cortical organization and connectivity [2]. The underlying genetic causes of neurodevelopmental disorders implicate numerous genetic variants critical for normal brain development and subsequently responsible for various neuropsychiatric dysfunctions. The inheritance pattern may be autosomal dominant (e.g., *PAFAH1B1, TUBA1A, TUBB2B,* and *TUBB3*), autosomal recessive (e.g., *RELN, VLDLR, TUBA8*), or X-linked (e.g., *DCX, ARX, FLNA*).

Malformations of cortical development

Malformations of cortical development (MCD) are an important group in which the clinical manifestation is dominated by epilepsy and cognitive impairment. More than 100 variants responsible for MCD are reported [3,4]. The large number of candidate genes for malformations of cortical development makes WES the test of choice. In addition, successively performed studies based on WES methodology confirmed the hypothesis that MCD-related genes are involved at multiple developmental stages, and neuronal proliferation and migration are genetically and functionally interdependent [5]. This increases the difficulty of selecting a gene for Sanger sequencing. Moreover, exome sequencing identified several new gene candidates. Bilgüvar et al. found recessive mutations in the *WDR62* gene responsible for a wide spectrum of severe MCD and cerebellar abnormalities [6]. Furthermore, WES of trio samples revealed a *de novo TUBA1A* mutation in a sporadic case [7]. Willemsen et al. have shown that *de novo* mutations in *DYNC1H1* can be linked to variable phenotypes, including severe developmental disability with variable neuronal migration defects and peripheral neuropathy [8].

Great interest has been paid to the group of diseases called ciliopathies with primary cilium dysfunction (a nonmotile cilium is a subcellular organelle playing a key role in embryonic development and contributing to tissue maintenance and regeneration). Ciliopathies relevant to neurology include Joubert and Meckel-Gruber syndromes.

- Joubert syndrome, a clinically and genetically heterogeneous group of disorders, is characterized by cerebellar vermis aplasia/hypoplasia and other brain and spinal malformations. A character-istic MRI finding in this condition is the "molar tooth"—an enlarged fourth ventricle secondary to abnormalities of the vermis and superior cerebellar peduncles. The clinical symptoms include developmental delay, hypotonia, ataxia, oculomotor abnormality (oculomotor apraxia), and breathing disturbances (hyperpnea/hypopnea) [9].
- Meckel-Gruber syndrome is a common neural tube defect characterized by a classic triad of clini-cal features as occipital encephalocele, polycystic kidneys, and liver fibrosis.

NGS has revolutionized gene identification and genome diagnostics in the ciliopathy field. Muta-tions in more than 100 ciliopathy genes have been identified, including more than 20 genes relevant to neurological disorders with autosomal recessive or X-linked inheritance. A very broad spectrum of phenotypic severity and variability in the same homozygous mutation in siblings remains an unan-swered question [10,11]. The panel of genes related to Joubert/Meckel-Gruber syndrome and orofa-ciodigital syndrome includes *AHI1, ARL13B, B9D1, B9D2, C5orf42, CC2D2A, CEP290, CEP41, CSPP1, INPP5E, KIF7, MKS1, NPHP1, NPHP3, PDE6D, OFD1, RPGRIP1L, TCTN1, TCTN2, TCTN3, TMEM67, TMEM138, TMEM216, TMEM231, TMEM237,* and *TTC21B*.

Inherited metabolic disorders

Inherited metabolic disorders are commonly present with neurological symptoms.

The main categories of metabolic disorders include disorders of amino acid and organic acid metab-olism, lysosomal and peroxisomal disorders, mitochondrial diseases, disorders of glucose and creatine metabolism, and disorders of neurotransmission.

Mitochondrial diseases

Mitochondrial diseases can be caused by mutations in about 1200 nuclear genes and in 37 mitochon-drial genes with a maternal inheritance pattern [12]. Therefore, selecting a suitable gene for Sanger sequencing may encounter substantial difficulties [12]. Calvo et al. applied a MitoExome sequencing panel of the entire mitochondrial DNA and exons of 1034 nuclear genes encoding mitochondrial pro-teins in 42 infants with clinical and biochemical evidence of mitochondrial oxidative phosphorylation disease [13]. Ten patients (24%) had mutations in genes previously linked to a disease and 13 patients (31%) had mutations in nuclear genes never linked to any disease, and the causative variant was not found in approximately half of the 42 sequenced patients. The lack of molecular diagnosis in 19 patients could be due to the fact that the causative variant was located in a nontargeted intronic or regulatory region [13]. Accordingly, Nishri et al. described a patient suspected to have a Leigh-like syndrome according to clinical and MRI findings; however, exome sequencing revealed a heterozygotic mutation in *GFAP* consistent with Alexander syndrome [14]. Furthermore, Fraser et al. described siblings with Leigh-like disease. The extensive laboratory examinations were nondiagnostic; only urine organic acids demonstrated elevations in lactic, α-ketoglutaric, and fumaric acids. WES demonstrated a homozygous thiamine pyrophosphokinase mutation. As thiamine pyrophosphate is a cofactor in many enzymatic reactions, including mitochondrial enzymes, its defects causes progressive encephalopathy, which can be prevented by early cofactor supplementation. This example clearly illustrates the importance of exome sequencing as a fast and reliable diagnostic tool in unresolved cases [15].

Congenital glycosylation disorders

Congenital glycosylation disorders (CGDs) are a group of genetic defects in glycans synthetic pathways expressed as a variety of phenotypes [16]. At least 2% of the human genome encodes glycan biosynthesis and recognition proteins [17]. To date more than 100 CDGs have been identified according to mutations in relevant glycosylation-related genes [17]. With the advances in NGS, there has been important progress in the diagnosis of these neurometabolic disorders. New CDGs have been described based on WES and WGS findings. It is remarkable that in 2013 a new CDG was reported, on average, every 17 days [17].

Lysosomal storage disorders

Lysosomal storage disorders (LSDs) are a group of over 50 inborn errors of metabolism characterized by the accumulation of specific substrates as a result of mutations in genes encoding lysosomal enzymes. Most of the affected individuals have neurological symptoms. The early diagnosis is of great importance because specific therapy for some LSDs is available. The diagnosis may be difficult, because of variable signs and symptoms, especially in the late-onset juvenile and adult forms [18]. Because of phenotype diversity and limitations of traditional laboratory methods, the diagnostic process may be slow [18]. Even though the diagnosis can be frequently confirmed on the basis of current biochemical assays, genetic testing should also be performed to provide counseling for the family. Fernández-Marmiesse et al. applied NGS testing in a group of 84 early-onset neurodegenerative disorders patients (potentially LSDs), including 18 patients with previously identified mutations and 66 unclassified subjects [18]. In the first group, all but one previously known mutations were detected by NGS. The missing one was not covered by hybridization baits, owing to its location close to repetitive regions on exon 1 of *NAGLU*—mucopolysaccharidosis IIIB (Sanfilippo type B). In the remaining group, the genetic diagnosis of LSD was obtained for 26 patients [18]. The performed analysis confirmed the effectiveness of NGS in the diagnostic investigation of neurometabolic disorders. The further advantage of NGS was the ability to detect gross deletions and insertions (macrodeletions were found in two genes) [18].

The diagnosis of neurometabolic diseases is often difficult because of their complex phenotypes. To diagnose these conditions the patient has to be subjected to an integrated set of clinical, neuroimaging, and sophisticated biochemical tests. The analysis of their results together with genetic testing is warranted to obtain a diagnosis in difficult cases. Genetic testing provides the final confirmation of clinical suspicions or constitutes an essential addition to prior examinations performed to reach a clinical diagnosis [19].

Table 1 presents genes included in diagnostic panels for metabolic disorders.

EPILEPSY

Epilepsy is defined as a disorder of the brain characterized by an enduring predisposition to seizures [20]. It is believed that genetic causes play a role in at least 70% of epileptic patients [21]. Inherited epilepsies can be classified as

- monogenic epilepsies (channelopathies caused by defects in genes regulating voltage or ligand-gated ion channels) and
- complex epilepsies, usually with no family history of seizures (defects in the function of excitatory or inhibitory neurotransmitters caused by the interaction of a few or several genetic variants) [21–23].

The highly heterogeneous genetic background of epilepsies and phenotypic variability make the preselection of an appropriate gene for sequencing difficult [24].

Table 1 Genes Included in Diagnostic Panels for Metabolic Disorders	
Lysosomal disorders	*AGA, ARSB, CTSA, FUCA1, GALC, GALNS, GLB1, GNPTAB, GNPTG, GNS, GUSB, HEXA, HEXB, HGSNAT, HYAL1, IDS, IDUA, MCOLN1, NAGLU, NEU1, NPC1, NPC2, PSAP, SGSH, SLC17A5, SMPD1, SUMF1*
Congenital disorders of glycosylation	*ALG1, ALG2, ALG3, ALG6, ALG8, ALG9, ALG11, ALG12, ALG13, B4GALT1, COG1, COG4, COG5, COG6, COG7, COG8, DDOST, DOLK, DPAGT1, DPM1, DPM2, DPM3, MAN1B1, MGAT2, MOGS, MPDU1, MPI, PGM1, PMM2, RFT1, SLC35A1, SLC35A2, SLC35C1, SRD5A3, TMEM165*
Peroxisomal disorders	*ABCD1, AMACR, PEX1, PEX2, PEX3, PEX5, PEX6, PEX7, PEX10, PEX11B, PEX12, PEX13, PEX14, PEX16, PEX19, PEX26*
Mitochondrial DNA-encoded mitochondriopathies	*MT-ND4, MT-ATP6, MT-ATP8, MT-CO1, MT-CO2, MT-CO3, MT-CYB, MT-DLOOP, MT-ND1, MT-ND2, MT-ND3, MT-ND4L, MT-ND5, MT-ND6, MT-RNR1, MT-RNR2, MT-TA, MT-TC, MT-TD, MT-TE, MT-TF, MT-TG, MT-TH, MT-TI, MT-TK, MT-TL1, MT-TL2, MT-TM, MT-TN, MT-TP, MT-TQ, MT-TR, MT-TS1, MT-TS2, MT-TT, MT-TV, MT-TW, MT-TY*
Nuclear-encoded mitochondriopathies	*AARS2, ABCB7, ACAD9, ACADM, ACADS, ACADVL, ACO2, ADCK3, AFG3L2, AGK, AIFM1, ALAS2, AMPD1, APTX, ATL1, ATP13A2, ATP5E, ATPAF2, AUH, BCS1L, BOLA3, C10orf2, C12orf62, C12orf65, CISD2, COA5, COQ2, COQ6, COQ9, COX10, COX14, COX15, COX20, COX4I2, COX6B1, CPT1A, CPT2, D2HGDH, DARS2, DECR1, DGUOK, DLAT, DLD, DNAJC19, DNM1L, EARS2, EIF2AK3, ETFA, ETFB, ETFDH, ETHE1, FASTKD2, FH, FOXRED1, FXN, GAMT, GARS, GATM, GDAP1, GFER, GFM1, GLRX5, HADH, HADHA, HADHB, HARS2, HSPD1, IDH2, ISCU, KARS, KIF5A, L2HGDH, LAMP2, LIAS, LRPPRC, MARS2, MFN2, MPC1, MPV17, MRPS16, MRPS22, MTFMT, MTO1, MTPAP, NDUFA1, NDUFA10, NDUFA11, NDUFA12, NDUFA2, NDUFA9, NDUFAF1, NDUFAF2, NDUFAF3, DUFAF4, NDUFAF5, NDUFAF6, NDUFB3, NDUFB9, NDUFS1, NDUFS2, NDUFS3, NDUFS4, NDUFS6, NDUFS7, NDUFS8, NDUFV1, NDUFV2, NFU1, NUBPL, OPA1, OPA3, OXCT1, PANK2, PC, PDHA1, PDHB, PDHX, PDP1, PDSS1, PDSS2, PFKM, PYGM, PNPT1, POLG, POLG2, PUS1, RARS2, REEP1, RMND1, RRM2B, SARS2, SCO1, SCO2, SDHA, SDHAF1, SDHAF2, SDHB, SDHC, SDHD, SERAC1, SLC19A2, SLC19A3, SLC22A5, SLC25A12, SLC25A19, SLC25A20, SLC25A3, SLC25A38, SLC25A4, SLC33A1, SLC6A8, SPAST, SPG20, SPG7, SUCLA2, SUCLG1, SURF1, TACO1, TAZ, TIMM8A, TK2, TMEM126A, TMEM70, TPK1, TRMU, TSFM, TTC19, TUFM, TYMP, UQCRB, UQCRC2, UQCRQ, WFS1, XPNPEP3, YARS2*

The familial Mendelian inheritance form of epilepsy has been attributed to mutations in

- the potassium channel genes, such as *KCNQ2* and *KCNQ3,* in benign familial neonatal seizures;
- *PRRT2* gene (encoding a transmembrane protein and an ion channel regulator) in benign familial neonatal epilepsy;
- *SCN1A, SCN2A,* and *SCN1B* genes (encoding voltage-gated sodium channel containing α and β subunits) in benign familial neonatal infantile seizures [25,26].

Table 2 Genes Related to Selected Epileptic Syndromes

Progressive myoclonic epilepsy	*SCARB2, PRICKLE2, KCTD7, COL6A2, CERS1, CERS2*
Benign familial neonatal convulsions	*KCNQ3, KCNQ2, SCN2A*
Generalized epilepsy with febrile seizures plus	*SCN1A, SCN1B, GABRD, GABRG2*
Dravet syndrome	*SCN1A, SCN1B, GABRD, GABRG2, GABRA1*
Malignant migrating partial seizures of infancy	*TBC1D24, PLCB1, SCN1A, KCNT1*
Epileptic encephalopathies	*KCNQ2, SCN2A, CDKL5, KCTD7, ARHGEF9, CHD2, SYNGAP1, MBD5, STXBP1, SCN8A, ALG13, HNRNPU, GNAO1, SLC25A22, SLC35A3, IQSEC2, GRIN1, HDAC4, HCN1, GABRB3, TBC1D24*

In 2012, it was found that mutations in *KCNQ2* underlie early-onset epileptic encephalopathy with a suppression-burst electroencephalography (EEG) pattern [27,28].

The spectrum of clinical symptoms secondary to a mutation in *SCN1A* comprises

- generalized epilepsy with febrile seizures plus,
- simple febrile seizures,
- benign familial neonatal–infantile seizures,
- severe myoclonic epilepsy in infancy and related syndromes (Dravet syndrome), myoclonic–astatic epilepsy (Doose syndrome) [25,29].

The identification of an *SCN1A* pathogenic variant may influence the choice of drugs in patients with treatment-resistant epilepsy. Additionally, a mutation in this gene is described in familial hemiplegic migraine (migraine with aura characterized by hemiparesis during the episodes) [30]. *De novo* mutations in genes encoding nicotinergic cholinergic receptors, ion channels, and other molecular pathways have been identified as causing autosomal dominant nocturnal frontal lobe epilepsy, malignant migrating partial seizures in infancy, and epileptic encephalopathies. Carvill et al. performed WES on 13 patients and targeted resequencing on 67 patients, all with *SCN1A*-negative Dravet syndrome. They detected disease-causing mutations in two novel genes for Dravet syndrome, *GABRA1* and *STXBP1* [31]. They also found three patients with previously undetected *SCN1A* mutations, despite screening for point mutations using denaturing high-performance liquid chromatography or bidirectional sequencing and a lack of large copy number variants [31].

Table 2 presents genes related to selected epileptic syndromes.

The diagnostic panels for epilepsy include genes encoding ion channels, neurotransmitter receptor subunits, cellular pathway proteins, and genes specific for many well-defined diseases for which epilepsy is a characteristic symptom. Exome sequencing has enabled the rapid detection of *de novo* point mutations in case–parents trios. It is likely that *de novo* mutagenesis plays a large role in both Mendelian and polygenic epilepsies [21]. At present the diagnostic panels for epilepsy including up to 400 variants (CeGaT GmbH) are commercially available [23].

The large group of inherited metabolic diseases is often accompanied by epilepsy. This group includes glucose transporter 1 deficiency (*SLC2A1*), disorders of B6 metabolism (*ALDH7A1, PNPO*), biotinidase deficiency (*BTD*), serine deficiency (*PHGDH*), and neurodegeneration due to cerebral folate transporter deficiency (*FOLR1*). All these genes should be incorporated in NGS epilepsy/seizures panels, as these diseases require early and precise diagnosis and prompt introduction of specific therapy to avoid irreversible damage of the brain. On the other hand, there is no specific therapy for many inherited metabolic diseases and accompanying epilepsy is drug resistant [23,32].

Epilepsy is also common in neurodevelopmental disorders. NGS technologies have enabled the detection of specific variants of *MECP2, FOXG1,* and *CDKL5* responsible for neurodevelopmental disorders with concomitant epilepsy [23]. Ng et al. applied WES to 10 unrelated individuals with Kabuki syndrome to discover underlying mutations in the *MLL2* gene [33].

Rett syndrome (OMIM 312750) is an X-linked neurodevelopmental condition with mutations in the gene encoding methyl-CpG-binding protein 2 (*MECP2*). The atypical variants of Rett syndrome may be due to mutations in the cyclin-dependent kinase-like 5 gene (*CDKL5*) (early seizure variant), in forkhead box G1 (*FOXG1*) (congenital variant), and also in *MECP2* [34,35]. Mutation in *CDKL5* underlies an early-onset epileptic encephalopathy with severe neurological impairment and a female-to-male ratio of about 12:1 [36,37]. Grillo et al. described two pairs of sisters carrying the same mutation in *MECP2*, but in both pairs one of the siblings presented with classical Rett syndrome, while the other child presented with the Zappella variant [38]. Whereas this discrepancy could be due to X-chromosome inactivation (XCI) status, it was shown that all the girls had balanced XCI, indicating that other factors contributed to the phenotypic outcome [38]. To provide insight into expression variability WES was performed, demonstrating that each subject had multiple mutations resulting in functional variants that could exacerbate or ameliorate the final clinical outcome [38]. This study highlights the leading role of NGS in understanding the diversity of phenotypes caused by the same pathogenic mutation.

Table 3 presents genes related to selected metabolic and structural epilepsies.

Table 3 Genes Related to Selected Metabolic and Structural Epilepsies [23,32]	
Pyridoxine-dependent epilepsy	*ALDH7A1*
Pyridoxal 5′-phosphate-dependent epilepsy	*PNPO*
Glucose transporter type 1 deficiency syndrome	*SLC2A1*
Glycine encephalopathy	*AMT, GCSH, GLDC*
Creatine deficiency syndromes	*GAMT, GATM*
Adenylosuccinate lyase deficiency	*ADSL*
POLG-related disorders (i.e., Alper–Huttenlochner syndrome)	*POLG*
Infantile neuronal ceroid lipofuscinosis 1, late infantile neuronal ceroid lipofuscinosis	*PPT1, TPP1*
Unverricht-Lundborg disease	*CSTB*
Brain malformations	*ACTB, ACTG1, AHI1, ARFGEF2, ARX, CASK, CC2D2A, CEP290, CEP41, CHMP1A, DCX, EOMES, EXOSC3, FKRP, FKTN, FLNA, GPR56, KIAA1279, KIF7, LAMC3, LARGE, MKS1, NPHP1, OPHN1, PAX6, PEX7, PAFAH1B1, POMGNT1, POMT1, POMT2, PQBP1, RAB3GAP1, RAB3GAP2, RARS2, RELN, RPGRIP1L, RTTN, SRPX2, TMEM138, TMEM216, TMEM237, TMEM67, TSEN2, TSEN34, TSEN54, TUBA1A, TUBA8, TUBB2B, TUBB3, VLDLR, VRK1, WDR62*
Tuberous sclerosis	*TSC1, TSC2*
Rett syndrome	*MECP2*
Mental retardation, autistic symptoms, ataxia, epilepsy	*MEF2C*

Much attention has been paid to the roles of intracellular proteins and their interactions, signaling cascades, and feedback regulation in the development of epilepsy. Special emphasis has been placed on the mammalian target of rapamycin (mTOR) pathway, which plays an essential role in cell growth, differentiation, proliferation, metabolism, protein synthesis, and autophagy [39]. The mTOR is a serine–threonine protein kinase that forms two complexes: rapamycin-sensitive mTORC1 and the relatively rapamycin-insensitive mTORC2 [39]. Mutations in the genes for tuberous sclerosis complex 1 protein (TSC1) or TSC2 lead to abnormal cellular differentiation, proliferation, and growth [40]. The tuberous sclerosis complex is a consequence of mTOR pathway dysregulation. Tuberous sclerosis is associated with hamartoma or tumor formation in multiple organs, including subependymal giant cell astrocytomas and renal angiomyolipomas. The mTOR pathway is also implicated in the pathogenesis of other brain tumors, particularly some types of gliomas. Rivière et al. performed WES on patients with early postnatal megalencephaly (megalencephaly–polymicrogyria–polydactyly–hydrocephalus and megalencephaly–capillary malformation syndromes). They identified *de novo* germ-line and somatic mutations in components of the PI3K/AKT/mTOR pathway [41]. Lee et al. performed WES on DNA from resected brain tissue and blood of five patients with hemimegalencephaly and proved that somatic mutations in components of the PI3K/AKT/mTOR pathway limited to the brain are responsible for this malformation [42]. The TSC and megalencephaly/hemimegalencephaly syndromes are often associated with treatment-resistant epilepsy.

Familial focal epilepsy with variable foci (FFEVF) is an autosomal dominant epilepsy characterized by focal seizures arising from different cortical regions in different family members (OMIM 604364). Causative mutations underlying FFEVF were found in *DEPDC5* by WES [43]. Further, Scheffer et al. confirmed this etiology, emphasizing the clinicoradiological phenotypes with malformations in some patients varying from focal cortical dysplasia to subtle band heterotopia in MRI [44]. *DEPDC5* mutations overactivate mTOR signaling and therefore FFEVF belongs to a group of "mTO-Ropathies" [44]. These studies emphasize an important role for the mTOR pathway in focal epilepsies.

NGS screening panels for epilepsy are fast and relatively inexpensive tools in the diagnostics of monogenic epilepsies. NGS epilepsy/seizures panels differ in the selection of genes, and some of the panels include infrequent and thus not routinely screened variants [24] as well as genes responsible for nonepileptic paroxysmal disorders such as hyperekplexia (usually caused by *GLRA1* mutation) or migraine. The impact of secondary or modulatory genes is an important issue in the genetic study of epilepsy and requires further research. The influence of *SCN9A*, *CACNB4*, and *CACNA1A* genes on the *SCN1A*-related Dravet syndrome is an example of such relationship [45]. An important problem is whether single variants of strong effect or a combination of multiple variants with weaker effects underlie susceptibility to epilepsy in a patient [21].

ATAXIA

Ataxia is defined as a lack of movement coordination (irregular character of movement or posture, dysmetria, dyssynergia, dysdiadochokinesia, dysartria, oculomotor disturbances). Hereditary ataxia results from neurodegeneration of the spinal cord (Friedreich ataxia) or cerebellum (ataxia–telangiectasia, Niemann Pick disease, Marinesco–Sjögren syndrome), with most cases featuring both to some extent (spinocerebellar ataxia) [46]. Gait ataxia may also be secondary to peripheral sensory dysfunction [46]. Cerebellar ataxia occurs frequently in neurodegenerative and inherited metabolic disorders.

The inherited cerebellar ataxias (CAs) are a genetically heterogeneous group of neurodegenerative disorders including at least 37 autosomal dominant CAs, more than 20 autosomal recessive CAs, X-linked ataxias, and several forms of ataxia associated with mitochondrial defects [47,48]. Several mutation types have been identified—repeat expansions in either coding (CAG—polyglutamine) or noncoding (CTG, CAG, TGGAA, ATTCT, GGCCTG) parts of genes, missense mutations, deletions, duplications, and splice and truncating mutations [48]. Causative mutations for cerebellar atrophy and ataxia have been found in more than 150 genes [49]. The inherited CAs may present as an isolated cerebellar syndrome or be associated with a spectrum of neurological symptoms (pyramidal or extrapyramidal syndrome, peripheral neuropathy, cognitive dysfunction, seizures) [47].

Ataxia may also accompany inherited metabolic disorders (e.g., Niemann Pick type C, Wilson disease, Refsum disease, congenital disorders of glycosylation) [50]. Ataxia–telangiectasia is caused by a defect in the *ATM* gene, which encodes a protein detecting DNA double-strand breaks, being responsible for DNA repair. *ATM* is a huge gene with 66 exons. Because of the overlapping phenotypes the appropriate diagnosis of ataxia-related disease is difficult. NGS is a powerful diagnostic tool for neurological disorders with concomitant ataxia; however, it has limitations due to its poor ability to sequence repetitive DNA expansions such as the polyglutamine repeats that are common in spinocerebellar ataxias [51]. As an illustration, Friedreich ataxia is caused by expanded *GAA* repeats in the first intron of the *FXN* gene, which is located on chromosome 9 and encodes the protein frataxin. The mutation causes gene silencing and leads to frataxin functional deficiency. Such repeat expansions can be easily tested using standard PCR methods [52]. Németh et al. used NGS to diagnose 50 patients with ataxia who had failed to be previously diagnosed [52]. The exonic and 25-bp intronic flanking sequences of 118 genes were selected. The detection rate was 18% and varied from 8.3% in those with an adult-onset progressive disorder to 40% in those with a childhood or adolescence onset [52]. The undiagnosed group included patients with CNVs and larger deletions/insertions [52]. Ohba et al. performed WES on 25 patients from 23 families with cerebellar and/or vermis atrophy of unknown origin [49]. They identified 15 pathological mutations in seven genes within nine families. An atypical mild phenotype of Zellweger syndrome with no white matter abnormalities, caused by *PEX16* mutations, was reported in one patient [49]. Lines et al. performed WES on three adult siblings with slowly progressive, juvenile-onset cerebellar atrophy and ataxia accompanied by intellectual disability, hearing loss, hypogonadism, and demyelinating sensorimotor neuropathy [53]. The course of this disease is usually severe, with onset in infancy. The elevation of very long chain fatty acids is a very characteristic feature of this disorder. In the examined patients compound heterozygous causative mutations in *HSD17B4*, encoding peroxisomal D-bifunctional protein, were identified by WES [53]. The above reports prove that NGS is a method of choice for neurometabolic diseases with an atypical course. However, the selection of a large panel of genes and WES or even WGS in ataxias of unknown origin is warranted [49,52].

Most autosomal dominant ataxias belong to the group of spinocerebellar ataxias (SCAs) and episodic ataxias. SCAs are clinically and genetically heterogeneous. A common feature is dysfunction of the cerebellum [54]. The classification of SCAs is based on genetic findings (SCA1 to SCA38 with vacant numbers 9 and 24) [55]. Other autosomal dominant disorders with ataxia are dentatorubropallidoluysian atrophy, vanishing white matter, Alexander disease, and Huntington disease. SCA35 was the first dominant ataxia to be identified through WES. Missense mutations in the cerebral transglutaminase gene (*TGM6*) were reported in two families [56]. Autosomal recessive CAs include Friedreich ataxia, ataxia–telangiectasia, Wilson disease, ataxia with vitamin E deficiency, abetalipoproteinemia, ataxia with oculomotor apraxia (AOA1 and AOA2), spastic ataxia of Charlevoix-Saguenay, spinocerebellar ataxia with epilepsy, ataxia associated with coenzyme Q10 deficiency, ataxia in lysosomal storage disorders, CDG1A, and Niemann–Pick type C disease.

In countries where large families are rare, autosomal recessive CAs often present as sporadic cases [57]. The same phenotype of autosomal recessive (AR) cerebellar ataxia may be due to different variants; on the other hand, mutations in the same AR CA gene may be responsible for distinct phenotypes; both conditions cause diagnostic difficulties [57]. By combining SNP array-based linkage analysis and targeted resequencing of relevant sequences in the linkage interval with the use of NGS, Vermeer et al. identified a mutation in the *ANO10* gene in patients with AR CA, downbeat nystagmus, and involvement of lower motor neurons [58]. Doi et al., using WES, identified a homozygous missense mutation in *SYT14* encoding synaptotagmin XIV in two patients with AR CA [59]. They also analyzed the expression of *SYT14* mRNA in human fetal and adult brain tissue and showed that *SYT14* is localized in Purkinje cells of the cerebellum [59]. Furthermore, WES allowed identification of a *STUB1* mutation in two patients with Gordon Holmes syndrome—a rare neurodegenerative disorder with ataxia and hypogonadism [61]. So far, this disorder has been associated with mutation in the E3 ligase gene *RNF216* and deubiquitinase gene *OTUD4* [60]. These authors demonstrated that the *STUB1* mutation causes a loss of function of CHIP (C-terminus of HSC70-interacting protein) acting as a molecular cochaperone, autonomous chaperone, and ubiquitin E3 ligase [61]. Four research groups found *STUB1* variants in families with AR CA with and without hormonal disturbances [62]. These studies show a significant role for WES in understanding the pathomechanisms of neurodegeneration.

Ataxia can also be associated with certain X-linked disorders. This group includes fragile X syndrome, tremor ataxia syndrome, and adrenomyeloneuropathy [46].

The most frequent ataxias associated with mitochondrial DNA mutations are mitochondrial encephalomyopathy, lactic acidosis, and stroke-like episodes; myoclonic epilepsy with ragged-red fibers; neuropathy, ataxia, and retinitis pigmentosa; and Kearns–Sayre syndromes. One of the most common mitochondrial disorders with ataxia of the first years of life is Leigh syndrome. Genetic heterogeneity in Leigh syndrome includes subunits of the mitochondrial electron transport chain complex, *SURF1* mutation, deficits in the coenzyme Q10 and pyruvate dehydrogenase complex (OMIM 256000), and Leigh-like syndrome with *SERAC1* mutation (OMIM 614739).

According to the above evidence, the selection of a gene panel for targeted sequencing in ataxias is not easy because of their heterogeneity. Furthermore, NGS does not screen for trinucleotide repeat expansions that underlie the majority of SCAs, autosomal recessive Friedreich ataxia, and a variety of other hereditary forms of ataxia. Undoubtedly, an NGS panel should include metabolic diseases with ataxia, mitochondrial diseases, and genes expressed in the cerebellum. Genetic testing guidelines for patients with ataxias have been proposed by the European Federation of Neurological Societies [63].

HEREDITARY SPASTIC PARAPLEGIAS

Hereditary spastic paraplegias (HSPs) are a group of clinically and genetically heterogeneous neurodegenerative disorders characterized by progressive lower limb spasticity and weakness. Other manifestations may occur, including cognitive impairment, epilepsy, ataxia, and extrapyramidal symptoms, in addition to the "pure" phenotype [64]. Nonneurological features include optic atrophy and retinal degeneration, cataract, and deafness [64]. HSPs are classified according to the spastic paraplegia genetic variant (SPG) and the mode of inheritance. HSP can be inherited in an autosomal dominant,

Table 4 Examples of Diagnostic Panels—Movement Disorders (www.dnalabsindia.com/dna_test.php)

Cerebellar ataxia panel	*ADCK3, APTX, COQ2, COQ9, DNMT1, FXN, PDSS1, PDSS2, POLG, SACS, SETX, SYNE1, TTPA, VLDLR*
Spastic paraplegia, autosomal dominant NGS panel	*ATL1, BSCL2, HSPD1, KIAA0196, KIF5A, NIPA1, REEP1, RTN2, SLC33A1, SPAST, ZFYVE27*
Spastic paraplegia, autosomal recessive NGS panel	*AP4M1, CYP2U1, CYP7B1, FA2H, GJC2, KIF1A, PNPLA6, SPG11, SPG20, SPG21, SPG7, ZFYVE26*
Parkinson disease NGS panel	*SNCA, LRRK2, VPS35, PARK2, PINK1, PARK7, ATP13A2, PLA2G6, FBXO7, DNAJC6*
Dystonia NGS panel	*TOR1A, THAP1, GCH1, TH, SPR, SLC2A1, CIZ1, PRRT2, PNKD*

autosomal recessive, X-linked, or maternally inherited manner with high genetic penetrance. NGS has allowed the identification of many new disease-causative genes. At least 72 SPG loci and 55 genes related to HSPs are known so far [65]. Conventional techniques of molecular diagnosis are impractical in this group of diseases because of phenotypic and genetic heterogeneity and inter- and intrafamilial variability. Both large NGS panels and smaller panels in preselected cases are in use for this condition. Large panels using targeted NGS SureSelect run on the HiSeq system that include an extensive panel of genes that can be tested in ataxia patients. This may be done when the case/family history has not been assessed in detail or there is a need to differentiate between HSP and Charcot-Marie-Tooth disease. On the other hand, smaller MiSeq/Ion Torrent panels are less expensive, but fewer genes can be tested and it is easier to interpret the results. This is preferred when the case is clinically well defined.

Table 4 presents examples of diagnostic panels for movement disorders.

MOVEMENT DISORDERS

The genome-wide association studies (GWAS) and NGS technologies allowed the discovery of new genetic risk factors in movement disorders, including parkinsonism, dystonia, and paroxysmal dyskinesias.

Parkinsonism is defined by clinical criteria such as a combination of hypokinesia (decrease in the number of movements), bradykinesia (slowness of movements), rigidity (increased resistance to passive movements), tremor, and postural instability. In children the term "hypokinetic–rigid syndrome" is more accurate [66]. In children parkinsonism is often incomplete, atypical, and complex, usually without resting tremor [66]. The cause of parkinsonism in children and adolescents may by acquired (e.g., infection, hypoxic–ischemic encephalopathy, immunological disorders) or hereditary (inborn errors in metabolism, neurodegeneration). Parkinson disease (PD) is a neurodegenerative disorder with motor and nonmotor symptoms due to decreased striatal dopamine levels, arising from selective loss of dopaminergic cells within the substantia nigra pars compacta and locus ceruleus of the midbrain. Defects in α-synuclein protein are considered the main pathogenic factor. The autosomal dominant form of PD is linked to several genes. Three well-defined autosomal recessive genes have been identified, PARK2 (parkin), PARK6 (DJ-1), and PARK7 (PINK1), causing autosomal recessive, early-onset parkinsonism with typical clinical signs [67]. Autosomal recessive, very early onset atypical parkinsonism with accompanying symptoms may be due to a mutation in several genes. As an illustration, *PLA2G6* gene

(OMIM 603604) mutations are linked with three phenotypes: PARK14 (OMIM 612953), neurodegeneration with brain iron accumulation 2B (OMIM 610217), and infantile neuroaxonal dystrophy (OMIM 256600). Mutations in the *ATP13A2* gene (OMIM 610513) were found in patients with Kufor–Rakeb syndrome (PARK9 OMIM 606693) and neuronal ceroid lipofuscinosis-12 (OMIM 606693). F-box-only protein 7 gene *FBXO7* (OMIM 605648) mutations cause PARK15 (OMIM 260300) with pyramidal signs. Additionally, the α-synuclein gene (*SNCA*—OMIM 163890; PARK 1—OMIM 168601) and leucine-rich repeat kinase 2 gene (*LRRK2*—OMIM 609007; PARK 8—OMIM 607060) are associated with this disorder. In 2011, a novel gene was identified by NGS—vacuolar protein sorting 35 (*VPS35*—OMIM 601501; PARK 17—OMIM 614203) [68,69]. Another causative variant, the eukaryotic translation initiation factor 4γ1 gene (*EIF4G1*—OMIM 600495; PARK 18—OMIM 614251), was identified in 2013 [67]. On the other hand, Nuytemans et al., using WES, found no evidence for an overall contribution of genetic variability in *VPS35* or *EIF4G1* to PD development [70]. This study is an example of the complexity of the problem. Mutations in another two genes linked to PD, *DNAJC6*—PARK 19 (OMIM 615528) and *SYNJ1*—PARK 20 (OMIM 615530), were identified using WES technology [71,72]. The majority of PD patients do not have any defined genetic background. Early onset and positive family history increase the chance of identifying a Mendelian genetic factor. Parkinsonism associated with dystonia may be due to conditions such as dopa-responsive dystonia, Wilson disease, or primary dystonia (*DYT1*). Parkinson disease associated with pallidopyramidal syndromes may be linked to *PANK2, ATP13A2, PLA2G6,* and *SPG11* gene variants and may be associated with dementia, eye movement disorder, and spasticity. Neuroimaging showing brain iron accumulation may be a clue pointing to one of these disorders.

Dystonia is defined as a movement disorder characterized by sustained or intermittent muscle contractions causing abnormal, often repetitive, movements, postures, or both. Dystonic movements are typically patterned and twisting and may be tremulous [73]. A scheme for classification of inherited dystonias is based on the DYTn coding system taking into account gene loci defined by linkage analyses [73]. Genes are known for some of these gene loci. Additionally, in many neurometabolic and neurodegenerative diseases, dystonia may be the dominant feature or concomitant with other symptoms. According to the inheritance pattern, hereditary dystonia is classified into four forms: autosomal dominant (torsin family 1 member A, *TOR1A*; ATPase α3 polypeptide, *ATP1A3*), autosomal recessive (tyrosine hydroxylase, *TH*), X-linked recessive (such as TATA box-binding protein-associated factor, *TAF1*), and mitochondrial (Leigh syndrome) [74]. So far, approximately 20 genes have been identified as causing dystonia. For children with dopa-responsive dystonia (Segawa dystonia), mutation screening for genes encoding products involved in dopamine metabolic pathways, such as TH or sepiapterin reductase, should be performed. Pantothenate kinase-associated neurodegeneration is a neurodegenerative disorder with initial symptoms of dystonia or parkinsonism due to iron accumulation in the brain. Deletion or missense mutations of the *PANK2* gene lead to the deficiency of pantothenate kinase 2, resulting in cell damage. Targeted NGS may be a useful and cost-effective method to screen for mutations in multiple genes associated with dystonia. Grundmann et al. analyzed a cohort of 27 unselected dystonia patients, using NGS. They identified disease-causing mutations in two patients (7%) and six possible disease-causing mutations including mutations in *ATP7B, PLA2G6, PARK2, FBCO7, VPS13A,* and *GCDH* [75].

Childhood- and juvenile-onset parkinsonism, dystonia, and ataxia are the phenotypes that may co-occur in inherited metabolic disease as well as in inherited neurodegenerative disorders. Extensive MRI and laboratory examinations usually direct the diagnosis. NGS enables rapid diagnosis, bypassing the need for numerous painful and complicated procedures.

NEUROMUSCULAR DISORDERS

Neuromuscular diseases (NMDs) are a heterogeneous group of disorders caused by injury to or dysfunction of the peripheral nervous system: anterior horn cells of the spinal cord, peripheral nerve, neuromuscular junction, or skeletal muscle. Most of the NMDs are hereditary. They are transmitted as autosomal dominant, autosomal recessive, or X-linked traits; mitochondrial DNA mutations are also responsible for some cases. NMDs present with a wide range of symptoms. The patient may present with severe congenital flaccid paresis, progressing early to respiratory insufficiency. Others have delayed or normal motor milestones with various degrees of skeletal muscle weakness. Some NMDs are marked by episodic symptoms such as periodic paralysis, recurrent rhabdomyolysis, or life-threatening malignant hyperthermia after general anesthesia. Cardiomyopathy may develop late in the course of some NMDs or may be a presenting feature, preceding symptomatic skeletal muscle involvement. In neuromuscular junction diseases (myasthenia or myasthenic syndromes) weakness may be fluctuating, with marked fatigability or worsening during the day and frequent involvement of extraocular muscles. In some families the onset of symptoms is in mid- or late adulthood; often striking intrafamilial variability is observed. NMDs meet the epidemiological criteria of rare diseases. Some are extremely rare and have been described in single families worldwide. The rarity of NMDs, a high incidence of *de novo* or private mutations, and the lack of definite biological markers of many NMDs make the diagnosis challenging.

In only a few of the hereditary NMDs is clinical and genetic diagnosis straightforward. Clinical presentation and results of simple diagnostic tests such as serum creatine kinase, electromyography (EMG), or nerve conduction studies (NCS) are sufficient to qualify the patient for targeted genetic testing for, e.g., Duchenne/Becker muscular dystrophy (DMD, OMIM 310200), proximal spinal muscular atrophy 5q (SMA, OMIM 253300, 253550, 253400, 271150), myotonic dystrophy type 1 and 2 (DM1, OMIM 160900, DM2 OMIM 602668), or the most frequent of hereditary neuropathies: Charcot-Marie-Tooth 1A (CMT1A, OMIM 118220). In SMA and DMD genetic testing became the first step of the diagnostic algorithm, right after clinical evaluation. EMG and muscle biopsy are no longer required in those patients before genetic testing. Such approach cannot be used for the majority of NMD patients when high genetic heterogeneity is linked to a given phenotype. Genetic counseling is hindered by lack of informative genetic results in many patients with congenital myopathies, muscular dystrophies, congenital myasthenic syndromes, or neuropathies. Molecular diagnosis may aid in differentiating some highly treatable hereditary congenital myasthenic syndromes from congenital myopathy patients. As of this writing, mutations in over 200 genes in the nuclear genome have been attributed to various NMDs [76].

Several approaches addressing the genetic diagnosis of NMD are currently employed. All are based on the clinical diagnosis combined with epidemiological data or identification of founder mutations predominating in a given population. They range from Sanger-based mutation screening of individual genes to tests of gene panels dedicated to a subgroup of NMD, for example, CMT neuropathy, congenital muscular dystrophy, and limb-girdle muscular dystrophy (other than DMD). The choice of gene panels may be guided by muscle biopsy findings. For example, nemaline myopathy is diagnosed when muscle pathology reveals the presence of rods in the cytoplasm of muscle fibers. Several hereditary types have been identified so far (NEM1–NEM9), with mutations in the α-tropomyosin 3 gene (*TPM3*), chromosome 1q21; nebulin (*NEB*), 2q23; α-actin (*ACTA1*), 1q42; β-tropomyosin (*TPM2*), 9p13; troponin T1 (*TNNT1*), 19q13; *KBTBD13*, 15q22; cofilin-2 (*CFN2*), 14q13; *KLHL40,* 3p22; and *KLHL41*, 2q31. Similar structures have also been described in families with mutations in the ryanodine receptor (*RYR1,* 19q13). It is of interest that sporadic nemaline myopathy responding to pharmacological treatment has been reported in HIV-positive and HIV-negative patients [77,78].

Testing a single gene in rare NMD yields the highest success rate in populations in which founder mutations have been linked to a specific phenotype. A good example comes from genetic testing of congenital myasthenic syndrome (CMS) patients. So far mutations in 15 genes have been identified as a cause of CMS. In a 2012 study of a cohort of 680 patients suspected of CMS, using panel testing, disease-causing mutations were identified in 299 (44%) of them [79]. On the other hand, in Romani-origin patients, 30% of cases of CMS are due to a mutation in exon 12 of the acetylcholine receptor ε subunit (epsilon1267delG, OMIM 100725.0012) [80].

Clinical diagnosis of hereditary NMD combines information about clinical presentation; family history; results of laboratory tests including EMG, NCS, repetitive nerve stimulation test, or single-fiber electromyography; muscle pathology; or imaging. Physical examination has to be conducted with attention to the presence and distribution of skeletal muscle weakness and atrophy (proximal, distal, facial, or oculomotor muscle weakness), sensory abnormalities, and dysmorphic features. In patients with episodic muscle symptoms such as malignant hyperthermia or rhabdomyolysis, information regarding the course of the episodes (as reflected by medical history and medical documentation) before, during, and after the episode needs to be noted. Family history and evaluation of oligosymptomatic family members may be valuable for clinical diagnosis. Whenever an autosomal recessive or *de novo* variant of autosomal dominant disease is suspected, the availability of DNA of trios (proband and both parents) will greatly aid in interpretation of genetic test results.

WGS and WES have proven to be powerful methods to identify genes causing a variety of hereditary diseases. NGS seems to be a very good tool for investigating NMDs, with their complexity as described above. In most cases NGS reveals high numbers of variants compared with reference sequence; some of them will have unknown significance. Both the "phenotype-down" and the "genotype-up" approaches need to be employed in NMD: looking for sequence variants in the genes most appropriate for a given phenotype (phenotype-down) or looking for certain phenotype features with respect to identified sequence variants in "neuromuscular" genes (genotype-up). Interpretation of NGS may be straightforward, if a previously described mutation linked to a specific phenotype is identified. It was demonstrated that targeted NGS was superior in clinical yield compared with sequential Sanger sequencing of a number of genes associated with congenital muscular dystrophy, a group of infantile-onset NMDs with severe weakness and contractures [81], although in some series, despite good sequence coverage, NGS leaves over 90% of patients with hereditary NMD without molecular confirmation [82]. With currently employed techniques and analyzing tools clinical WES and WGS identify causative mutation in up to 25% of tested patients [83].

NGS allowed identification of a number of new NMD genes. Novel, potentially pathogenic mutations need validation by linkage or segregation analysis or by functional studies possible in research laboratories. Therefore NGS cannot be viewed as a routine NMD diagnostic method yet, and the clinicians need to understand its current limitations [83,84].

Faced with new challenges associated with increasing availability of NGS, the neuromuscular community initiated joint efforts to define phenotypes of various rare NMDs. Integration of clinical presentation and the results of laboratory tests, including muscle pathology and muscle imaging, should guide the interpretation of new sequence variants identified with NGS [85].

NGS can offer explanation for some cases presenting with striking intrafamilial variability. Patients with phenotypes resulting from "double-trouble," or mutations of two different genes, were occasionally identified with Sanger sequencing in previous years. In a series of 250 patients tested with WES, surprisingly four (6%) patients had more than one pathological mutation: three had both an autosomal dominant and an autosomal recessive disorder and one was diagnosed with concomitant autosomal recessive and an X-linked disease, leading to a complex clinical presentation [83]. When mutations of two different genes affect the same pathway or the same effector they may lead to an unexpectedly severe phenotype. Several such patients with Charcot-Marie-Tooth neuropathy were reported [86,87].

In Duchenne muscular dystrophy the *LTBP4* (latent transforming growth factor-β binding protein 4) genotype acts as a phenotype modifier both in a murine disease model and in humans. In DMD boys the IAAM haplotype of *LTBP4* is related to almost 2 years older age at loss of ambulation than in boys heterozygous or homozygous for the VTTT haplotype [88]. Possibly future NGS studies will reveal more mutations outside the causative gene, acting as disease modifiers in other NMDs.

NGS in autoimmune NMD patients may serve as a research tool to uncover complex genetic backgrounds leading to disease development or influencing the disease course in large cohorts of patients. GWAS were undertaken in myasthenia gravis (MG), an IgG-mediated disease of the neuromuscular junction. In early-onset MG (age <50 years) the association of TNIP1 was demonstrated, indicating a role for ubiquitin-dependent dysregulation of NF-κB signaling in disease development [89]. Another group of researchers employed a 35-candidate gene approach implicating *VAV1* and *BAFF* as genetic risk factors of early-onset MG. Both *VAV1* and *BAFF* play important roles in B-cell activation and proliferation [90]. Similar studies in late-onset myasthenia gravis are ongoing.

In conclusion, NGS offers an attractive option for genetic diagnosis of many rare NMDs, with a current success rate in confirming causative mutations approaching approximately 25%. It may identify genetic modifiers of the clinical course or response to treatment in many hereditary and acquired diseases. NGS/GWAS testing may lead to better understanding of the pathophysiology and predisposition to acquired NMD or improve the approach to pharmacotherapy and pharmacogenetics.

NEURODEGENERATIVE DISEASES WITH DEMENTIA

Dementia is usually a chronic syndrome characterized by various symptoms of cognitive deterioration, especially memory, but also a decline in other cognitive abilities (*International Classification of Diseases,* 10th revision—ICD-10). The global prevalence of dementia is about 24 million [91]. According to the *Diagnostic and Statistical Manual of Mental Disorders,* 5th revision (DSM-V) the term "dementia" has been replaced by major or minor neurocognitive disorder. Genetic factors are important in all forms of dementia but especially in early-onset dementia [92].

Alzheimer disease (AD) is the most common cause of dementia. The neuropathological findings are neuronal and synaptic loss with amyloid-β peptide plaques and neurofibrillary tangles formed by abnormally hyperphosphorylatedmicrotubule-associated protein tau. Disease is caused by the combination of genetic and environmental factors. There are two forms of the disease. The early-onset AD (1–5% of all cases) that affects people before 65 years is usually characterized by Mendelian inheritance [93]. The autosomal dominant highly penetrant genetic pattern of inheritance is observed in families with early-onset disease caused by mutations in three genes, *APP, PSEN1,* and *PSEN2* [92].

Late-onset AD, after 65 years of age, is more common than the early one; it has no consistent mode of transmission and the genes involved increase disease risk in a non-Mendelian fashion [91,93]. The major late-onset Alzheimer susceptibility gene is *APOE ε4*—encoding apolipoprotein E. The most common genotype is *APOE ε3ε3*. The *APOE ε4* allele increases the risk of AD by two or three times [91]. The risk of AD for *APOE ε4ε4* carriers at the age of 60–69 years is much higher than for *APOE ε3ε3* carriers [94].

GWAS have led to the identification of other genes linked to the risk of late-onset AD, including *BIN1, CLU, PICALM, CR1, CD2AP, CD33, EPHA1, ATXN1, ABCA7*, the *MS4A4A/MS4A4E/MS4A6E* cluster, and *EPHA1* [91,93]. Variants in these genes are linked to significant differences in risk of AD, but their effects are much smaller than that of *APOE*. NGS enabled detection of a rare variant of *TREM2* carrying risk comparable with that of *APOE ε3ε4*. Two WES studies identified mutations in *NOTCH3* and in *SORL1*, the latter in the group of patients with early-onset AD [94]. The Alzheimer's Disease Sequencing Project is currently under way. Using WGS, 582 subjects from 111 families will be examined with the aim to identify new genomic variants contributing to increased disease risk as well as a protective constellation of genes (https://www.niagads.org/adsp/content/home). Moreover, several research teams have joined the International Genomics of Alzheimer's Project with the goal of carrying out the largest genetics study of AD to date and providing further insights into the inheritance of the condition (www.alz.org).

Frontotemporal dementia (FTD), in its familial (approximately 50% of cases) and sporadic forms, is characterized by gradual damage of neurons in the frontal and temporal lobes with slowly progressive dementia and behavioral changes and, at later stage, by psychomotor slowing and apathy. The two causal genes *MAPT* and *PGRN* account for 10–20% of familial FTD. Mutations in *TDP43, FUS, UBQLN2, VCP*, and *CHMP2B* are associated with some familial FTD cases. Most recently, a large intronic hexanucleotide (GGGGCC) expansion in the *C9orf72* gene was associated with 5% of cases of familial as well as some sporadic forms of FTD. The hexanucleotide GGGGCC expansion ranges from zero to around 30 copies in unaffected individuals to several thousand repeats in mutation carriers. The detection of such a repeat expansion by current NGS technologies is not possible [95,96].

Amyotrophic lateral sclerosis (ALS) is a devastating neurodegenerative disease in which upper and lower motor neurons are lost, accounting for respiratory muscle paralysis within 3–5 years after the disease onset. ALS has familial (10%) and sporadic (90%) forms. Missense mutations in the gene encoding the antioxidant enzyme Cu/Zn superoxide dismutase, *SOD1*, together with mutations in *C9orf72, TDP43*, and *FUS*, account for approximately 65% of the familial ALS cases [95,96]. Other rare genes linked to familial ALS include *MAPT, PGRN, VCP, UBQLN2*, and *CHMP2B*. Furthermore, it has been established that symptoms of FTD and ALS can co-occur in the same patient. Interestingly, the most convincing direct molecular link between ALS and FTD has been the identification of a hexanucleotide GGGGCC expansion in *C9orf72* in families with ALS, FTD, and overlapping ALS–FTD syndrome. This expansion accounts for approximately 40% of familial ALS, 10% of sporadic ALS, 5% of sporadic FTD, and up to 80% of familial ALS–FTD cases; hence, it makes GGGGCC repeat expansion the most common cause of ALS and FTD. Accordingly ALS and FTD are supposed to be linked genetically and have overlapping phenotypes, and each disease should now be recognized as representative of a continuum of a broad neurodegenerative disorder, supporting the hypothesis about a spectrum of neurodegenerative diseases, further confirmed by the coexistence of various neurodegenerative diseases in relatives of patients with ALS [95,96].

Huntington disease (HD) is a rare progressive autosomal dominant neurodegenerative condition associated with the accumulation of intracellular pathogenic proteins. The course of the disease is variable, and dementia can occur at any stage. HD is caused by a mutation in the *HTT* gene. The diagnosis of HD is based on the estimation of the CAG repeat length at the *HTT* locus. The number of trinucleotide CAG repeats in the normal *HTT* gene is reported to be less than 27. Individuals affected with HD typically have at least one *HTT* allele containing a CAG repeat size of 40 or greater. Currently such analysis cannot be performed using standard NGS technology.

NEUROPSYCHIATRIC DISORDERS
INTELLECTUAL DISABILITY

Both the DSM-V and the ICD-10 have defined intellectual disability (ID) as (1) a significant limitation in intellectual functioning; (2) a significant limitation in adaptive behavior, such as social and practical skills; and (3) an onset before the age of 18 years [97]. This meets the current standards of the American Association on Intellectual and Developmental Disabilities and for the sake of this publication also includes such capacious terms as mental retardation and intellectual impairment. At the milder end of this spectrum are individuals whose IQ is around 70, and a chance of finding a genetic cause of their phenotype is estimated to be only up to 30% owing to the multifactorial nature of the problem [98]. In contrast, severely affected patients with IQ below 35 mostly represent chromosomal or monogenic forms of the condition. The likelihood of unraveling the etiological factors of their phenotype is much higher.

The overall estimates of prevalence of ID differ depending on the study and symptom severity but are reported to be between 0.3 and 0.5% for the severe end and up to 7–8% for mild forms [99]. Finding the genetic causes of ID enables reliable genetic counseling, including recurrence risks and prenatal testing, as well as optimization of treatment and anticipation of possible late negative sequelae. It also ends the "diagnostic odyssey," saving precious energy and money of the caretakers.

Microscopically visible CNVs account for up to 16% of all causes of ID [100]. The introduction of various methods of submicroscopic analysis, such as FISH, MLPA, or aCGH, has significantly increased these numbers [101]. According to the latest data, another 12% of pathogenic CNVs are identified at the submicroscopic level with the use of the aCGH technique [102]. Monogenic X-linked forms of ID have been relatively easy to pick up, being limited to up to 80 genes so far, but other single-gene defects are diagnosed with variable efficiency reflecting the heterogeneous nature of ID [98]. The total yield when utilizing all the above techniques in a single cohort of ID individuals has been estimated as 40% in the pre-NGS era [103].

NGS has been appreciated as a diagnostic tool in uncovering the causality of ID since the early studies released in 2009. These were divided according to the selection criteria used to characterize patients by Topper et al. into exome sequencing in (1) syndromes, (2) nonsyndromic sporadic cases, and (3) familial cases [104].

The "phenotype first" approach allowed for identification of the causes of syndromic conditions after application of WES. Thus, relatively few, yet clinically well-selected, individuals provided enough power to filter against insignificant variants in patients' exomes. In such cases the phenotype must be convincing in favor of a certain diagnosis. Early examples of such an approach, such as Schinzel–Giedion, Kabuki, and Miller syndromes, have been followed by much more genetically heterogeneous conditions such as Kallmann and Noonan syndromes [105–109]. In all, more than 30 studies have proved decisive in recognizing the cause of over 20 syndromic forms of ID [110].

"Genotype first" attempts have been fruitful when certain syndromic conditions had a well-known molecular cause picked up by WES but could not initially be recognized clinically. Herein, the utility of WES has been highlighted by Classen et al., who questioned CNVs detected in their patients as the sole cause of their phenotypes and identified critical single nucleotide variants (SNVs) in the *FBN1*, *EYA1*, and *BLM* genes linked with syndromic diagnoses [111].

Most cases of intellectual disability have no associated features and family history is not relevant. This nonspecific or nonsyndromic ID group provides no phenotypic clues as to possible causative defects other than the degree of severity of their impairment. Here, one can either compare the number and type of SNVs per exome of such individuals against controls or directly look for *de novo* pathogenic changes compatible with the hypothesis that they are pathogenic in most cases of ID. Vissers et al. looked at the exomes of 10 children and found *de novo* pathogenic mutations in six of them [112]. Two genes were known as causative from previous works, and the other four were critical for central nervous system activity. Another group carefully selected their patients against family history, inbreeding, any congenital anomaly or dysmorphic feature, or CNVs detected by array [113]. Nine trios (parents + child) were sequenced with a median 31-fold coverage. Three pathogenic variants in known ID genes, and another three genes, were identified as candidates, thus potentially explaining the cause of the phenotype in most patients.

The most promising results have been obtained at the more severe end of the nonsyndromic ID spectrum. Rauch et al. chose 51 such trios (sporadic cases + parents) with IQ less than 60, for which CNVs had been excluded by molecular karyotyping [114]. They focused on *de novo* variants and identified them at a rate of about 1.41 mutations per patient. More importantly, they then compared their findings with a control cohort. In the latter, fewer loss-of-function variants were observed, the mutation rate was lower (1.15), and there were more synonymous changes. The authors suggested that the clinical bias exists that favors looking for specific syndromic diagnoses. Thus, WES should be a first-tier test for patients without a recognizable condition. About the same time de Ligt et al. reported results of a study in 100 patients with IQ below 50 [115]. They found pathogenic mutations in 16 patients, including 13 *de novo* and three inherited variants.

Familial cases of intellectual disability strongly suggest recessive variants as key etiological factors. In the Nijmegen ID database cases of two or more affected sibs account for just 6%. In 2013 they investigated 19 families with two to five affected siblings, identifying pathogenic mutations in one novel, two known, and five candidate syndromic and nonsyndromic ID genes [116]. The authors highlight the need to explain the causes of mental impairment in these particular families due to high risk of phenotype recurrence. Other studies focus on inbred families (usually nonsyndromic) to identify underlying defects, such as a familial mutation in the *TECR* gene, that could not be picked up by linkage and homozygosity mapping alone [117]. Another example of a similar approach targeting specific exons instead of the exome has been done in consanguineous families from Iran [118]. Of interest are attempts at combining multiple nonconsanguineous families with similar phenotypes to uncover causative mutations behind recessive syndromes. Successful studies have thus far been done for the Dubowitz and CHIME syndromes [119,120].

So-called nongenic DNA segments, including DNA translating into noncoding RNAs, escape WES. Thus, a more powerful WGS approach has to be applied [121]. In an elegant work by Gilissen et al. causative mutations were found in 21 of 50 patients with IQ less than 50 using WGS [122]. This included 20 dominant *de novo* variants in well-known or candidate genes and one compound heterozygous change. It is noteworthy that the WGS study group was carefully selected after array and WES were applied with diagnostic yields of 12 and 27%, respectively. Their final estimate of WGS effectiveness in reaching genetic diagnosis was about 60%.

ID is an extremely heterogeneous symptom (rather than a condition itself). In most cases, where no specific syndrome is recognized clinically, array is applied as the first diagnostic step followed by WES. In some instances WGS can be done as well, albeit thus far mostly on a research basis. New methods of data analysis empower WES/WGS to detect CNVs as well. However, the more sensitive the test is, the more challenges it poses in results interpretation. Some atypical syndromic presentations picked up by WES broaden the clinical knowledge but also necessitate significant changes in the genetic counseling process. A similar question is low-level somatic mosaicism detected by the WES/WGS technique that could not initially be seen on Sanger sequencing [123]. Good examples in the context of an ID phenotype are Cornelia de Lange syndrome, in which mosaics may constitute up to 25% of all cases, and cerebral cortical, malformations with the rate up to 30% when targeted WES is used [124,125]. Another challenge is interpreting *de novo* vs inherited mutations. A *de novo* paradigm for mental retardation was put forward early on by Vissers et al. [112]. In part this theory was based on the fact of constant proportion of mentally disabled individuals within the general population despite the reduced fecundity and increased lethality of this group. However, the *de novo* mutation rate is relatively high in healthy people. Moreover, the role of inherited variants may be underestimated, based on their possible reduced penetrance. Last, it need not be one particular variant but more changes fulfilling the definition of an oligo- or polygenic model, especially in nonspecific ID phenotypes.

AUTISM

In the new classification of the American Psychiatric Association DSM-V, a category called autistic spectrum disorders (ASDs) has been introduced, replacing autistic disorder, Asperger syndrome, Rett syndrome, childhood disintegrative disorder, and pervasive developmental disorder not otherwise specified. ASDs are defined by two basic psychopathological dimensions, communication disturbances and stereotyped behaviors, and the diagnosis is complemented by the assessment of language and intellectual development. The prevalence of ASDs in the general population is about 1%, and lifetime psychiatric comorbidity is observed in the majority of patients.

Two overlapping working etiological hypotheses have been accepted for ASDs and/or other neuropsychiatric conditions, such as schizophrenia [126]. The common disease–common variant theory assumes that common DNA changes of relatively small individual impact, that is, SNVs or chromosomal rearrangements (CNVs), cause clinical symptoms of the condition. In practice, this is seen as a variable (milder or more severe) disease expression in the relatives of the affected persons, where the closer the relation, the higher the risk of symptoms developing in a given individual. The hypothesis of common disease–rare variant (CD–RV) suggests that the presence of a single rare defect (mutation) of significant impact can be held responsible for the symptoms. Important arguments in favor of the CD–RV hypothesis include early onset of disease, frequent co-occurrence of ID, and large differences in concordance rates among mono- and dizygotic twins.

CNVs have been linked with both simplex and syndromic autism [127,128]. Specifically, *de novo* unique (i.e., rare compared to an available database of variants) events identified by microarray technology are believed to be causal for the substantial proportion of autism cases [127,129]. Data generated by Sebat et al. showed a 10-times excess of mainly small *de novo* CNVs in sporadic autism vs controls [130]. Yet, better association at the clinical level has been established for larger recurrent microdeletions or microduplications, such as the 16p11.2 deletion/duplication or 7q11.23 duplication [127,131]. For these CNVs, a number of candidate dose-sensitive genes have been proposed. However, it is currently unknown how they mechanistically contribute to the disease pathogenesis. Moreover, it seems that higher array probe density does not lead to more pathogenic CNVs being identified. Thus, while rare high-risk microdeletions or microduplications are causal, common low-risk variants may not be [129].

NGS technologies are expected to identify both rare and common SNVs that are likely to fall within more than 70% of autism cases, the pathogenesis of which is currently unknown [132]. These methods allow reading and interpreting hundreds to thousands of megabases of nucleotide sequence in a single run [126]. However, the sole identification of SNVs/CNVs is only the beginning in the process of uncovering the causality of ASDs. It is hoped that eventually modern techniques will allow the establishment of connections at a protein level between products of genes implicated as pathogenic.

The early published studies applying WES to autism were based mainly on sporadic (i.e., nonfamilial) cases. The basic rationale behind such analyses has been that sporadic autism is simply much more prevalent than familial and that *de novo* events are most likely to be responsible for the former. In the approach of O'Roak et al. 20 parent/child trios with idiopathic ASDs were carefully selected after microarray had failed to identify pathogenic CNVs [133]. Eighteen coding and three noncoding *de novo* variants were identified, of which 11 were predicted to alter protein function. Despite their deleteriousness, they resided in different genes, which precluded any further conclusions as to their causality. However, four candidates were eventually put forward as probably pathogenic: *FOXP1, GRIN2B, SCN1A,* and *LAMC3.* The authors speculated about a multihit or oligogenic ASD model based on the observation that the *FOXP1* mutation carrier also had a *CNTNAP2* variant. Another such case consisting of two *de novo* truncating mutations in *CHD8* and *CUBN* was revealed by the same group a year later [134]. In this study, the authors added another 189 trios from the Simons Complex Collection of cases (see Autism databases). The cases were more severe and a significant proportion had overlapping phenotypes of ID. In total, 248 *de novo* changes, including SNVs and CNVs, were identified in 60 genes. Of these, 120 variants were severe and nonsynonymous, with nearly one-third being truncating mutations. The authors noted that a number of the most severe *de novo* variants mapped to interconnected pathways of proteins, including β-catenin and p53 signaling cascades. The only recurrently mutated genes were *CHD8* and *NTNG1.* The added value of the paper was the inclusion of the control cohort consisting of 50 healthy sibs.

Further studies by Iossifov et al., Sanders et al., and Neale et al. had significantly more power owing to both larger study and larger control groups, in which the latter consisted of healthy siblings [135–137]. The Iossifov team's aim was to specifically compare the incidence of *de novo* mutations between patients and their sibs. They studied 343 families of probands with high-functioning autism. No recurrent mutations or recurrently changed genes were found; however, two overlaps (*NRXN1, PHF2*) with their previous CNV study existed. The significant subset of their 59 genes translates into mRNAs that may be controlled by the fragile X mental retardation gene product (FMRP) [138]. Sanders et al. applied the WES methodology to 238 families, that is, 928 individuals with ASDs, identifying 279 nonsynonymous *de novo* SNVs, of which nonsense and splice-site mutations were particularly associated with the ASD phenotype [136]. Two of these nonsense variants affected the same gene, *SCN2A* (sodium channel, voltage-gated, type II, α subunit). This, according to the authors, may point to a significant overlap between epilepsy and autism phenotypes [139]. While both Iossifov and Sanders highlighted a causative role of single *de novo* mutations in autism pathogenesis, Neale et al. developed a polygenic model, in which the role of even severe disruptive *de novo* variants in coding genes may be limited [137]. This hypothesis was based on the observation of an only slightly higher overall mutation rate in affected persons, the sheer number of affected genes, as well as their reduced penetrance. Nevertheless, Neale's group put forward two candidate genes, *KATNAL2* and *CHD8*, in which *de novo* mutations were unlikely to occur by chance. Furthermore, additional variants in these genes were found in the replication cohort.

A substantial proportion of autism cases are from multiplex families, in which at least one other close or distant relative is affected. Here, identity-by-descent filtering allows for tracking pathogenic inherited mutations. Cukier et al. applied this technique to 40 families with autism (i.e., 164 individuals) [140]. They identified 36 genes with segregating SNVs in at least two families. They noticed that their SNVs were also observed in other studies in families with schizophrenia, ID, bipolar disorder, or major depression. Multiple, damaging, validated variants in the *ABHD14A*, *FAT1*, *OFCC1*, and *PRICKLE1* genes overlapped with epilepsy, Tourette syndrome, or bipolar disorder in other publications. This overlap together with the vast complexity of autism pathogenesis were the main conclusions of this study.

Two studies applying WES specifically focused on the possible damaging recessive variants. Chahrour et al. identified segregating homozygous, potentially pathogenic mutations in four of 16 apparently nonconsanguineous families [141]. Transcription of the genes hosting these mutations is regulated by neuronal depolarization, suggesting their role in neurons. Homozygosity analysis in three independent sibpairs led to identification of *HERC2* as a potential candidate in a small study by Puffenberger et al. [142]. The phenotype of gait instability, autistic behavior, and cognitive delay showed some overlap with Angelman syndrome.

Candidate variants that may elude identification by WES reside in nongenic DNA segments, including DNA transcribed into noncoding RNAs. On the other hand, applying the WGS approach requires that the study has significantly more power, with the cost of about 2.5 times higher than conventional WES.

Although small sample size prevented far-reaching conclusions, Jiang et al. suggested that a significant increase in the yield of pathogenic variants may be achieved by a more uniform coverage [143]. They studied 32 families with ASDs and in six identified *de novo* mutations, while in 10 they found inherited X-linked or autosomal variants. The latter corresponds well with high heritability of the autism phenotype.

Important conclusions may be drawn from monozygotic discordant twins, where revealing a nonshared variant in an affected sib strongly implies its pathogenicity. However, such families are rare and difficult to find. Michaelson et al. looked for germ-line inherited pathogenic mutations presenting in concordant twins [144]. Of 668 germ-line mutations identified, 565 (87%) were *de novo* and 53 (9%) were true-positive SNVs inherited from one of the parents. Mutability issues were put forward by the authors and clusters of variants uncovered in the genome.

Seven candidate genes, of which the most likely is *ANK3*, were suggested by Shi et al. after sequencing genomic DNA in a single multiplex family [145]. Rightly so, the authors bring back the idea of searching for responsible variants in large multiplex pedigrees.

Protein–protein interaction networks and several other models, including NETBAG, have allowed translation of massive parallel sequencing data at the molecular level [146,147]. One of the first such interconnected protein networks included β-catenin/chromatin remodeling and p53 pathways, with almost half of the changed genes' products mapping to these pathways [134,148]. In turn, a quarter of the genes interrupted by mutations in Iossifov's study were found to be regulated by FMR1 protein [135]. The complex model of autism protein networks based on identified CNVs called NETBAG was worked out by Gilman's group [147]. Every gene pair in this network was assigned a score proportional to the log of the ratio of the likelihood that the two genes participate in the same genetic phenotype to the likelihood that they do not. NETBAG was instrumental in revealing *WNT*, postsynaptic complexes, and some dendritic spine proteins as key pathways in ASD pathogenesis. The *WNT* pathway and chromatin remodeling network appear to be connected by the same strong candidate CHD8 protein, first put forward as altered in autism by O'Roak et al. and Neale et al. [134,137]. Now, the *CHD8* gene product is speculated to be responsible for the macrocephaly phenotype of autism. Novel approaches to protein interactions in autism have been proposed by Krumm et al., who noticed that a central network can be formed from proteins altered in overlapping phenotypes of autism and ID [146].

Earlier studies based their assumptions on the paradigm of the crucial role of *de novo* mutations in neuropsychiatric disease [112]. However, high genetic variation is present in the exomes of perfectly healthy individuals [149]. Thus, each autistic and nonautistic person carries per se close to one disruptive *de novo* variant. This problem can theoretically be addressed by counting and comparing the numbers of per-gene *de novo* mutations in patients and controls [150]. In practice, recurrent mutations in a single gene or multiple independent *de novo* SNVs in the same gene in unrelated patients have the potential to identify clear risk alleles [134,136,148]. Such events within and across studies of autism are shown in Table 5. It is estimated that lifting the power of the future studies by increasing sample sizes to thousands of families should identify candidate genes only if three or more *de novo* disruptive mutations in these genes are present in probands [150].

Given the complex nature of neuropsychiatric disease, a role of *de novo* changes cannot be overestimated [151]. Poly- or oligogenic models of autism have been proposed based on the fact that no locus contributes to more than 1% of ASD cases [132]. Moreover, there is significant overlap between autism and other diseases, such as schizophrenia, ID, or major depression. These and other questions may successfully be tackled by creating functional protein networks of neuropsychiatric genes.

Table 5 Examples of Key Autism Genes Hit At least Twice by the Same or Different *De Novo* **Disruptive Mutations within the Same or Across Different WES Autism Studies**

Gene	Function	Study
SCN2A	Controlling electrical excitability in neurons; sodium channel component	Sanders, 2012; Neale, 2012
CHD8	Regulation of Wnt signaling via β-catenin; CHD7 binding	Neale, 2012; O'Roak, 2012A; O'Roak, 2012B
DYRK1A	Calcineurin signaling; neuronal differentiation regulation	Iossifov, 2012; O'Roak, 2012A; O'Roak, 2012B
GRIN2B	NMDA receptor; synaptic plasticity	O'Roak, 2011; O'Roak, 2012B
KATNAL2	Unknown; high expression in central nervous system	Neale, 2012; Sanders, 2012; O'Roak, 2012A
POGZ	Heterochromatin formation; mitotic progression	Iossifov, 2012
NRXN1	Synaptic signaling	Iossifov, 2012
PHF2	Transcriptional regulation	Iossifov, 2012
NTNG1	Axon guidance; growth of thalamocortical axons	O'Roak, 2012A
CUL3	Chromatin modification; mitotic progression; completion of cytokinesis	O'Roak, 2012A; Kong, 2012
PTEN	Cell cycle progression; cell migration	O'Roak, 2012B
TBL1XR1	Recruitment of proteasome complex; nuclear receptor-mediated gene activation	O'Roak, 2012B
TBR1	Transcriptional regulation	O'Roak, 2012B

INTELLECTUAL DISABILITY AND AUTISM—A PRACTICAL APPROACH

The clinical diagnostic process of an autistic child or ID individual has to take into account the various phenotypes overlapping in a single patient, including autistic features, ID, or epilepsy, together with other nonneurodevelopmental physical characteristics, for example, dysmorphism. This should allow for selection of those affected for whom the chance of establishing a genetic cause of the disorder using available genetic techniques, including array CGH and NGS, is the highest.

Phenotypic variables of diagnostic importance (biomarkers) increasing the likelihood of diagnosis are disordered physical development, neurological problems, natural history of the disorder, and family history data [152]. The most important symptoms belonging to the first category are dysmorphism and micro-/macrocephaly, as well as congenital defects, including CNS anomalies. The combination of dysmorphic features found in children with ASDs and/or ID should help establish the final clinical diagnosis (e.g., long, triangular face, and long earlobes in fragile X syndrome). No one dysmorphic feature stands out as characteristic of the entire population of ASDs/ID; however, the examiner may want to focus on the face and distal limb segments, as these are the most informative.

Seizures are present in about 25% of the autistic population, whereas abnormal EEG concerns 50% of the affected. These numbers are most likely similar for ID. Further points to consider are the disease onset (the earlier, the higher the likelihood of the presence of a rare *de novo* mutation) and accompanying developmental regression. Family history information should include the presence of the same phenotype as well as other neuropsychiatric conditions (especially epilepsy and schizophrenia but also bipolar affective disorder or alcoholism).

A clinically useful division of neurodevelopmental phenotypes highlights autism/ID without other accompanying features and autism/ID (i.e., autism/ID simplex) with other clinical symptoms (i.e., autism/ID+). An especially high chance of identification of the genetic defect, up to 25%, concerns the latter subgroup. Comparative characteristics of the clinical features of autism/ID and autism/ID+ are shown in Table 6.

GENETIC TESTING IN AUTISM AND ID—A PROPOSAL

The revolutionary progress in laboratory medicine seems to have always anticipated the clinics. Therefore, a genotype-first approach is now advocated by most clinical groups in diagnostics of autism or ID. Yet, the phenotypic description of an affected individual by a skilled clinical geneticist renders interpretation of vast array/NGS data much easier. Another crucial aspect of care is the availability of

Table 6 Comparison of Alternative Types of Autism/ID (Modified after [153])		
	Autism/ID COMPLEX, More Severe, with Genetic Etiology (Autism/ID+)	**Only Autism or ID, of Unknown Etiology**
Etiology	Point mutations or aberrations, *de novo* or inherited	Unknown mutations, unknown inheritance
Clinical heterogeneity	Significant	Possibly less
Environmental contribution	Less important	More important
Paternal age effect	Relevant	Less relevant
Dysmorphism	Frequent	Usually absent
Head circumference	Microcephaly or extreme macrocephaly	Mild microcephaly possible
Developmental regression	Rare	More common

different techniques and cost coverage by the insurers. As shown in Figure 1, various choices can be made depending on the phenotype, family history, and financial coverage. A general comparison of diagnostic NGS targeted gene panels and WES is presented in Table 7. Two main specific targeted panels used currently in autism diagnostics are shown in Table 8. However, it still needs to be pointed out that, according to the most recent data [122], even at the severe end of neurodevelopmental phenotypes, close to one-third of individuals remain undiagnosed after application of array, WES, and WGS.

Figure 1 presents a diagnostic flowchart for ID and/or autism phenotypes.

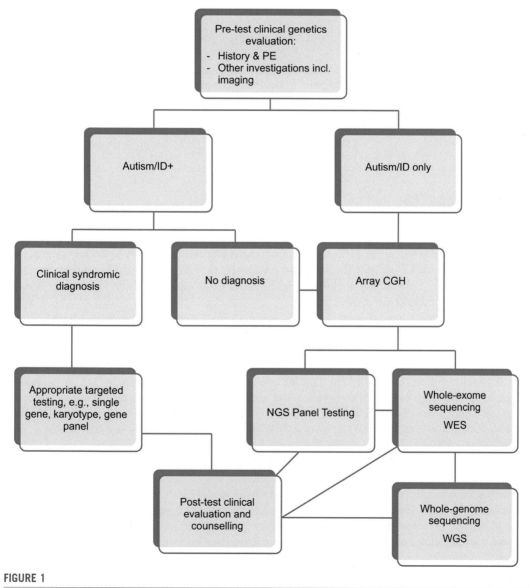

FIGURE 1

Suggested diagnostic flowchart for ID and/or autism phenotypes.

Table 7 General Comparison of NGS Gene-Targeted Panels Versus WES

	Panel Testing	WES
Performance	Del/dup included; exons of genes fully screened	Not all exons screened; no del/dup analysis
Coverage	Only genes on the panel	All coding genes
Pretest clinical assessment	Necessary to choose the right panel	Not needed
Inheritance pattern	Not needed	Parents to be screened for variant interpretation
Incidental findings	Absent	Can be present, including medically actionable

Table 8 Main Autism NGS Gene-Targeted Panel Performance

	TruSight Illumina	Emory Genetics
Number of genes	101	62
Coverage	Mean 100×; minimum 20×	Average 168×
% Exons covered	≥95%	97.6%

AUTISM DATABASES

Large databases that have been created by extracting data from peer-reviewed scientific publications and integrating information about diverse autism risk genes represent an important clinical as well as a research tool. They serve as the advanced resource addressing the complexity of autism genetics for health care providers. All available databases are continually optimized as user-friendly tools for clinicians and the research community, supplying standardized information. Moreover, the databases can be searched to recruit patients for clinical trials.

The Simons simplex collection

The Simons simplex collection (SSC) of more than 2700 simplex autism families has been created with the help of The Simons Foundation Autism Research Initiative (SFARI) (http://sfari.org/resources/simons-simplex-collection, [154]). The families are recruited from 12 autism centers in the United States, as the emphasis is primarily on the careful clinical evaluation of the affected family members given the highly heterogeneous nature of the disorder. Probands are examined by a psychologist and a clinical geneticist with the use of ADOS and ADI-R tools. Those under 4 years or over 18 years as well as the patients whose evaluation might have been compromised by other conditions were excluded from the resource. The aim of the SSC is to facilitate chip-based searches for both CNVs and SNVs identified by high-throughput sequencing. Main sequencing efforts that have been based upon the SSC include those of Sanders et al., Iossifov et al., and O'Roak et al. [134–136].

In addition to the clinical data, SFARI investigators developed a software platform called SFARI Base to support data acquisition, validation, and distribution. In turn, the SFARI Gene database has become a unique source of both rare and common variation data. Users will find a thoroughly annotated list of genes that have been studied in the context of autism, with information on the genes themselves, relevant references from the literature, and the nature of the evidence. SFARI Gene can be accessed through https://gene.sfari.org/autdb/Welcome.do.

SFARI Gene is the latest modification of the Autism Database (AutDB)—the first widely available genetic database for ASDs [155]. AutDB includes all known genes that have been implicated in ASDs, together with risk-conferring candidates associated with these conditions. The genetic information includes data from linkage and association studies, cytogenetic abnormalities, and specific mutations associated with ASDs.

A National Institutes of Health-funded initiative called National Database for Autism Research (NDAR) is one of the largest databases of autism so far (http://ndar.nih.gov/). Its aim is to integrate multiple resources at both the phenotypic and the genotypic level. Moreover, it combines the function of a data repository, which holds genetic, phenotypic, clinical, and medical imaging data, and the function of a scientific community platform, which defines the standard tools and policies to integrate the computational resources developed by scientific research institutions, private foundations, and other federal and state agencies supporting ASD research. Furthermore, NDAR is working to develop the means to connect relevant repositories together through data federation.

The autism genetic resource exchange

The autism genetic resource exchange (AGRE) includes both a DNA repository and a family registry housing a database of genotypic and phenotypic information (http://agre.autismspeaks.org/site/, [156]). It is the world's largest private shared open-access resource for the study of autism and related disorders and was developed as a joint effort of the Cure Autism Now Foundation and the Human Biological Data Interchange. AGRE houses a collection of more than 2000 well-characterized multiplex and simplex families available to the greater scientific community. This resource is now used by more than 150 research groups worldwide.

The autism genetic database

The autism genetic database was developed as a comprehensive list of susceptibility genes and CNVs found to have a relationship to autism integrated with known noncoding RNAs (snoRNA, miRNA, and piRNA) and chemically induced fragile sites (http://wren.bcf.ku.edu, [157]). Furthermore, genomic information about human fragile sites and noncoding RNAs was also downloaded from miRBase, snoRNA-LBME-db, piRNABank, and the MIT/ICBP siRNA database. Such a unique autism genetic database will facilitate the evaluation of autism susceptibility factors in relation to known human noncoding RNAs and fragile sites having an impact on human diseases.

The autism chromosome rearrangement database

The autism chromosome rearrangement database (ACRD) stores information about break points and other genomic variants related to autism and downloaded from publicly available literature: databases and unpublished data (http://projects.tcag.ca/autism/, [158]). The database is continually updated with information from in-house experimental data as well as data from published research studies. ACRD enables integration of molecular data with cytogenetic findings collected over the years.

The autism genome project database

The autism genome project database (AGP), the world's largest research project on identifying genes associated with risk for autism, contains genetic linkage and CNV data that connect autism to genetic loci [159]. Established by the Autism Speaks community, AGP supports using autism-related subphenotypes linked to susceptible loci. The Autism Genome Project brings together a large group of leading researchers to collaborate and build the largest data set of DNA from families affected by autism.

The researchers, who represent over 50 institutions worldwide, assembled a database of about 2500 families, most of which contained at least two members affected with autism. This project is the largest collaboration to study the genetics of autism. The methods used by the AGP involved studying both the entire genome and regions of interest within all chromosomes.

Interactive autism network

Interactive autism network (IAN) is an Internet-based registry founded by the Centers for Disease Control and Prevention (http://www.iancommunity.org/cs/about_ian). IAN requires parents to self-register their child and fill out forms. IAN pools data retrieved from electronic medical records to create one of the largest autism databases to date. Using these resources, clinicians have identified 20,000 subjects with ASDs. Unlike similar networks, this database contains standardized information that will aid researchers in their efforts. IAN is funded by Autism Speaks, the Simons Foundation, and the National Institutes of Health.

MAJOR PSYCHIATRIC DISORDERS

Virtually all major psychiatric disease entities such as schizophrenia (SZ), bipolar disorder (BPD), major depression disorder (MDD), obsessive–compulsive disorder (OCD), attention-deficit hyperactivity disorder (ADHD), and anxiety disorders are genetically complex. This complexity is accounted for by numerous determinants, including multiple common and rare variants (SNVs) and CNVs, in various proportions or by different factors in different individuals [160]. Until recently, large-scale approaches to such complex traits yielded limited results, and consequently only a few variants, usually of individually minor effect, were identified.

Currently, for common disorders in psychiatry a more systemic view of biological networks is emerging, promising new possibilities of clinical genetic testing [161,162]. As of this writing, multiple projects are being undertaken to uncover a genetic framework of mental disorders and to identify the genomic contributions to risk for severe psychiatric entities, including ongoing projects of the National Institute of Mental Health (www.nimh.nih.gov/) and the Psychiatric Genomics Consortium (http://www.med.unc.edu/pgc).

Potential application of genetic testing in major psychiatric disorders

Genetic tests in routine psychiatry could potentially be used in several ways, including:

- risk stratification of developing a particular disorder,
- support for establishing and confirming a diagnosis,
- identification of treatment targets,
- optimal selection and dosing of a drug. (http://www.psychiatrictimes.com/neuropsychiatry/genetic-testing-psychiatric-disorders-its-current-role-clinical-psychiatric-practice#sthash.V9UzZ1pL.dpuf)

To date, routine guidelines for clinical genetic screening have been established for some neurodevelopmental disorders, but not for major psychiatric disorders. Even increasing division of mental health phenotypes into smaller and better defined categories is not helping to find a genetic test supporting the specific diagnosis. Moreover, it is unlikely that a DSM-defined mental disorder can be considered as a gold standard for genetically based stratification of psychiatric entities.

The genetic predisposition to a psychiatric disorder tends to cross categories as well; moreover, the majority of genetic findings in psychiatry are of a small or moderate effect and consecutive large-scale GWAS have identified only variants associated with small effects on risk [163–166].

Databases of genetic variants in mental diseases

A large panel of known genetic associations in SZ, BPD, ADHD, MDD, and OCD is compiled in large databases extracting information from peer-reviewed scientific publications and integrating information about diverse risk genes. All databases serve as a resource for standardized information for health care providers. Several databases are available. The clinical and genetic information gathered on several thousand affected individuals produced the schizophrenia database SZGene with over 1700 studies of 1000 genes and 8000 polymorphisms leading to hundreds of statistically significant associations in SZ (http://www.szgene.org). Schizophrenia Gene Resource (SZGR) is the largest database collecting and categorizing all genetic data from various sources, including major genetic studies and all other available surveys, that is, association studies, linkage analyses, gene expression studies, gene ontology annotations, gene networks, cellular and regulatory pathways analyses, and microRNAs and their target sites (http://bioinfo.mc.vanderbilt.edu/SZGR/). SZGR is linked to many other public databases such as NCBI and the SchizophreniaGene database.

The Sullivan Lab Evidence Project was developed for the linkage and association evidence of genes or loci based on data from genome-wide linkage analyses, GWAS, and microarray studies for ADHD, ASD, BPD, eating disorders, MDD, nicotine dependence, and SZ (http://slep.unc.edu).

BDgene is a genetic database that aims to integrate multiple genetic factors of BPD from published genetic studies. This project is targeted to help reveal the genetic basis of BPD and its shared mechanism with SZ and MDD from cross-disorder studies for shared diseases etiology research. The results include gene prioritization analysis, pathway-based analysis for GWAS data, intersection analysis of candidate genes, functional annotation, and pathway enrichment analysis for BPD core genes and BPD shared genes (www.bdgene.psych.ac.cn). The most common genes associated with increased risk of psychiatric disorders include *SLC6A4, SULT4A1, CACNA1C, ANK3, COMT, DRD1, DRD2, DRD4, DAT1, MTHFR, SHANK, CHD8, GRM3*, and *GRIN2A*.

Current application of diagnostic genetic tests in major psychiatric disorders

As of this writing the only real opportunity for psychiatry is to use genetics to supplement the diagnostic process in particular cases, rather than replacing the symptom-driven diagnosis. At present genetic testing in psychiatry primarily addresses only several clinical situations:

1. Predicting the response or adverse effects of a drug, including antipsychotic agents, serotonin and serotonin/norepinephrine reuptake inhibitors, and medications used in SUD (substance-use disorder), such as disulfiram. Expert panels have published guidelines for the use of CYP450 testing in psychiatry [167]. These guidelines do not recommend genetic testing of every patient, but provide guidance if genotype data are already available. For more information see Chapter 11, "Next Generation Sequencing in Pharmacogenomics."
2. Excluding disorders mimicking psychiatric conditions or causing symptoms specific for mental diseases (i.e., certain neurometabolic or neurodegenerative diseases such as gangliosidoses and other storage disorders, porphyrias).

3. Early identification of genetic variants associated with a higher risk for major psychiatric disorders (controversial—applicable only in specific situations). It is not known how the results of various tests alone or in combination may help to determine overall risk, nor how they may influence treatment decisions and health outcomes. For major psychiatric phenotypes single genetic variants are not sufficient to cause disease, and there are no genetic tests that can establish a diagnosis or predict individual risk. The association between genetic variants and disease risk appears to be relatively weak and not consistently demonstrated across studies. At present, the study of *de novo* rare disruptive loss-of-function point mutations in genes critical for the biology of the brain seems the most powerful approach.

4. As of this writing only very few tests are used to confirm the diagnosis or support the selection of a therapy (fragile X syndrome, *22q11* deletion syndrome). In the future such testing may open up the possibility of personalized care for patients with certain rare mutations.

Several panels of genetic tests including genes relevant to mental health have been developed and proposed for use in the management of mental disorders. To date it has not been possible to estimate the clinical sensitivity and specificity of those tests. The studies of clinical validity and utility in defined groups of patients were also very limited. The International Society of Psychiatric Genetics elaborated preliminary recommendations for the medical community with growing attention given to clinical genetic testing and the utility of such testing in psychiatry (http://ispg.net/genetic-testing-statement/). This statement based on the published evidence will be updated periodically. The guidelines include the description of available genetic tests including prenatal and postnatal analysis.

NGS in major psychiatric disorders

NGS contributes to the global effort to find genes affecting brain function and paves a new way to the identification of novel variants, helping to unravel this genetic complexity. Several large projects of WGS/WES are ongoing, with some preliminary data already reported [168,169]. WES may offer advantages over traditional molecular diagnostic approaches in complex psychiatric traits, as it ana-lyzes simultaneously the coding regions of thousands of genes, including genes relevant to the mental diseases.

In comparison with traditional genetic testing, NGS is regarded as a perfect tool to find the genetic background of complex psychiatric conditions, because of the amount and diversity of variants this technology can reveal. NGS testing could be justified in:

- neurodevelopmental disorders,
- early-onset neuropsychiatric disorders,
- patients without family history of any comorbid psychiatric disorder (search for *de novo* variants),
- members of a family with extensive history of mental illnesses (search for transmitted variants).

The sequencing of the exome of a patient, comparing it with a normal reference sequence, and further relating to the patient phenotype may help to unravel the genetic determinants of a disorder. WES uncovers genetic defects associated with various complex traits, particularly when combined with sequencing exome-flanking noncoding regions to unravel mutations in noncoding sectors of the genome [170,171]. WES together with the search for CNVs is regarded as the key to understanding psychiatric disorders of multivariate genetic origin [171]. It has become evident that a considerable fraction of exomic mutations and CNVs may occur as *de novo* events and contribute to a significant fraction of cases of mental diseases such as SZ [172–174]. However, known CNVs lack disease-related specificity as a given CNV may increase the risk for a range of psychiatric illnesses, intellectual disabilities, or SUDs. Large population-based studies are needed to establish the lifetime risk for psychiatric disorder in individuals who carry specific CNVs. Identification of CNVs may also help to diagnose rare identified psychiatric disorders that have further implications for individual patients and for family counseling, such as *22q11* deletion syndrome increasing around 30 times the SZ risk or Phelan–McDermid syndrome. The analysis of NGS testing results should refer to the available resources including the above-mentioned databases. According to the reported evidence, the analysis should focus on gene networks in specific pathways contributing to the genetic architecture of mental disorders, rather than single genes. The analysis of gene networks with greater emphasis on the existence of interactions among genes that are overlooked when gene variants are examined separately seems to be an optimal approach. Such analysis requires advanced expertise and understanding of the biological significance of the findings to corroborate the outcomes. The application of NGS technology could help identify many variants that may be simultaneously involved in the various spectrum conditions. Furthermore, there is mounting evidence that many neuropsychiatric disorders share a panel of common genetic risk factors. The Cross-Disorder Group of the Psychiatric Genomics Consortium aims to conduct cross-disorder studies and meta-analyses to find the common genetic background of major psychiatric traits [175]. Data from family and twin genetic studies confirm that there are shared genetic risk factors between major psychiatric disorders. Lotan et al. searched for genetic commonality across major neuropsychiatric disorders including SZ, BPD, ADHD, ASD, MDD, and anxiety disorder [176]. The study was based on the list of genes included in the National Human Genome Research Institute catalog. Of 180 analyzed genes, 22% overlapped two or more disorders. The most widely shared genes common to five of six disorders included *ANK3, AS3MT, CACNA1C, CACNB2, CNNM2, CSMD1, DPCR1, ITIH3, NT5C2, PPP1R11, SYNE1, TCF4, TENM4, TRIM26*, and *ZNRD1*. The authors noticed that many of the shared genes are implicated in postsynaptic density. Shared genes are related to CNS development, neural projections, and synaptic transmission or implicated in various intracellular structures and cellular metabolism, transport, and binding. Taken together these genetic components account for 20–30% of the genetic load. The remaining risk is conferred by variants specific for each disorder. These data contribute to the ongoing debate concerning the new taxonomy of mental disorders on the basis of genetic markers. In this respect, Lotan et al. hypothesized that a shared genetic architecture may be necessary to induce a primary susceptibility to mental illness. Further separate molecular processes, which build up on top of this common infrastructure, ultimately lead, in certain patients, to the development of a specific neuropsychiatric disorder.

Reverse phenotyping of psychiatric phenotypes

A new, promising opportunity for psychiatry is reverse phenotyping, stratifying patients according to genetic biomarkers but not according to phenotypes. Such approach resulted from the observation that similar phenotypes appear when specific genes are disrupted. This approach bypasses the debates about revision of the psychiatric taxonomy of all diseases and the precise diagnostic limits, does not need any golden standard, and does not require a complete understanding of etiology. Moreover, stratification of patients or phenotypes to predict prognosis becomes possible. Thus, such system could coexist alongside the conventional diagnostic systems. As an illustration, this approach identified *MBD5*, a new candidate gene for mental disorders such as BPD and SZ [177]. Accordingly biological psychiatry may need to change the way in which studies are done and reported. In this situation, it would be more efficient to start with genotyping, followed by detailed phenotyping based on genotype information [178, 179]. In a 2014 report of the Molecular Genetics of Schizophrenia Consortium, the analysis of genome-wide SNP profiles for thousands of individuals with or without schizophrenia enabled them to define sets of risk variants that co-occur. Ultimately eight sets of risk variants that consistently coincided with the presence or lack of certain SZ symptoms and served to define eight SZ subtypes were found. Those defined subtypes were subsequently replicated in more than 1000 additional cases enrolled through the Clinical Antipsychotic Trial of Intervention Effectiveness and the Psychiatric Genomics Consortium projects [180]. These subgroups included individuals with varying types of symptoms and severity, from hallucinations to disorganized speech and behavior, which were linked to SNP sets associated with SZ risk to varying degrees. Such approach may identify homogeneous populations for new targeted therapy, thus moving beyond descriptive psychiatry and toward a nosology related to disease etiology [181].

REFERENCES

[1] Huang Y, Yu S, Wu Z, Tang B. Genetics of hereditary neurological disorders in children. Transl Pediatr 2014;3:108–19.

[2] Flores-Sarnat L, Sarnat HB. Axes and gradients of the neural tube and gradients for a morphological molecular genetic classification of nervous system malformations. In: Sarnat HB, Curatolo P, editors. Malformations of the nervous system. Handbook of clinical neurology, vol. 87. Elsevier; 2008. 3rd ed.

[3] Barkovich AJ, Kuzniecky RI, Jackson GD, Guerrini R, Dobyns WB. A developmental and genetic classification for malformations of cortical development. Neurology 2005;65:1873–87.

[4] Guerrini R, Dobyns WB. Malformations of cortical development: clinical features and genetic causes. Lancet Neurol 2014;13:710–26.

[5] Poirier K, Lebrun N, Broix L, et al. Mutations in TUBG1, DYNC1H1, KIF5C and KIF2A cause malformations of cortical development and microcephaly. Nat Genet 2013;45(6). http://dx.doi.org/10.1038/ng.2613.

[6] Bilgüvar K, Öztürk AK, Louvi A, et al. Whole-exome sequencing identifies recessive WDR62 mutations in severe brain malformations. Nature 2010;467:207–10.

[7] Shimojima K, Narita A, Maegaki Y, et al. Whole-exome sequencing identifies a *de novo* TUBA1A mutation in a patient with sporadic malformations of cortical development: a case report. BMC Res notes 2014;7:465.

[8] Willemsen MH, Vissers LE, Willemsen MA, et al. Mutations in DYNC1H1 cause severe intellectual disability with neuronal migration defects. J Med Genet 2012;49:179–83.

[9] Davis EE, Katsanis N. The ciliopathies: a transitional model into systems biology of human genetic disease. Curr Opin Genet Dev 2012;22:290–303.

[10] Akizu N, Silhavy JL, Rosti RO. Mutations in CSPP1 Lead to classical joubert syndrome. Am J Hum Genet 2014;94:80–6.

[11] Tuz K, Bachmann-Gagescu R, O'Day DR, et al. Mutations in CSPP1 cause primary cilia abnormalities and joubert syndrome with or without jeune asphyxiating thoracic dystrophy. Am J Hum Genet 2014;94: 62–72.

[12] Carroll CJ, Brilhante V, Suomalainen A. Next-generation sequencing for mitochondrial disorders. Br J Pharmacol 2014;171:1837–53.

[13] Calvo SE, Compton AG, Hershman SG, et al. Molecular diagnosis of infantile mitochondrial disease with targeted next-generation sequencing. Sci Transl Med 2012;4:118ra10.

[14] Nishri D, Edvardson S, Lev D, et al. Diagnosis by whole exome sequencing of atypical infantile onset Alexander disease masquerading as a mitochondrial disorder. Eur J Paediatr Neurology 2014;18:495–501.

[15] Fraser JL, Vanderver A, Yang S, et al. Thiamine pyrophosphokinase deficiency causes a Leigh disease like phenotype in a sibling pair: identification through whole exome sequencing and management strategies. Mol Genet Metab Rep 2014;1:66–70.

[16] Ng BG, Freeze HH. Human genetic disorders involving glycosylphosphatidylinositol (GPI) anchors and glycosphingolipids (GSL). Journal of inherited metabolic disease. J Inherit Metab Dis 2015;38:171–8.

[17] Freeze HH, Chong JX, Bamshad MJ, Ng BG. Solving glycosylation disorders: fundamental approaches reveal complicated pathways. Am J Hum Genet 2014;94:161–75.

[18] Fernández-Marmiesse A, Morey M, Pineda M, et al. Assessment of a targeted resequencing assay as a support tool in the diagnosis of lysosomal storage disorders. Orphanet J Rare Dis 2014;9:59.

[19] Szymańska K, Szczałuba ŁA, et al. The analysis of genetic aberrations in children with inherited neurometabolic and neurodevelopmental disorders. BioMed Res Int 2014;2014, Article ID 424796, 8 p. http://dx.doi.org/10.1155/2014/424796.

[20] Panayiotopoulos CP. The new ILAE report on terminology and concepts for the organization of epilepsies: critical review and contribution. Epilepsia 2012;53:399–404.

[21] Hildebrand MS, Dahl H-HM, Damiano JA, et al. Recent advances in the molecular genetics of epilepsy. J Med Genet 2013;50:271–9.

[22] Deng H, Xiu X, Song Z. The molecular biology of genetic-based epilepsies. Mol Neurobiol 2014;49:352–67.

[23] Garofalo S, Cornacchione M, Di Costanzo A. From genetics to genomics of epilepsy. 2012. http://dx.doi.org/10.1155/2012/876234. Volume.

[24] Lemke JR, Riesch E, Scheurenbrand T, et al. Targeted next generation sequencing as a diagnostic tool in epileptic disorders. Epilepsia 2012;53:1387–98.

[25] Pandolfo M. Pediatric epilepsy genetics. Curr Opin Neurol 2013;26:137–45.

[26] Helbig I, Lowenstein DH. Genetics of the epilepsies: where are we and where are we going? Curr Opin Neurol 2013;26:179–85.

[27] Weckhuysen S, Korff CM. Epilepsy: old syndromes, new genes. Curr Neurol Neurosci Rep 2014;14:447.

[28] Kato M, Yamagata T, Kubota M, et al. Clinical spectrum of early onset epileptic encephalopathies caused by KCNQ2 mutation. Epilepsia 2013;54:1282–7.

[29] Tang S, Pal DK. Dissecting the genetic basis of myoclonic-astatic epilepsy. Epilepsia 2012;53:1303–13.

[30] Dichgans M, Freilinger T, Eckstein G, et al. Mutation in the neuronal voltage-gated sodium channel SCN1A in familial hemiplegic migraine. Lancet 2005;366:371–7.

[31] Carvill GL, Weckhuysen S, McMahon JM, et al. GABRA1 and STXBP1: novel genetic causes of Dravet syndrome. Neurology 2014;82:1245–53.

[32] Rahman S, Footitt EJ, Varadkar S, Clayton PT. Inborn errors of metabolism causing epilepsy. Dev Med Child Neurol 2013;55:23–36.

[33] Ng SB, Buckingham KJ, Lee C, et al. Exome sequencing identifies the cause of a Mendelian disorder. Nat Genet 2010;42:30–5.

[34] Neul JL, Kaufmann WE, Glaze DG. Rett syndrome: revised diagnostic criteria and nomenclature. Ann Neurol 2010;68:944–50.

[35] Guerrini R, Parrini E. Epilepsy in Rett syndrome, and CDKL5- and FOXG1-gene–related encephalopathies. Epilepsia 2012;53:2067–78.

[36] Veeramah KR, Johnstone L, Karafet TM, et al. Exome sequencing reveals new causal mutations in children with epileptic encephalopathies. Epilepsia 2013;54:1270–81.

[37] Mei D, Darra F, Barba C, et al. Optimizing the molecular diagnosis of CDKL5 gene–related epileptic encephalopathy in boys. Epilepsia 2014;55(11):1748–53. http://dx.doi.org/10.1111/epi.12803.

[38] Grillo E, et al. Revealing the complexity of a monogenic disease: Rett syndrome exome sequencing. PloS one 2013;8:e56599.

[39] Wong M. Mammalian target of rapamycin (mTOR) pathways in neurological diseases. Biomed J 2013;36:40–50.

[40] Child ND, Benarroch EE. mTOR its role in the nervous system and involvement in neurologic disease. Neurology 2014;83:1562–72.

[41] Rivière J-B, Mirzaa GM, O'Roak BJ, et al. *De novo* germline and postzygotic mutations in AKT3, PIK3R2 and PIK3CA cause a spectrum of related megalencephaly syndromes. Nat Genet 2012;44:934–40.

[42] Lee JH, Silhavy JL, Kim S, et al. *De novo* somatic mutations in components of the PI3K-AKT3-mTOR pathway cause hemimegalencephaly. Nat Genet 2012;44(8):941–5.

[43] Dibbens LM, de Vries B, Donatello S, et al. Mutations in DEPDC5 cause familial focal epilepsy with variable foci. Nat Genet 2013;45(5):546–51.

[44] Scheffer IE. Epilepsy genetics revolutionizes clinical practice. Neuropediatrics 2014;45:70–4.

[45] Ohmori I, Ouchida M, Kobayashi K, et al. CACNA1A variants may modify the epileptic phenotype of Dravet syndrome. Neurobiol Dis 2013;50:209–17.

[46] Manto MU. Cerebellar disorders: a practical approach to diagnosis and management. Cambridge University Press; 2010.

[47] Mancuso M, Orsucci D, Siciliano G, Bonuccelli U. The genetics of ataxia: through the labyrinth of the Minotaur, looking for Ariadne's thread. J Neurol 2014;261(Suppl. 2):S528–41.

[48] Smeets CJLM, Verbeek DS. Cerebellar ataxia and functional genomics: identifying the routes to cerebellar neurodegeneration. Biochim Biophys Acta 2014;1842:2030–8.

[49] Ohba C, Osaka H, Iai C, et al. Diagnostic utility of whole exome sequencing in patients showing cerebellar and/or vermis atrophy in childhood. Neurogenetics 2013;14:225–32.

[50] Hersheson J, Haworth A, Houlden H. The inherited ataxias: genetic heterogeneity, mutation databases, and future directions in research and clinical diagnostics. Hum Mutat 2012;33:1324–32.

[51] Sailer A, Houlden H. Recent advances in the genetics of cerebellar ataxias. Curr Neurol Neurosci Rep 2012;12:227–36.

[52] Németh AH, Kwasniewska AC, Lise S, et al. Next generation sequencing for molecular diagnosis of neurological disorders using ataxias as a model. Brain 2013;136:3106–18. http://dx.doi.org/10.1093/brain/awt236.

[53] Lines MA, Jobling R, Brady L, et al. Peroxisomal D-bifunctional protein deficiency three adults diagnosed by whole-exome sequencing. Neurology 2014;82:963–8.

[54] Matilla-Dueñas A. The ever expanding spinocerebellar ataxias. Editorial. Cerebellum 2012;11:821–7.

[55] Rossi M, Perez-Lloret S, Doldan L, et al. Autosomal dominant cerebellar ataxias: a systematic review of clinical features. Eur J Neurol 2014;21:607–15.

[56] Wang JL, Yang X, Xia K, et al. TGM6 identified as a novel causative gene of spinocerebellar ataxias using exome sequencing. Brain 2010;133:3510–8.

[57] Anheim M, Tranchant C, Koenig M. The autosomal recessive cerebellar ataxias. N. Engl J Med 2012;366:636–46.

[58] Vermeer S, Hoischen A, Meijer RP, et al. Targeted next-generation sequencing of a 12.5 Mb Homozygous region Reveals *ANO10* mutations in patients with autosomal-recessive cerebellar ataxia. Am J Hum Genet 2010;87:813–9.

[59] Doi H, Yoshida K, Yasuda T, et al. Exome sequencing reveals a homozygous SYT14 mutation in adult-onset, autosomal-recessive spinocerebellar ataxia with psychomotor retardation. Am J Hum Genet 2011;89:320–7.

[60] Margolin DH, Kousi M, Chan Y-M, et al. Ataxia, dementia, and hypogonadotropism caused by disordered ubiquitination. N. Engl J Med 2013;368:1992–2003.

[61] Shi CH, Schisler J, Rubel CE, et al. Ataxia and hypogonadism caused by the loss of ubiquitin ligase activity of the U box protein CHIP. Hum Mol Genet 2014;23:1013–24.

[62] Heimdal K, Sanchez-Guixé M, Aukrust I, et al. STUB1 mutations in autosomal recessive ataxias–evidence for mutation-specific clinical heterogeneity. Orphanet J Rare Dis 2014;9(1):146.

[63] Gasser T, Finsterer J, Baets J, et al. EFNS guidelines on the molecular diagnosis of ataxias and spastic paraplegias. Eur J Neurol 2010;17:179–88.

[64] Finsterer J, Löscher W, Quasthoff S, et al. Hereditary spastic paraplegias with autosomal dominant, recessive, X-linked, or maternal trait of inheritance. J Neurol Sci 2012;318:1–18.

[65] Giudice TL, Lombardi F, Santorelli FM, et al. Hereditary spastic paraplegia: clinical-genetic characteristics and evolving molecular mechanisms. Exp Neurol 2014;261:518–39.

[66] Garcia-Cazorla A, Duarte ST. Parkinsonism and inborn errors of metabolism. J Inherit Metab Dis 2014;37(4):627–42. http://dx.doi.org/10.1007/s10545-014-9723-6.

[67] Bonifati V. Genetics of Parkinson's disease – state of the art 2013. Parkinsonism Relat Disord 2014;20(Suppl. 1):S23–8.

[68] Vilarino-Guell C, Wider C, Ross OA, et al. VPS35 mutations in Parkinson disease. Am J Hum Genet 2011;89:162–7.

[69] Zimprich A, Benet-Pages A, Struhal W, et al. A mutation in VPS35, encoding a subunit of the retromer complex, causes late-onset Parkinson disease. Am J Hum Genet 2011;89:168–75.

[70] Nuytemans K, Bademci G, Inchausti V, et al. Whole exome sequencing of rare variants in EIF4G1 and VPS35 in Parkinson disease. Neurology 2013;80:982–9.

[71] Edvardson S, Cinnamon Y, Ta-Shma A, et al. A deleterious mutation in DNAJC6 encoding the neuronal-specific clathrin-uncoating co-chaperone auxilin, is associated with juvenile parkinsonism. PLoS One 2012;7(5):e36458.

[72] Krebs CE, Karkheiran S, Powell JC, et al. The Sac1 domain of SYNJ1 identified mutated in a family with early-onset progressive Parkinsonism with generalized seizures. Hum Mutat 2013;34:1200–7.

[73] Albanese A, Bhatia K, Bressman SB, et al. Phenomenology and classification of dystonia: a consensus update. Mov Disord 2013;28:863–73.

[74] Lohmann K, Klein C. Genetics of dystonia: what's known? What's new? What's next? Mov Disord 2013;28:899–905.

[75] Grundmann K, Söhn A, Sturm M, Riess O, Bauer P. Diagnosing dystonia using a next-generation-sequencing gene panel [abstract]. Mov Disord 2014;29(Suppl. 1):1371.

[76] Kaplan JC, Hamroun D. The 2014 version of the gene table of monogenic neuromuscular disorders (nuclear genome). Neuromuscular Disord 2013;23:1081–111.

[77] Milone M, Katz A, Amato AA, et al. Sporadic late onset nemaline myopathy responsive to IVIg and immunotherapy. Muscle Nerve 2010;41:272–6. http://dx.doi.org/10.1002/mus.21504.

[78] Romero NB, Sandaradura SA, Clarke NF. Recent advances in nemaline myopathy. Curr Opin Neurol 2013;26(5):519–26. http://dx.doi.org/10.1097/WCO.0b013e328364d681.

[79] Abicht A, Dusl M, Gallenmüller C, Guergueltcheva V, Schara U, Della Marina A, et al. Congenital myasthenic syndromes: achievements and limitations of phenotype-guided gene-after-gene sequencing in diagnostic practice: a study of 680 patients. Hum Mutat 2012;33(10):1474–84. http://dx.doi.org/10.1002/humu.22130. Epub June 27, 2012.

[80] Abicht A, Stucka R, Karcagi V, Herczegfalvi A, Horváth R, Mortier W, et al. A common mutation (epsilon1267delG) in congenital myasthenic patients of Gypsy ethnic origin. Neurology 1999;53(7):1564–9.

[81] Valencia CA, Ankala A, Rhodenizer D, et al. Comprehensive mutation analysis for congenital muscular dystrophy: a clinical PCR-based enrichment and next-generation sequencing panel. PLoS One 2013;8:e53083. http://dx.doi.org/10.1371/journal.pone.0053083.

[82] Ylikallio E, Johari M, Konovalova S, et al. Targeted next-generation sequencing reveals further genetic heterogeneity in axonal Charcot-Marie-Tooth neuropathy and a mutation in HSPB1. Eur J Hum Genet 2014;22:522–7. http://dx.doi.org/10.1038/ejhg.2013.190.

[83] Yang Y, Muzny DM, Reid JG, et al. Clinical whole-exome sequencing for the diagnosis of mendelian disorders. N Engl J Med 2013;369:1502–11.

[84] Biesecker LG, Green RC. Diagnostic clinical genome and exome sequencing. N Engl J Med June 19, 2014;370(25):2418–25. http://dx.doi.org/10.1056/NEJMra1312543.

[85] North KN, Wang CH, Clarke N, et al. International Standard of Care Committee for Congenital Myopathies. Approach to the diagnosis of congenital myopathies. Neuromuscular Disord 2014;24:97–116. http://dx.doi.org/10.1016/j.nmd.2013.11.003. Epub November 18, 2013.

[86] Cassereau J, Casasnovas C, Gueguen N, Malinge MC, Guillet V, Reynier P, et al. Simultaneous MFN2 and GDAP1 mutations cause major mitochondrial defects in a patient with CMT. Neurology 2011;76:1524–6.

[87] Kostera-Pruszczyk A, Kosinska J, Pollak A, Stawinski P, Walczak A, Wasilewska K, et al. Exome sequencing reveals mutations in MFN2 and GDAP1 in severe Charcot-Marie-Tooth disease. J Peripher Nerv Syst 2014;19(3):242–5. http://dx.doi.org/10.1111/jns.12088.

[88] Flanigan KM, Ceco E, Lamar KM, Kaminoh Y, Dunn DM, Mendell JR, et al. LTBP4 genotype predicts age of ambulatory loss in Duchenne muscular dystrophy. United Dystrophinopathy Project Ann Neurol 2013;73(4):481–8. http://dx.doi.org/10.1002/ana.23819.

[89] Gregersen PK, Kosoy R, Lee AT, Lamb J, Sussman J, McKee D, et al. Risk for myasthenia gravis maps to a (151) Pro→Ala change in TNIP1 and to human leukocyte antigen-B*08. Ann Neurol 2012;72(6):927–35. http://dx.doi.org/10.1002/ana.23691.

[90] Avidan N, Le Panse R, Harbo HF, et al. VAV1 and BAFF, via NFκB pathway, are genetic risk factors for myasthenia gravis. Ann Clin Transl Neurol 2014;1:329–39.

[91] Reitz C, Mayeux R. Alzheimer disease: epidemiology, diagnostic criteria, risk factors and biomarkers. Biochem Pharmacol 2014;88:640–51.

[92] Cohn-Hokke PE, Elting MW, Pijnenburg YAL, van Swieten JC. Genetics of dementia: update and guidelines for the clinician. Am J Med Genet Part B Neuropsychiatr Genet 2012;159:628–43.

[93] Tanzi RE. The genetics of Alzheimer disease. Cold Spring Harbor Perspect Med 2012;2(10):a006296.

[94] Bettens K, Sleegers K, Van Broeckhoven C. Genetic insights in Alzheimer's disease. Lancet Neurol 2013;12(1):92–104.

[95] DeJesus-Hernandez M, et al. Expanded GGGGCC hexanucleotide repeat in noncoding region of C9ORF72 causes chromosome 9p-linked FTD and ALS. Neuron 2011;72(2):245–56.

[96] Verma A. Tale of two diseases: amyotrophic lateral sclerosis and frontotemporal dementia. Neurol India 2014;62(4):347.

[97] Schalock RL, Borthwick-Duffy SA, Bradley VJ, et al. Intellectual disability: definition, classification and system of supports, vol. 11. Washington DC: American Association on Intellectual and Developmental Disabilities; 2010.

[98] Ropers HH. Genetics of early-onset cognitive impairment. Annu Rev Genomics Hum Genet 2010;11:161–87.

[99] Roeleveld N, Zielhuis GA, Gabreels F. The prevalence of mental retardation: a critical review of recent literature. Dev Med Child Neurol 1997;39:125–32.

[100] Rauch A, Hoyer J, Guth S, et al. Diagnostic yield of various genetic approaches in patients with unexplained developmental delay or mental retardation. Am J Med Genet 2006;140A:2063–74.

[101] Bartnik M, Wiśniowiecka-Kowalnik B, Nowakowska B, et al. The usefulness of array comparative genomic hybridization in clinical diagnostics of intellectual disability in children. Dev Period Med 2014; 18:307–17.

[102] Moeschler JB, Shevell M. Comprehensive evaluation of the child with intellectual disability or global development al delays. Pediatrics 2014;134:e903–18.

[103] Willemsen MH, Kleefstra T. Making headway with genetic diagnostics of intellectual disabilities. Clin Genet 2014;85:101–10.

[104] Topper S, Ober C, Das S. Exome sequencing and the genetics of intellectual disability. Clin Genet 2011;80(2):117–26.

[105] Hoischen A, van Bon BW, Gillisen C, et al. De novo mutations of SETBP1 cause Schinzel-Giedion syndrome. Nat Genet 2010;42:483–5.

[106] Ng SB, Bigham AW, Buckingham KJ, et al. Exome sequencing identifies MLL2 mutations as a cause of Kabuki syndrome. Nat Genet A 2010;42:790–3.

[107] Ng SB, Buckingham KJ, Lee C, et al. Exome sequencing identifies the cause of a mendelian disorder. Nat Genet B 2010;42:30–5.

[108] Kotan LD, Hutchins BI, Ozkan Y, et al. Mutations in FEZF1 cause Kallmann syndrome. Am J Hum Genet 2014;95:326–31.

[109] Bertola DR, Yamamoto GL, Almeida TF, et al. Further evidence of the importance of RIT1 in Noonan syndrome. Am J Med Genet A 2014;164A(11):2952–7. Epub August 13, 2014.

[110] Rabbani B, Tekin M, Mahdieh N. The promise of whole-exome sequencing in medical genetics. J Hum Genet 2014;59:5–15.

[111] Classen CF, Riehmer V, Landwehr C, et al. Dissecting genotype of syndromic intellectual disability using whole exome sequencing in addition to genome-wide copy number analysis. Hum Genet 2013;132:825–41.

[112] Vissers L, de Ligt J, Gilissen C, et al. A *de novo* paradigm for mental retardation. Nat Genet 2010;42:1109–12.

[113] Athanasakis E, Licastro D, Faletra F, et al. Next generation sequencing in nonsyndromic intellectual disability: from a negative molecular karyotype to a possible causative mutation detection. Am J Med Genet 2013;164A:170–6.

[114] Rauch A, Wieczorek D, Graf E, et al. Range of genetic mutations associated with severe non-syndromic sporadic intellectual disability: an exome sequencing study. Lancet 2012;380:1674–82.

[115] de Ligt J, Willemsen MH, Bregje WM, et al. Diagnostic exome sequencing in persons with severe intellectual disability. N Engl J Med 2012;367:1921–9.

[116] Schuurs-Hoeijmakers JHM, Vulto-van Silfhout AT, Vissers LELM, et al. Identification of pathogenic variants in small families with intellectually disabled siblings by exome sequencing. J Med Genet 2013;50:802–11.

[117] Caliskan M, Chong JX, Uricchio L, et al. Exome sequencing reveals a novel mutation for autosomal recessive non-syndromic mental retardation in the TECR gene on chromosome 19p13. Hum Mol Genet 2011;20:1285–9.

[118] Najmabadi H, Hu H, Garshasbi M, et al. Deep sequencing reveals 50 novel genes for recessive cognitive disorders. Nature 2011;478:57–63.

[119] Martinez FJ, Lee JH, Lee JE, et al. Whole exome sequencing identifies a splicing mutation in NSUN2 as a cause of a Dubowitz-like syndrome. J Med Genet 2012;49:380–5.

[120] Ng BG, Hackmann K, Jones MA, et al. Mutations in the glycosylphosphatidylinositol gene PIGL cause CHIME syndrome. Am J Hum Genet 2012;90:685–8.

[121] Drmanac R, et al. Human genome sequencing using unchained base reads on self-assembling DNA nanoarrays. Science 2010;327:78–81.

[122] Gilissen C, Hehir-Kwa JY, Thung DT, et al. Genome sequencing identifies major causes of severe intellectual disability. Nature 2014;511:344–7.

[123] Pagnamenta AT, Lise S, Harrison V, et al. Exome sequencing can detect pathogenic mosaic mutations present at low allele frequencies. J Hum Genet 2012;57:70–2.

[124] Huisman SA, Redeker EJ, Maas SM, et al. High rate of mosaicism in individuals with Cornelia de Lange syndrome. J Med Genet 2013;50:339–44.

[125] Jamuar SS, Lam ATN, Kircher M, et al. Somatic mutations in cerebral cortical malformations. N Engl J Med 2014;371:733–43.

[126] O'Roak BJ, State MW. Autism genetics: strategies, challenges, and opportunities. Autism Res 2008;1:4–17.

[127] Sanders SJ, Ercan-Sencicek AG, Hus V, et al. Multiple recurrent CNVs, including duplications of 7q11.23 Williams syndrome region, are strongly associated with autism. Neuron 2011;70(5):863–5.

[128] Wiśniowiecka-Kowalnik B, Kastory-Bronowska M, Bartnik M, et al. Application of custom-designed oligonucleotide array CGH in 145 patients with autism spectrum disorders. Eur J Hum Genet 2013;21(6):620–5.

[129] Levy D, Ronemus M, Yamrom B, et al. Rare *de novo* and transmitted copy-number variation in autistic spectrum disorders. Neuron 2011;70(5):886–97.

[130] Sebat J, Lakshmi B, Malhotra D, et al. Strong association of *de novo* copy number mutations with autism. Science 2007;316(5823):445–9.

[131] Koolen DA, de Vries BBA. Newly recognized microdeletion/duplication syndromes. In: Knight SJL, editor. Genetics of mental retardation: an overview encompassing learning disability and intellectual disability. Oxford: Karger; 2010. p. 101–13.

[132] Schaaf CP, Sabo A, Sakai Y, et al. Oligogenic heterozygosity in individuals with high-functioning autism spectrum disorders. Hum Mol Genet 2011;20(17):3366–75.

[133] O'Roak BJ, Deriziotis P, Lee C, et al. Exome sequencing in sporadic autism spectrum disorders identifies severe *de novo* mutations. Nat Genet 2011;43(6):585–9.

[134] O'Roak BJ, Vives L, Girirajan S, et al. Sporadic autism exomes reveal a highly interconnected protein network of *de novo* mutations. Nature 2012;485:246–50.

[135] Iossifov I, Ronemus M, Levy D, et al. *De novo* gene disruptions in children on the autistic spectrum. Neuron 2012;74:285–99.

[136] Sanders SJ, Murtha MT, Gupta AR, et al. *De novo* mutations revealed by whole-exome sequencing are strongly associated with autism. Nature 2010;485:237–41.

[137] Neale BM, Kou Y, Liu L, et al. Patterns and rates of exonic *de novo* mutations in autism spectrum disorders. Nature 2012;485:242–5.

[138] Darnell JC, Van Driesche SJ, Zhang C, et al. FMRP stalls ribosomal translocation on mRNAs linked to synaptic function and autism. Cell 2011;146:247–61.

[139] Meisler MH, O'Brien JE, Sharkey LM. Sodium channel gene family: epilepsy mutations, gene interactions and modifier effects. J Physiol 2010;588(11):1841–8.

[140] Cukier HN, Dueker ND, Slifer SH, et al. Exome sequencing of extended families with autism reveals genes shared across neurodevelopmental and neuropsychiatric disorders. Mol Autism 2014;5(1).

[141] Chahrour MH, Yu TW, Lim ET, et al. Whole-exome sequencing and homozygosity analysis implicate depolarization-regulated neuronal genes in autism. PLoS Genet 2012;8(4):e1002635.

[142] Puffenberger EG, Jinks RN, Wang H, et al. A homozygous missense mutation in HERC2 associated with global developmental delay and autism spectrum disorder. Hum Mutat 2012;33(12):1639–46.

[143] Jiang YH, Yuen RK, Jin X, et al. Detection of clinically relevant genetic variants in autism-spectrum disorder by whole-genome sequencing. Am J Hum Genet 2013;93(2):249–63.

[144] Michaelson JJ, Shi Y, Gujral M, et al. Whole-genome sequencing in autism identifies hot spots for *de novo* germline mutation. Cell 2012;151(7):1431–42.

[145] Shi L, Zhang X, Golhar R, et al. Whole-genome sequencing in an autism multiplex family. Mol Autism 2013;4(1):8.

[146] Krumm N, O'Roak BJ, Shendure J, Eichler EE. A *de novo* convergence of autism genetics and molecular neuroscience. Trends Neurosci 2014;37(2):95–105.

[147] Gilman SR, Iossifov I, Levy D, et al. Rare *de novo* variants associated with autism implicate a large functional network of genes involved in formation and function of synapses. Neuron 2011;70: 898–907.

[148] O'Roak BJ, Vives L, Fu W, et al. Multiplex targeted sequencing identifies recurrently mutated genes in autism spectrum disorders. Science 2012;338(6114):1619–22.

[149] Bamshad MJ, et al. Exome sequencing as a tool for Mendelian disease gene discovery. Nat Rev Genet 2011;12:745–55.

[150] Gratten J, Visscher PM, Mowry BJ, Wray NR. Interpreting the role of *de novo* protein-coding mutations in neuropsychiatric disease. Nat Genet 2013;45(3):234–8.

[151] Vermeesch JR, Balikova I, Schrander-Stumpel C, Fryns JP, Devriendt K. The causality of *de novo* copy number variants is overestimated. Eur J Hum Genet 2011;19:1112–3.

[152] Miles JH. Autism spectrum disorders—a genetics review. Genet Med 2011;13:278–94.

[153] Beaudet AL. Preventable forms of autism? Science 2012;338:342–3.

[154] Fischbach GD, Lord C. The simons simplex collection: a resource for identification of autism genetic risk factors. Neuron 2010;68:192–5.

[155] Basu SN, Kollu R, Banerjee-Basu S. AutDB: a gene reference resource for autism research. Nucleic Acids Res 2009;37:D832–6.

[156] Geschwind DH, Sowinski J, Lord C, et al. The autism genetic resource exchange: a resource for the study of autism and related neuropsychiatric conditions. Am J Hum Genet 2001;69:463–6.

[157] Matuszek G, Talebizadeh Z. Autism genetic database (AGD): a comprehensive database including autism susceptibility gene-CNVs integrated with known noncoding RNAs and fragile sites. BMC Med Genet 2009;10:102.

[158] Marshall CR, Noor A, Vincent JB, et al. Structural variation of chromosomes in autism spectrum disorder. Am J Hum Genet 2008;82:477–88.

[159] Hu-Lince D, Craig DW, Huentelman MJ, et al. The autism genome project. Am J PharmacoGenomics 2005;5:233–46.

[160] Schizophrenia Working Group of the Psychiatric Genomics Consortium. Biological insights from 108 schizophrenia-associated genetic loci. Nature 2014;511:421–7.

[161] Schadt EE, Buchanan S, Brennand KJ, Kalpana M. Merchant evolving toward a human-cell based and multiscale approach to drug discovery for CNS disorders. Front Pharmacol 2014;5:252. http://dx.doi.org/10.3389/fphar.2014.00252.

[162] Califano A, Butte AJ, Friend S, Ideker T, Schadt E. Leveraging models of cell regulation and GWAS data in integrative network-based association studies. Nat Genet 2012;44:841–7. http://dx.doi.org/10.1038/ng.2355.

[163] Stefansson H, Ophoff RA, Steinberg S, Andreassen OA, Cichon S, Rujescu D, et al. Genetic risk and outcome in psychosis (GROUP) common variants conferring risk of schizophrenia. Nature 2009;460:744–7.

[164] Shi J, Levinson DF, Duan J, Sanders AR, Zheng Y, Pe'er I, et al. Common variants on chromosome 6p22.1 are associated with schizophrenia. Nature 2009;460:753–7.

[165] Purcell SM, Wray NR, Stone JL, Visscher PM, O'Donovan MC, Sullivan PF, et al. International schizophrenia consortium common polygenic variation contributes to risk of schizophrenia and bipolar disorder. Nature 2009;460:748–52.

[166] Purcell SM, et al. A polygenic burden of rare disruptive mutations in schizophrenia. Nature 2014;506:185–90.

[167] Hicks JK, Swen JJ, Thorn CF, et al. Clinical pharmacogenetics implementation consortium guideline for CYP2D6 and CYP2C19 genotypes and dosing of tricyclic antidepressants. Clin Pharmacol Ther 2013;93:402–8.

[168] Kato T. Whole genome/exome sequencing in mood and psychotic disorders. Psychiatry Clin Neurosci October 16, 2014;69(2):65–76. http://dx.doi.org/10.1111/pcn.12247.

[169] Schreiber M, Dorschner M, Tsuang D. Next-generation sequencing in schizophrenia and other neuropsychiatric disorders. Am J Med Genet B Neuropsychiatr Genet 2013;162B:671–8.

[170] Lehne B, Lewis CM, Schlit T. Exome localization of complex disease association signals. BMC Genomics 2011;12:92.

[171] Raffan E, Semple RK. Next generation sequencing–implications for clinical practice. Br Med Bull 2011;99:53–71.

[172] Xu B, Roos JL, Dexheimer P, Boone B, Plummer B, et al. Exome sequencing supports a *de novo* mutational paradigm for schizophrenia. Nat Genet 2011;43:864–8.

[173] Girard SL, Gauthier J, Noreau A, Xiong L, Zhou S, et al. Increased exonic *de novo* mutation rate in individuals with schizophrenia. Nat Genet 2011;43:860–3.

[174] Xu B, Ionita-Laza I, Roos JL, Boone B, Woodrick S, et al. *De novo* gene mutations highlight patterns of genetic and neural complexity in schizophrenia. Nat Genet 2012;44:1365–9.

[175] Cross-Disorder Group of the Psychiatric Genomics Consortium. Identification of risk loci with shared effects on five major psychiatric disorders: a genome-wide analysis. Lancet 2013;381(9875):1371–9.

[176] Lotan A, Fenckova M, Bralten J, Alttoa A, Dixson L, Williams RW, et al. Neuroinformatic analyses of common and distinct genetic components associated with major neuropsychiatric disorders. Front Neurosci November 6, 2014;8:331. http://dx.doi.org/10.3389/fnins.2014.00331.

[177] Hodge JC, Mitchell E, Pillalamarri V, et al. Disruption of MBD5 contributes to a spectrum of psychopathology and neurodevelopmental abnormalities. Mol Psychiatry 2014;19:368–79.

[178] Stessman HA, Bernier R, Eichler EE. A genotype-first approach to defining the subtypes of a complex disease. Cell 2014;156:872–7.

[179] Schulze TG, McMahon FJ. Defining the phenotype in human genetic studies: forward genetics and reverse phenotyping. Hum Hered 2004;58:131–8.

[180] Arnedo J, Svrakic DM, Del Val C, et al. Uncovering the hidden risk architecture of the schizophrenias: confirmation in three independent genome-wide association studies. Am J Psychiatry September 15, 2014;172(2):139–53. http://dx.doi.org/10.1176/appi.ajp.2014.14040435.

[181] Kapur S, Phillips AG, Insel TR. Why has it taken so long for biological psychiatry to develop clinical tests and what to do about it&quest. Mol Psychiatry 2012;17.12:1174–9.

NEXT GENERATION SEQUENCING IN DYSMORPHOLOGY

Robert Smigiel[1], Urszula Demkow[2]

[1]Department of Social Pediatrics, Wroclaw Medical University, Wroclaw, Poland; [2]Department of Laboratory Diagnostics and Clinical Immunology of Developmental Age, Medical University of Warsaw, Warsaw, Poland

CHAPTER OUTLINE

DYSMORPHOLOGY—PAST AND PRESENT

Dysmorphology, the study of human congenital malformations and syndromes, is quite a young discipline in clinical genetics. Dysmorphic syndrome (DS) includes a particular set of developmental anomalies that create a recognizable and consistent pattern of abnormalities. DS has a known or assumed single etiology. The term dysmorphic is used to describe children whose physical features, particularly facial features, and also features of the entire body (stature, neck, limbs, hands, feet, trunk, and genitalia) are not usually found in children of the same age and/or ethnic group. Dysmorphology attempts to interpret the patterns of human growth and structural defects. These include malformation, disruption, deformation, and dysplasia [1–3]. Many dysmorphic children have significant internal or external malformations or developmental delay or some combination of these.

Congenital malformations represent one of the most frequent and important reasons for genetic counseling. Children with a congenital defect and/or DS represent 2–3% of live births [4]. Many multidefect syndromes and DSs are very rare (even 1:50,000 to 1:200,000 live births), but with a large number of individual syndromes (about 7000 currently recognized syndromes described by OMIM and Orphanet) [5–7], they are collectively frequent, affecting millions of people worldwide, accounting for a major proportion of serious disorders, disability, hospitalizations, and early deaths in children [8–12]. Moreover, they account for rising costs of health care in every country and society [9–11].

Dr David Smith, a pediatrician, was the founder of clinical dysmorphology. He formed, in 1960, in the United States (Madison, WI, and Seattle, WA), a clinical and scientific group of specialists devoted to the study of congenital malformations [13]. A few years later, Dr Jon Aase, a former student of Dr Smith, elaborated a detailed concept of dysmorphology, which "as a scientific discipline combines concepts, knowledge, and techniques from the fields of embryology, clinical genetics and pediatrics. As a medical subspecialty, dysmorphology deals with people who have congenital abnormalities and with their families" [3,13].

From 1968 to 1974, Victor A. McKusick organized annual conferences on birth defects and subsequently published new discoveries in this field. Victor McKusick understood that congenital malformations overlapped with a wider group of genetic disorders that were progressively becoming clinically and scientifically defined. In the 1970s and after, clinical dysmorphology rapidly developed owing to the common efforts of numerous researchers, including Robert Gorlin, Michael M. Cohen, John Opitz, Judith Hall, Jon Aase, David Rimoin, Robin Winter, Dian Donnai, John Carey, Giovanni Neri, and many others [13]. At that time, clinicians and dysmorphologists attempted to interpret the patterns of human structural defects based on their own knowledge and experience. Owing to the difficulties in providing a definitive diagnosis, further frustrating the family of the affected child, some criticism arose toward dysmorphology as a discipline. However, around 1990, clinical dysmorphology began rapidly advancing in parallel with the enormous progress in cytogenetics and molecular genetics [1,2]. The accurate delineation of DS at the molecular level increases the possibilities for patient management and is essential to provide genetic counseling to the families [1–3].

DIAGNOSTIC PROCESS IN DYSMORPHOLOGY

The classical clinical approach in dysmorphology consists of collecting the family pedigree (comprising a three-generation family history and consanguinity) and a medical history (pregnancy and delivery history, neonatal history, developmental milestones and current intellectual and social development, other medical symptoms and signs), physical and behavioral examinations, as well as imaging, biochemical, and metabolic findings (Table 1). After collecting all this information, dysmorphologists attempt to pinpoint a specific diagnosis of a single disorder or a group of disorders, including differential diagnosis [1–3].

A detailed collection of the personal history and accurate physical examinations of the patient are an essential part of the diagnostic process [14]. Furthermore, the complexity of dysmorphology diagnosis was the reason for the development of computerized dysmorphological databases. The most used of these databases is the *London Medical Database*, known as the *London Dysmorphology Database* (created by Robin Winter and Michael Baraitser), and the Australian database *Possum* [15,16]. The ability to place patients' photographs in these databases and the progressive simplification of search procedures have made them increasingly important tools in the daily practice of the medical geneticist. These databases, after about 30 years of development, are now the main source of clinically delineated cases. On the other hand, many DSs are of unknown origin (about 40–50% of all congenital defects), and a number of human malformation phenotypes remain yet to be defined [1–3].

Table 1 Diagnostic Algorithm in Dysmorphology
Family history and pedigree including at least three generations (and consanguinity)
Personal history (prenatal, perinatal period including delivery, postnatal history, morbidity)
Physical examination (dysmorphological examination) with anthropometric measurements and standard pictures
Additional consultations and other additional tests (i.e., neurological, cardiological, ophthalmological, metabolic, immunological, endocrine, gastroenterological, audiological, psychological, brain imaging, skeletal imaging)
Clinical and differential diagnosis
Genetic, metabolic, and biochemical tests
Diagnostic synthesis

The DYSCERNE project was launched in 2007 by the European Commission Public Health Executive Agency to share expertise in rare dysmorphic disorders (www.dyscerne.org). As part of the DYSCERNE program, the Dysmorphology Diagnostic System was initiated as a web-based tool to submit difficult-to-diagnose cases for further support in establishing a diagnosis of unknown condition associated with physical, developmental, and behavioral disturbances. More than 30 designated experts in dysmorphology from 28 centers of expertise across the EU countries are involved in the project. Every case is being reviewed by several experts (at least five) to reach, if possible, a consensus clinical diagnosis. Genetic testing is being suggested for most of the submitted cases [17]. A further tool supporting the diagnostic process of patients with congenital malformations is the Database of Genomic Variants and Phenotype in Humans using Ensembl Resources (DECIPHER—www.decipher.sanger.ac.uk). The aim of this database is to assist treating doctors and geneticists in the interpretation of candidate genes in the light of known and suspected pathogenic loci and, using Ensembl tools, to annotate detected variants by comparing sequence and structural variants with the latest functional annotation of the current human reference genome, with the ultimate goal of improving the diagnostic process and management of rare human diseases.

GENETIC TESTING IN DYSMORPHOLOGY

A genetic etiology of dysmorphism in children should be suspected especially if associated with:

- Congenital anomalies (e.g., major anomaly or several minor anomalies),
- Growth deficit (e.g., short stature or failure to thrive),
- Developmental delay and intellectual disability or developmental regression,
- Failure to develop secondary sexual characteristics,
- Ambiguous genitalia.

Until now, to determine the genetic etiology of a congenital malformation, very laborious diagnostic and/or research efforts were necessary (linkage mapping, candidate gene analysis, multiple sequencing of single genes, etc.). Currently, for patients with unexplained multiple congenital anomalies, the international consensus proposes a chromosomal microarray as a first-line test [18]. Microarray testing for copy number variations (CNVs) is recommended as the initial evaluation of patients with multiple defects not specific to a well-delineated genetic syndrome. The test detects whole extra or missing chromosomes and deletions/duplications of fragments of chromosomes. Various next generation

Table 2 The Principal Elements of Genetic Counseling
Differential diagnosis and proposed clinical diagnosis
Confirmation of clinical diagnosis in independent biochemical, metabolic, cytogenetic, or molecular tests
Elucidation of etiology and genetic aspects of a recognizable syndrome, description of symptoms, and prognosis, as well as recommendations for further investigation and therapeutic options
Genetic risk estimation (recurrence and offspring risks), discussion of available prenatal testing
Communication with patient, whole family, and therapists
Support
Follow-up

sequencing (NGS)-based tailored gene panels constitute the second line of testing. Whole-exome sequencing (WES) will be considered if these first- and second-line tests cannot determine a definitive diagnosis. Such approach enables genetic diagnoses in 20–25% children with moderate to severe intellectual disability accompanied by malformations or dysmorphic features [19].

When the underlying pathogenic mechanism of a congenital anomaly is confirmed, family and prenatal counseling, as well as proper clinical management of the patient and family support, is possible (Table 2).

Progress in genomics technology enables finding the genetic background of clinically defined dysmorphic syndromes. The introduction of new cytogenetic and molecular testing allowed the identification of several new syndromes [20,21]. As an illustration, the recently initiated Deciphering Developmental Disorders Study is aimed at finding out the etiology of undiagnosed developmental disorders. The project involves 24 regional genetics services and scientists at the Wellcome Trust Sanger Institute providing genetic testing. The final aim is to collect DNA and clinical information from 12,000 undiagnosed children with developmental disorders and their parents (http://www.ddduk.org/). As of 2014 the study had enrolled 1133 children with severe, undiagnosed developmental disorders and enabled discovery of 12 novel genes associated with congenital malformations by means of the combination of exome sequencing and exome-focused array comparative genomic hybridization (exome—aCGH). The whole analysis was further supplemented by genome-wide genotyping of the trios (child plus parents) to identify deletions, duplications, uniparental disomy, and mosaic large chromosome rearrangements [22]. Altogether 28% (5317) of the children were classified and probable pathogenic variants were found. Most of these diagnoses involved de novo single-nucleotide polymorphisms, indels (insertions/deletions), or CNVs. Among the single-variant diagnoses, most genes associated with a developmental disorder were only detected once, although there were eight (*ANKRD11*, *ARID1B*, *DYRK1A*, *MED13L*, *SATB2*, *SYNGAP1*, *SCN1A*, and *STXBP1*) of which each accounted for 0.5–1% of conditions in the examined cohort. Novel genes with compelling evidence for a role in developmental disorders (*COL4A3BP*, *PPP2R5D*, *ADNP*, *POGZ*, *PPP2R1A*, *CHAMP1*, *BCL11A*, *PURA*, *DNM1*, *TRIO2*, *PCGF2*) were identified as well [22]. In two of the children with novel dysmorphic syndrome (similar facial appearance), identical Pro65Leu mutations in *PCGF2*, encoding a component of a Polycomb transcriptional repressor complex, were identified. Four other phenotypically similar patients had identical missense mutations in four of these novel developmental disorder-associated genes (*PCGF2*, *COL4A3BP*, *PPP2R1A*, and *PPP2R5D*), and for a fifth gene, *BCL11A*, nonidentical missense mutations were found. The three identical Ser132Leu mutations in *COL4A3BP*, which encodes an intracellular transporter of ceramide, probably result in an intracellular imbalance in ceramide and derangements of its metabolism. The two further patients had a mutation in *PPP2R1A*, which encodes the scaffolding A subunit of the protein phosphatase 2 complex [22].

NGS TESTING IN DYSMORPHOLOGY AND RARE MULTIPLE CONGENITAL DEFECTS SYNDROMES

NGS has changed the approach to rare dysmorphic and multidefects syndromes [23,24]. NGS comprises whole-genome sequencing, WES, and gene panel sequencing, being a tool for both diagnostic and research fields. It is likely that further progress in NGS techniques will establish a closer link between clinical and research genetic testing (a similar process was observed in detecting chromosomal aberrations using microarrays).

NGS has proven to be an accurate tool for mutations causing Mendelian disorders. Since 2012 the discovery of new genes in rare Mendelian syndromes using NGS techniques has increased rapidly— more than 250 novel genes involved in rare disorders, including dysmorphic syndromes, have been discovered in this way [23–25] and this number is expected to grow [23,24,26].

NGS-based discoveries include the following [23]:

- Novel genes linked to well-known dysmorphic syndromes.
- Novel genes causing previously unknown syndromes.
- Novel genes linked to new syndromes previously regarded as atypical/complex presentations of well-delineated disorders.

Unfortunately, there is still a gap between what has been discovered and what is available as a clinical test. Because of that there is a need for a systematic network that will cover a wide range of rare genetic disorders on a national or international level.

Patients with rare dysmorphic syndromes were among the first beneficiaries of NGS testing. Some of the first dysmorphic syndromes with a causative variant discovered by NGS in 2010 were Freeman–Sheldon syndrome (*MYH3* gene, autosomal dominant inheritance), Miller syndrome (*DHODH* gene, autosomal recessive inheritance), and Schinzel–Giedion syndrome (*SETBP1* gene, autosomal dominant inheritance) [27–29]. Using Freeman–Sheldon syndrome and Miller syndrome as a proof-of-concept, clinicians and scientists demonstrated for the first time that NGS analysis of a small number of unrelated affected individuals can identify a disease-causing gene [27,28]. Other known dysmorphic syndromes molecularly defined by the NGS approach include Weaver syndrome (*EZH2* gene), Floating-Harbor syndrome (*SRCAP* gene), Hajdu–Cheney syndrome (*NOTCH2* gene), Proteus syndrome (*AKT1* gene), and others [23–25,30–34]. The first report of an application of NGS testing to discover somatic *de novo* mutations as a cause of a genetic disorder was a case of Proteus syndrome [32]. NGS is extremely useful when the differential diagnosis includes several conditions with overlapping phenotypic features (e.g., Noonan, Costello, LEOPARD, and cardiofaciocutaneous syndromes) or when mutations in any one of many genes can cause the same syndrome or disorder. A classical example is Coffin–Siris syndrome, in which many identified genes such as *SMARCB1*, *SMARCA4*, *SMARCA2*, *SMARCE1*, and *ARID1A* are transmitted as autosomal dominant traits [35]. Mutations in any of these genes encoding subunits of a single complex are found to be causal for Coffin–Siris syndrome [35]. The Noonan spectrum disorders, also known as RASopathies, are a group of developmental syndromes characterized by extensive clinical and genetic heterogeneity but with a considerable phenotypic overlap [5,6]. Noonan spectrum disorders are caused by dysregulation of the RAS/mitogen-activated protein kinase (RAS/MAPK) signaling pathway (http://www.ncbi.nlm.nih.gov/pubmed/19467855). Thirteen gene variants have been detected in patients with Noonan and Noonan-like syndromes. Ten of these genes (*PTPN11, SOS1, RAF1, KRAS, HRAS, SHOC2, BRAF, NRAS, MAP2K1, MAP2K1*) encode components of the RAS/MAPK signaling pathway, while three others (*CBL, KAT6B*, and *RIT1*) encode regulatory proteins for this pathway (https://www.preventiongenetics.com/clinical-dna-testing/test/noonan-spectrum-disordersrasopathies-nextgen-sequencing-ngs-panel/2685). Although most causative mutations in Noonan spectrum disorders occur *de novo* (mainly missense and indels resulting in inframe alterations of the transcribed protein), familial cases have been reported as well, inherited in an autosomal dominant manner (https://www.preventiongenetics.com/clinical-dna-testing/test/noonan-spectrum-disordersrasopathies-nextgen-sequencing-ngs-panel/2685).

RASopathies include:

- Noonan syndrome,
- Noonan syndrome with multiple lentigines (known as LEOPARD syndrome),
- Cardiofaciocutaneous syndrome,
- Costello syndrome,
- Neurofibromatosis-1,
- Legius syndrome.

Noonan syndrome (NS) is characterized by dysmorphic facial appearance, short stature, congenital heart defects in up to 80% of patients (pulmonary valve stenosis, atrial septal defect, atrioventricular canal defect, and hypertrophic cardiomyopathy), and musculoskeletal abnormalities (chest deformity and short webbed neck). Intelligence is usually normal; however, learning difficulties can occur. NS is characterized by extensive clinical heterogeneity, even among members of the same family (http://www.ncbi.nlm.nih.gov/pubmed/4025385). NS, in approximately 50% of cases, is caused by missense mutations in the *PTPN11* gene, resulting in a gain of function of the nonreceptor protein tyrosine phosphatase SHP-2 protein.

Cardiofaciocutaneous syndrome is characterized by a distinctive facial appearance, congenital cardiac defects (pulmonary valve stenosis and atrial septal defects), ectodermal abnormalities (café au lait, erythema, keratosis, ichthyosis, eczema, sparse hair, and nail dystrophy), short stature, and neurological findings (seizures, hypotonia, macrocephaly, and various degrees of mental and cognitive delay) [36] (http://www.ncbi.nlm.nih.gov/pubmed/3789005).

Noonan syndrome with multiple lentigines (NSML) is known as LEOPARD syndrome (multiple lentigines, electrocardiographic-conduction abnormalities, ocular hypertelorism, pulmonary stenosis, abnormal genitalia, retardation of growth, sensorineural deafness). Other less common features include short stature, mild mental retardation, and abnormal genitalia. This syndrome is caused by at least 10 different missense defects in *PTPN11*, accounting for over 90% of all cases genotyped, and *RAF1* (rarely) [37]. Unlike NS, NSML-causative mutations in the *PTPN11* gene act through a dominant negative effect, which appears to disrupt the function of the wild-type gene product (SHP2 protein) [38] (http://www.ncbi.nlm.nih.gov/pubmed/15121796).

Costello syndrome (CS) is characterized by coarse facial features, thick and loose skin of the hands and feet, papillomata, heart defects (pulmonary valve stenosis), short stature, macrocephaly, and mild to moderate mental retardation. Most CS cases are sporadic, resulting from *de novo HRAS* mutations [39] (http://www.ncbi.nlm.nih.gov/pubmed/9934987).

Several NGS panels of 13 genes that have been implicated in NS and related disorders are available for patients with NS-like clinical features (https://www.preventiongenetics.com/clinical-dna-testing/ test/noonan-spectrum-disordersrasopathies-nextgen-sequencing-ngs-panel/2685/).

Holoprosencephaly is a structural anomaly of the developing forebrain, resulting from a failure of the prosencephalon to divide into hemispheres and associated with neurologic impairment and dysmorphism of the brain. Craniofacial anomalies include cyclopia, hypotelorism, anophthalmia or microophthalmia, bilateral cleft lip, absent nasal septum, flat nose, or single central incisor. Developmental delay is consistently observed. Holoprosencephaly occurs rather frequently (1:250 embryos); however, owing to a high rate of fetal loss, the birth prevalence is around 1:10,000 live births [40]. Holoprosencephaly classification ranges from the most severe to the least severe, depending on the degree of forebrain separation:

- Alobar variant
- Semilobar variant
- Lobar variant
- Middle interhemispheric variant
- Microform.

Around 25% of newborns affected by holoprosencephaly have a defined monogenic syndrome, including Smith–Lemli–Opitz syndrome (MIM 270400), Pallister–Hall syndrome (MIM 146510), and Rubinstein–Taybi syndrome (MIM 180849). Chromosomal anomalies have been implicated in 24–45% of live births, most frequently numeric anomalies in chromosomes 13, 18, and 21 and structural variations involving 13q, 18p, 7q36, 3p24–pter, 2p21, and 21q22.3 [40]. Intragenic mutations in several genes have also been found as increasing susceptibility to holoprosencephaly: *DISP1, FGF8, FOXH1, GLI2, NODAL, PTCH1, SHH, SIX3, TDGF1, TGIF1*, and *ZIC2*. All except *FGF8* are inherited in an autosomal dominant manner (http://ltd.aruplab.com/Tests/Pdf/407). Prenatal testing is also possible. About 25% of individuals with nonsyndromic holoprosencephaly have a mutation in one of these four genes: *SHH, ZIC2, SIX3*, or *TGIF1*. Because incomplete penetrance and variable expressivity are features of dominantly inherited holoprosencephaly, high intrafamilial phenotypic variability occurs and relatively normal facial appearance can be seen in individuals who have the causative variant and affected first-degree relatives. Thus, holoprosencephaly, like many other entities, is a spectrum disorder characterized by complex traits that are not reliably predicted by the presence of a single mutation [40].

WES studies enabled discovering novel growth disorders. IGSF1 deficiency described by Sun et al. is characterized by tall stature, central hypothyroidism, macroorchidism and delayed puberty, and severe skeletal abnormalities [41]. Hannema et al. reported the case of a patient with gigantism without other dysmorphic features, associated with a novel activating variant in *NPR2* [42]. Furthermore, a novel heterozygous *FGFR3* variant in dominantly transmitted proportionate short stature was published in 2015 [43], as well as a report on two siblings with disproportionate short stature caused by a compound heterozygous mutation of *PAPSS2* [44]. Nilsson et al. discovered an *AKAN* gene defect responsible for a rare syndrome of short stature and advanced bone age [45]. Additionally, a novel cause of primordial dwarfism (*NINEIN* mutation) was reported by Dauber et al. [46]. Several other groups have also used WES, for example, Nikkel et al., discovering the gene associated with the above-mentioned Floating-Harbor syndrome (*SRCAP* mutations) [47].

Scientists have begun to understand the complexity of some genetic disorders and congenital defects [23,48–50]. Genetic diseases seem to be rather (with few exceptions) randomly scattered among different genes involved in similar pathways. This heterogeneity directly challenges many studies trying to understand and correlate genotype with phenotype. Many genetic conditions can be suspected through a combination of clinical features, including physical appearance and family history. For example, in Holt–Oram syndrome, called heart–hand syndrome, mutations in *TBX5* cause congenital heart and limb malformations [51]. There are several possible explanations for the presence of a cluster of findings in a patient with a genetic syndrome. One common reason is pleiotropy of the multiple effects of a single variant on different organs or tissues. Another possible explanation for the presence of a cluster of findings is that the patient has a contiguous gene syndrome (deletions or duplications involving a certain part of a chromosome). Because all genes in the altered regions are affected, the involvement of many genes can result in a complicated clinical picture. A well-known example of a contiguous gene syndrome is the 22q11.2 deletion syndrome [52]. Furthermore, a single locus may be responsible for multiple phenotypes, and different disorders may result from mutations in the same gene. Various NGS-based studies have discovered novel genes involved in the etiology of a congenital syndrome with the same or similar phenotype, as well as single genes associated with different phenotypes or with an atypical form of a well-known syndrome delineated as a new syndrome [23,53]. For example, Rubinstein–Taybi syndrome is caused by mutations in the *CREBBP* and *EP300* genes (both genes function as transcriptional coactivators in the regulation of gene expression through various signal transduction pathways and both are potent histone acetyltransferases). On the other side, *COL4A1* gene mutations are responsible for neuronal migration disorders as well as cataract and other ophthalmological syndromes and many collagen type IV-associated disorders. The diagnostic difficulties in the above-mentioned conditions strongly point to a prominent role for NGS technologies as a potent diagnostic tool in dysmorphology. NGS technology helps us understand the pathogenesis of known disorders through delineation of pathways responsible for their pathogenesis. For example, NGS enabled the implication of the AKT/PI3K/mTOR pathway in overgrowth syndromes such as Proteus, megalencephaly–capillary malformation and megalencephaly–polymicrogyria–polydactyly–hydrocephalus syndrome as well as the chromatin modeling of the SWI/SNF pathway in Coffin–Siris syndrome and the RAS/MAPK pathway in RASopathies [5,6,23,35,54–56].

NGS-based research has a further potential to pinpoint new therapies [23,56,57]. Beaulieu et al. published a review discussing some issues related to new therapeutic approaches in rare diseases [56]. For instance, the identification of mutations in riboflavin transporter genes (*SLC52A2* and *SLC52A3*) causing infancy-onset sensorineural deafness and pontobulbar palsy provides an opportunity to use riboflavin as a therapeutic agent [58]. Additionally, in an infantile-onset movement disorder caused by a novel disease-causing gene (*SLC18A2* gene) that encodes the vesicular monoamine transporter, cytosolic dopamine and serotonin accumulating into synaptic vesicles implies monoamine agonists as a therapeutic agents [59].

NGS will probably become part of the standard assessment for most rare dysmorphic syndromes, as it spectacularly facilitates, accelerates, and shortens the diagnostic process [48,49]. A good example is the diagnostic process in intellectual disability patients with or without dysmorphic features (see chapter Next generation sequencing in neurology and psychiatry). Nowadays, there is a plethora of diagnostic companies offering diagnostic NGS panels of genes known to be causal and potentially important for a heterogeneous group of disorders such as intellectual disability, autism, epilepsy (epileptic encephalopathy), schizophrenia and bipolar disorders, deafness, leukoencephalopathy, and peroxisomal defects, as well as ataxia, macular dystrophy, ciliopathies, cardiomyopathy, myopathy and neuropathy, skeletal dysplasia syndromes and connective tissue syndromes, RASopathies, metabolic disorders, etc. (see respective chapters). Moreover, panels can be customized on request (www.genetests.com). The EuroGentest project (www.eurogentest.org) has prepared clinical utility guidelines covering diagnostic NGS approaches [60]. Structured information from NGS providers (about 30 laboratories) includes panel name (about 1000 tests) and variants tested (about 3000 genes), collected, and linked to various conditions (3460).

DILEMMAS

The NGS revolution and subsequent new gene discovery and diagnostics progress predicts significant transformations of the routine practice for medical geneticists and dysmorphologists. It is expected that NGS testing will change traditional phenotype-driven diagnosis. The inclusion of NGS into the diagnostic process will help clinicians to further improve the diagnostic process far beyond what has been possible after integration of clinical findings, including personal history collection and physical examinations [3,17].

Currently, dysmorphologists struggle to precisely define the main clinical problem of an affected patient, gathering clinical and laboratory data and analyzing the differential diagnosis, which facilitates the selection of targeted genetic tests to establish a molecular diagnosis and to confirm the clinical suspicion. In the future, dysmorphologists will probably perform more accurate phenotyping to facilitate the interpretation of the large amount of data (mutations) generated by NGS testing. The definition of phenotypes for genetic study is very challenging as phenotyping should meet consistently high standards of reproducibility and validity. Raoul Hennekam and Leslie Biesecker are convinced that NGS development demands next generation phenotyping and transformation of the profession of medical geneticists and dysmorphologists [17]. A good example of the new insight for dysmorphology in the NGS era is intellectual disability with or without dysmorphic features [17,61,62]. NGS performed in such patients is likely to provide a multitude of candidate mutations [61,62]. In such a case, the medical geneticist is an appropriate specialist for further analysis of the causative mutation or mutations [17]. Using an NGS assay, the diagnostic process is accelerated but phenotypic research is still essential and the skill and experience of dysmorphologists are still needed [17,23]. It is also possible that, in the future, clinical geneticists, in addition to genetic counseling, will recommend, participate in, and conduct the therapy for genetic disorders.

The main questions for clinicians, as well as patients and their parents, include clinical indications for NGS testing (Mendelian disorders versus multifactorial disorders, single defects versus multiple defects, and multisystem, pleiotropic disorders), informed consent, and incidental findings and the right of the patient, or her/his parents, to know or not to know. Professionalism of medical geneticists is needed to address these questions [17,23].

Usually the patient and her/his family want to establish short- and long-term prognoses of the disorder, as genetic disorders are characterized by phenotypic heterogeneity—interfamilial, intrafamilial, and interindividual variability in phenotype and severity of symptoms. This counseling needs a deep knowledge and the experience of dysmorphologists.

It is widely discussed that genomic medicine, especially massive NGS technology, by generating plenty of personal genetic data, produces significant ethical dilemmas for physicians, individual patients, and society as a whole (see the chapter Ethical issues) [48,50,63].

NGS assays identify more variants that are suspected to be disease-causing mutations [67,68]. The American College of Medical Genetics defined six categories of genetic variants in clinical diagnostic testing as disease causing, likely to be disease causing, possibly disease causing, not likely to be disease causing, not disease causing, and variant of unknown clinical significance [48]. It is recognized that there are thousands of variants of unknown significance in every person's genome. This strongly shows the difficulties with interpretation of genetic variants found in NGS testing.

Moreover, many mutations that are relevant to adult medical care may be identified by chance in the process of NGS use and then they are called the secondary findings (for example, mutations responsible for cancers or progressive neurological disorders starting in adulthood). There is no agreement or consensus on whether and how to share this information with a patient [50].

It is commonly known that the cost of a single NGS assay is high for hospitals, health institutions, and individual patients [48,49]. But the total cost of genome sequencing is generally lower than costs of standard Sanger sequencing assays for many individual genes included in the pathogenesis of a concrete genetic syndrome, making application of NGS economically feasible in plenty of rare disorders. Moreover, a targeted NGS approach (specific gene panels) based on the suspected syndrome maximizes mutation identification and may decrease the overall costs of genetic testing.

Despite the unquestionable advantages of NGS, it is clear that this technology, as well as genomic medicine in general, generates plenty of personal genetic data and thus produces significant ethical dilemmas for physicians, individual patients, and society as a whole [48,50,63]. Moreover, there is no agreement or consensus on whether and how to share information about secondary findings with patients [50].

REVERSE DYSMORPHOLOGY

NGS technology is currently leading to a paradigm shift in the field owing to the possibility of the identification of genetic syndromes by a process called *reverse dysmorphology*, that is, delineation of new syndromes primarily by genotype followed by description of phenotype [17,18]. Such approach can be a very useful element in diagnostics of genetic syndromes with heterogeneous *loci*, including congenital malformations, but also intellectual disability, autism spectrum disorders, and epilepsy. The new approach "genotype first" points to the phenotype shared by all patients with the same variant/genotype. Reverse dysmorphology as a diagnostic process was described in a large group of patients when high-resolution CGH studies were started to be used in diagnostics allowing the linkage of new critical chromosomal regions (CNVs) to new phenotypes and dysmorphic syndromes [17,18]. Consistently, NGS can identify a mutation in a gene known to cause disease, prompting the clinicians to reevaluate the phenotype and make the correct diagnosis, compatible with reverse phenotyping.

NGS AND SCREENING OF RARE DISORDERS IN NEWBORNS

The definition of the genetic test has changed as the new technological assays have evolved, including the NGS revolution. Applications of clinical genetic testing are extended to newborn screening, carrier testing for inherited disorders, predictive and presymptomatic testing for adult-onset and complex disorders, as well as pharmacogenetic testing. Newborn screening programs are used to detect specific treatable diseases before the onset of irreversible symptoms (for instance, PKU, CF, congenital hypothyroidism, congenital adrenal hyperplasia, galactosemia, and some metabolic disorders screened using the tandem mass spectrometry technique, such as fatty acid disorders, organic acid disorders, maple syrup urine disease, and homocystinuria). At present newborn screening projects in many countries are the standard of neonatal care, which naturally encourages the discussion of the possibility of transition to a broad NGS-based screening. It could be advocated that apart from the early identification of genetic disorders in children, the collection of mutational data for future health should result in better outcomes of potential treatments of genetic disorders with delayed symptoms. But as discussed before, NGS identifies all variants in the targeted genes, creating potential problems when variants predictive for adult-onset genetic disorders (for instance, cancers), behavioral predilections, or a carriership status of rare genetic syndromes are found [64–66]. It is suspected that at the present time, integrating NGS into newborn screening projects in an untargeted manner might reduce moral authority and public participation of this kind of screening program [64].

CONCLUSIONS

The correct diagnosis of abnormal phenotype at an early age is important to provide appropriate patient care and family counseling with regard to the recurrence risk. Some dysmorphic conditions can be recognized by detailed clinical examination accompanied by conventional karyotyping to identify balanced chromosomal rearrangements and aCGH to exclude genomic imbalance. Apparently such approach is not always sufficient. NGS is a potent tool to fully characterize break points of all types of balanced chromosomal rearrangements, confirming gene disruption that could account for the phenotype. It is proposed that dysmorphic patients carrying balanced chromosomal rearrangements should be subjected to NGS testing once a genomic imbalance has been ruled out by aCGH. Such procedure raises the diagnostic output up to 30%. NGS is also a potent tool enabling the interpretation of variants of uncertain significance or reduced penetrance, frequently found in congenital malformations. The combination of all new genomics tools and techniques in research and diagnostics including NGS is making dysmorphology a very exciting and dynamic discipline of clinical genetics. The parallel improvements in both phenotyping and genotyping and their reciprocal interaction can facilitate molecular diagnosis in dysmorphology as well as increasing the knowledge about the pathogenesis of a number of diseases. However, new models of clinical implementation of genomics should be developed and the ethical, legal, and social implications of NGS technology should be considered.

LIST OF ABBREVIATIONS

CF Cystic fibrosis
CNV Copy number variant
NGS Next generation sequencing
PKU Phenyloketonuria
WES Whole-exome sequencing

REFERENCES

[1] Jones KL, Jones MC, Del Campo Cesanelles M. Smith's recognizable patterns of human malformation. 7th ed. Philadelphia: W.B. Saunders; 2013.

[2] Hennekam RCM, Krantz ID, Allanson JE. Gorlin's syndromes of head and neck. 5th ed. New York: Oxford University Press; 2010.

[3] Aase JM. Diagnostic dysmorphology. New York and London: Plenum Medical Book Company; 1990.

[4] Latos-Bieleńska A, Materna-Kiryluk A. Wrodzone wady rozwojowe w Polsce w latach 2005–2006. Dane z Polskiego Rejestru Wrodzonych Wad Rozwojowych. Poznań: Ośrodek Wydawnictw Naukowych; 2010.

[5] OMIM – http://www.ncbi.nlm.nih.gov.

[6] Orphanet – http://www.orpha.net.

[7] Ayme S, Urbero B, Oziel D, Lecouturier E, Biscarat AC. Information on rare diseases: the orphanet project. Rev Med Interne 1998;19:376S–7S.

[8] Baird PA, Anderson TW, Newcombe HB, Lowry RB. Genetic disorders in children and young adults: a population study. Am J Hum Genet 1988;42:677–93.

[9] Yoon PW, Olney RS, Khoury MJ, Sappenfield WM, Chavez GF, Taylor D. Contribution of birth defects and genetic diseases to pediatric hospitalizations. A population-based study. Arch Pediatr Adolesc Med 1997;151:1096–103.

[10] McCandless SE, Brunger JW, Cassidy SB. The burden of genetic disease on inpatient care in a children's hospital. Am J Hum Genet 2004;74:121–7.

[11] Dye DE, Brameld KJ, Maxwell S, Goldblatt J, Bower C, Leonard H, et al. The impact of single gene and chromosomal disorders on hospital admissions of children and adolescents: a population-based study. Pub Health Genomics 2010;14:153–61.

[12] Cooper DN, Chen JM, Ball EV, Howells K, Mort M, Phillips AD, et al. Genes, mutations, and human inherited disease at the dawn of the age of personalized genomics. Hum Mutat 2010;31:631–55.

[13] Harper PS. A short history of medical genetics. New York: Oxford University Press; 2008.

[14] Van Karnebeek CD, Scheper FY, Abeling NG, Alders M, Barth PG, Hoovers JM, et al. Etiology of mental retardation in children referred to a tertiary care center: a prospective study. Am J Ment Retard 2005;110(4):253–67.

[15] Winter R, Baraitser M, Douglas JM. A computerised data base for the diagnosis of rare dysmorphic syndromes. J Med Genet 1984;21(2):121–3.

[16] Bankier A, Keith CG. POSSUM: the microcomputer laser-videodisk syndrome information system. Ophthalmic Paediatr Genet 1989;10(1):51–2.

[17] Douzgou S, Clayton-Smith J, Gardner S, Day R, Griffiths P, Strong K, DYSCERNE expert panel. Dysmorphology at a distance: results of a web-based diagnostic service. Eur J Hum Genet 2014;22(3):327–32.

[18] Miller DT, Adam MP, Aradhya S, Biesecker LG, Brothman AR, Carter NP, et al. Consensus statement: chromosomal microarray is a first-tier clinical diagnostic test for individuals with developmental disabilities or congenital anomalies. Am J Hum Genet 2010;86(5):749–64.

[19] Beaudet AL. The utility of chromosomal microarray analysis in developmental and behavioral pediatrics. Child Dev 2013;84(1). http://dx.doi.org/10.1111/cdev.12050.

[20] Hennekam RC, Biesecker LG. Next-generation sequencing demands next-generation phenotyping. Hum Mutat 2012;33(5):884–6.

[21] van Ravenswaaij-Arts CM, Kleefstra T. Emerging microdeletion and microduplication syndromes; the counseling paradigm. Eur J Med Genet 2009;52(2–3):75–6.

[22] Wright CF, Fitzgerald TW, Jones WD, Clayton S, McRae JF, van Kogelenberg M, on behalf of the DDD study, et al. Genetic diagnosis of developmental disorders in the DDD study: scalable analysis of genome-wide research data. Lancet 2014;385(9975):1305–14. http://dx.doi.org/10.1016/S0140-6736(14)61705-0.

[23] Boycott KM, Vanstone MR, Bulman DE, MacKenzie AE. Rare-disease genetics in the era of next-generation sequencing: discovery to translation. Nat Rev Genet 2013;14(10):681–91.

[24] Bamshad MJ, Ng SB, Bigham AW, Tabor HK, Emond MJ, Nickerson DA, et al. Exome sequencing as a tool for Mendelian disease gene discovery. Nat Rev Genet 2011;12:745–55.

[25] Veltman JA, Brunner HG. *De novo* mutations in human genetic disease. Nat Rev Genet 2012;13:565–75.

[26] Berg JS, Khoury MJ, Evans JP. Deploying whole genome sequencing in clinical practice and public health: meeting the challenge one bin at a time. Genet Med 2011;13:499–504.

[27] Ng SB, Turner EH, Robertson PD, Flygare SD, Bigham AW, Lee C, et al. Targeted capture and massively parallel sequencing of 12 human exomes. Nature 2009;461:272–6.

[28] Ng SB, Buckingham KJ, Lee C, Bigham AW, Tabor HK, Dent KM, et al. Exome sequencing identifies the cause of a mendelian disorder. Nat Genet 2010;42:30–5.

[29] Hoischen A, van Bon BW, Gilissen C, Arts P, van Lier B, Steehouwer M, et al. *De novo* mutations of *SETBP1* cause Schinzel–Giedion syndrome. Nat Genet 2010;42:483–5.

[30] Gibson WT, Hood RL, Zhan SH, Bulman DE, Fejes AP, FORGE Canada Consortium, et al. Mutations in EZH2 cause Weaver syndrome. Am J Hum Genet 2011;90:110–8.

[31] Hood RL, Lines MA, Nikkel SM, Schwartzentruber J, Beaulieu C, FORGE Canada Consortium, et al. Mutations in *SRCAP*, encoding SNF2-related CREBBP activator protein, cause Floating–Harbor syndrome. Am J Hum Genet 2012;90:308–13.

[32] Lindhurst MJ, Sapp JC, Teer JK, Johnston JJ, Finn EM, Peters K, et al. A mosaic activating mutation in *AKT1* associated with the Proteus syndrome. N Engl J Med 2011;365:611–9.

[33] Simpson MA, Irving MD, Asilmaz E, Gray MJ, Dafou D, Elmslie FV, et al. Mutations in *NOTCH2* cause Hajdu-Cheney syndrome, a disorder of severe and progressive bone loss. Nat Genet 2011;43(4):303–5.

[34] Majewski J, Schwartzentruber JA, Caqueret A, Patry L, Marcadier J, FORGE Canada Consortium, et al. Mutations in *NOTCH2* in families with Hajdu–Cheney syndrome. Hum Mutat 2011;32:1114–7.

[35] Tsurusaki Y, Okamoto N, Ohashi H, Kosho T, Imai Y, Hibi-Ko Y, et al. Mutations affecting components of the SWI/SNF complex cause Coffin–Siris syndrome. Nat Genet 2012;44:376–8.

[36] Reynolds JF, Neri G, Herrmann JP, Blumberg B, Coldwell JG, Miles PV, et al. New multiple congenital anomalies/mental retardation syndrome with cardio-facio-cutaneous involvement–the CFC syndrome. Am J Med Genet 1986;25:413–27.

[37] Pandit B, Sarkozy A, Pennacchio LA, Carta C, Oishi K, Martinelli S, et al. Gain-of-function RAF1 mutations cause Noonan and LEOPARD syndromes with hypertrophic cardiomyopathy. Nat Genet 2007;39:1007–12.

[38] Legius E, Schrander-Stumpel C, Schollen E, Pulles-Heintzberger C, Gewillig M, Fryns J-P. PTPN11 mutations in LEOPARD syndrome. J Med Genet 2002;39:571–4.

[39] van EAM, van GI, Hennekam RC. Costello syndrome: report and review. Am J Med Genet 1999;82:187–93.

[40] Raam MS, Solomon BD, Muenke M. Holoprosencephaly: a guide to diagnosis and clinical management. Indian Pediatr 2011;48(6):457–66.

[41] Sun Y, Bak B, Schoenmakers N, van Trotsenburg AS, Oostdijk W, Voshol P, et al. Loss-of-function mutations in IGSF1 cause an X-linked syndrome of central hypothyroidism and testicular enlargement. Nat Genet 2012;44:1375–81.

[42] Hannema SE, van Duyvenvoorde HA, Premsler T, Yang RB, Mueller TD, Gassner B, et al. An activating mutation in the kinase homology domain of the natriuretic peptide receptor-2 causes extremely tall stature without skeletal deformities. J Clin Endocrinol Metab 2013;98:E1988–98.

[43] Kant SG, Cervenkova I, Balek L, Trantirek L, Santen GW, de Vries MC, et al. A novel variant of FGFR3 causes proportionate short stature. Eur J Endocrinol 2015;172(6):763–70. pii:EJE-14-0945.

[44] Oostdijk W, Idkowiak J, Mueller JW, House PJ, Taylor AE, O'Reilly MW, et al. PAPSS2 deficiency causes androgen excess via impaired DHEA sulfation - in vitro and in vivo studies in a family harboring two novel PAPSS2 mutations. J Clin Endocrinol Metab 2015;100(4):E672–80. jc20143556.

[45] Nilsson O, Guo MH, Dunbar N, Popovic J, Flynn D, Jacobsen C, et al. Short stature, accelerated bone maturation, and early growth cessation due to heterozygous aggrecan mutations. J Clin Endocrinol Metab 2014;99:E1510–8.

[46] Dauber A, Lafranchi SH, Maliga Z, Lui JC, Moon JE, McDeed C, et al. Novel microcephalic primordial dwarfism disorder associated with variants in the centrosomal protein ninein. J Clin Endocrinol Metab 2012;97:E2140–51.

[47] Nikkel SM, Dauber A, de Munnik S, Connolly M, Hood RL, FORGE Canada Consortium, et al. The phenotype of Floating-Harbor syndrome: clinical characterization of 52 individuals with mutations in exon 34 of SRCAP. Orphanet J Rare Dis 2013;8:63.

[48] Katsanis SH, Katsanis N. Molecular genetic testing and the future of clinical genomics. Nat Rev Genet 2013;14:415–26.

[49] Kingsmore SF, Saunders CJ. Deep sequencing of patient genomes for disease diagnosis: when will it become routine? Sci Transl Med 2011;3:87ps23.

[50] van El CG, Cornel MC, Borry P, Hastings RJ, Fellmann F, Hodgson SV, ESHG Public and Professional Policy Committee, et al. Whole-genome sequencing in health care: recommendations of the European Society of Human Genetics. Eur J Hum Genet 2013;21(6):580–4.

[51] Basson CT, Bachinsky DR, Lin RC, Levi T, Elkins JA, Soults J, et al. Mutations in human *TBX5* cause limb and cardiac malformation in Holt-Oram syndrome. Nat Genet 1997;15(1):30–5.

[52] Kobrynski LJ, Sullivan KE. Velocardiofacial syndrome, DiGeorge syndrome: the chromosome 22q11.2 deletion syndromes. Lancet 2007;370(9596):1443–52.

[53] McDaniell R, Warthen DM, Sanchez-Lara PA, Pai A, Krantz ID, Piccoli DA, et al. *NOTCH2* mutations cause Alagille syndrome, a heterogeneous disorder of the notch signaling pathway. Am J Hum Genet 2006;79:169–73.

[54] Riviere JB, Mirzaa GM, O'Roak BJ, Beddaoui M, Alcantara D, Finding of Rare Disease Genes (FORGE) Canada Consortium, et al. *De novo* germline and postzygotic mutations in *AKT3*, *PIK3R2* and *PIK3CA* cause a spectrum of related megalencephaly syndromes. Nat Genet 2012;44:934–40.

[55] Lee JH, Huynh M, Silhavy JL, Kim S, Dixon-Salazar T, Heiberg A, et al. *De novo* somatic mutations in components of the PI3K-AKT3-mTOR pathway cause hemimegalencephaly. Nat Genet 2012;44:941–5.

[56] Beaulieu CL, Samuels ME, Ekins S, McMaster CR, Edwards AM, Krainer AR, et al. A generalizable preclinical research approach for orphan disease therapy. Orphanet J Rare Dis 2012;7:39.

[57] Nakamura A, Takeda S. Exon-skipping therapy for Duchenne muscular dystrophy. Neuropathol 2009;29:494–501.

[58] Johnson JO, Gibbs JR, Megarbane A, Urtizberea JA, Hernandez DG, Foley AR, et al. Exome sequencing reveals riboflavin transporter mutations as a cause of motor neuron disease. Brain 2012;135:2875–82.

[59] Rilstone JJ, Alkhater RA, Minassian BA. Brain dopamine-serotonin vesicular transport disease and its treatment. N Engl J Med 2012;368:543–50.

[60] Dierking A, Schnidtke J. The future of clinical utility gene card in the context of next-generation sequencing diagnostic panels. Eur J Hum Genet 2014;22:1247.

[61] Vissers LE, de Ligt J, Gilissen C, Janssen I, Steehouwer M, de Vries P, et al. A *de novo* paradigm for mental retardation. Nat Genet 2010;42:1109–12.

[62] de Ligt J, Willemsen MH, van Bon BW, Kleefstra T, Yntema HG, Kroes T, et al. Diagnostic exome sequencing in persons with severe intellectual disability. N Engl J Med 2012;367:1921–9.

[63] Ayuso C, Millán JM, Mancheño M, Dal-Ré R. Informed consent for whole-genome sequencing studies in the clinical setting. Proposed recommendations on essential content and process. Eur J Hum Genet 2013;21(10):1054–9.

[64] Christenhusz GM, Devriendt K, Dierickx K. To tell or not to tell? A systematic review of ethical reflections on incidental findings arising in genetics contexts. Eur J Hum Genet 2013;21(3):248–55.

[65] Green RC, Berg JS, Berry GT, Biesecker LG, Dimmock DP, Evans JP, et al. Exploring concordance and discordance for return of incidental findings from clinical sequencing. Genet Med 2012;14:405–10.

[66] Bell CJ, Dinwiddie DL, Miller NA, Hateley SL, Ganusova EE, Mudge J, et al. Carrier testing for severe childhood recessive diseases by next-generation sequencing. Sci Transl Med 2011;3(65):65ra4.

[67] Abecasis GR, Altshuler D, Auton A, Brooks LD, Durbin RM, Gibbs RA, et al. A map of human genome variation from population-scale sequencing. Nature 2010;467:1061–73.

[68] Altshuler DM, Gibbs RA, Peltonen L, Altshuler DM, Gibbs RA, Peltonen L, et al. Integrating common and rare genetic variation in diverse human populations. Nature 2010;467:52–8.

NEXT GENERATION SEQUENCING IN VISION AND HEARING IMPAIRMENT

8

Monika Ołdak

Department of Genetics, World Hearing Center, Institute of Physiology and Pathology of Hearing, Warsaw, Poland

CHAPTER OUTLINE

INTRODUCTION

Vision and hearing are commonly listed together as the most important of the human senses. Acute hearing and vision are clearly required to function in an environment, as they constitute a vital part of a wide range of processes, from communication through gathering of information, from mobility to risk avoidance. Vision and hearing problems may be attributable to environmental or genetic factors or a combination of both. In the Online Mendelian Inheritance in Man database, a comprehensive compendium of human genes and genetic phenotypes, 25–30% of diseases contain a vision-associated term in their description of clinical features. Aberrant visual and hearing phenotypes that were previously underappreciated are now being recognized in numerous syndromic diseases [1].

Advances in genetics over the past decades have provided insight into the molecular background of a number of ophthalmic and hearing disorders. Gene mutations have been detected in many inherited, usually early-onset, disorders that display a Mendelian mode of inheritance, and genetic variations are increasingly recognized to play a role in common, usually adult-onset, disorders that are inherited as complex traits. Some examples include glaucoma, myopia, age-related macular or corneal degeneration, and noise-induced and age-related hearing loss (HL) [2–4]. Identification of a genetic component of vision and hearing disorders has been significantly accelerated by the use of homozygosity mapping and high-throughput technologies, such as microarrays and next generation sequencing (NGS).

Currently, more than 500 genes that affect vision have been reported (http://neibank.nei.nih.gov/cgi-bin/eyeDiseaseGenes.cgi), while hereditary syndromic and nonsyndromic HL may result from mutations in about 150 different genes (http://hereditaryhearingloss.org/). Enormous genetic heterogeneity is particularly striking for nonsyndromic HL, with almost 90 different potentially causative genes; retinitis pigmentosa (RP), with about 100 genes; Leber congenital amaurosis (LCA) and Bardet–Biedl syndrome (BBS), with respectively 41 and 22 different genes responsible for each condition; and Usher syndrome (USH), a common cause of combined hereditary deaf–blindness (sensorineural HL and RP), with 17 different genes involved (http://www.hgmd.org/, accessed June 2014) [5,6].

Studies on the USH protein network revealed that at the molecular level it is linked to other ciliopathies [7]. The term ciliopathy encompasses diseases affecting the primary cilium, which is a hairlike, immotile cellular organelle protruding from almost all eukaryotic cells and involved in transducing extracellular signals. Approximately one-third of retinal dystrophies involve a defect in a gene encoding a ciliary protein. Depending on the type of photoreceptors that are primarily affected, retinal dystrophies can have different clinical presentations, such as RP, cone dystrophy, cone–rod dystrophy, LCA, or macular degeneration. These conditions are genetically heterogeneous and although described as distinct clinical entities, they are often difficult to distinguish [8]. Among syndromic ciliopathies are, for example, BBS, Senior–Loken, Alström, and USH syndromes, the last two also characterized by sensorineural HL [8].

Sequential gene screening in causally heterogeneous conditions with no clearly defined genotype–phenotype correlations is laborious, expensive, time-consuming, and inefficient. To circumvent this problem, other technologies such as resequencing arrays and single-primer extension microarrays have been applied. However, these methods focus only on previously reported mutations and do not provide a comprehensive screening of all possibly affected genes [9,10].

The major advance in the field was the introduction of NGS, which has become the premier tool in genetic and genomic analyses for vision and hearing disorders. NGS, alternatively named massive parallel sequencing, allows millions of sequencing to run in parallel and offers the advantage of simultaneous sequencing of multiple genes. With declining cost and increasing availability of benchtop sequencers, NGS is no longer out of reach and is already used by many laboratories for routine diagnostic purposes.

In the diagnostic process of hereditary vision and hearing disorders, one can perform whole-exome sequencing (WES) and subsequently apply filters for a set of genes related to these disorders and in this way "open" and interrogate only the part of the sequenced exome that is currently of interest. This approach is used by our laboratory as well as several others and it has proven utility for the diagnosis of hereditary deafness and blindness [11]. The advantage of this approach is the possibility of "sequencing once" and "interrogating many times" as novel genes related to a particular disorder are discovered. A drawback of this strategy resulting from an overall lower coverage of the regions of interest is discussed below.

NGS TESTS FOR VISION AND HEARING DISORDERS

An alternative to WES is the use of commercially available NGS diagnostic panels that are specifically designed to enrich and sequence only a particular set of selected genes. The main differences between the NGS diagnostic panels are the methods used for enrichment of the target sequences, the sequencing technology, and the number of genes tested (for details see Chapter 1).

Two different enrichment methods, that is, target-specific primer design followed by PCR and target hybridization followed by hybrid capture, achieved more than 99% specificity and sensitivity and were successfully applied for studying the genetic background of HL. However, both approaches differed considerably in the number of genes tested [12–14]. In a capture-based platform (OtoSCOPE) developed by Shearer et al., initially 54, and as of this writing 90, different HL genes could be tested [12]. Using a similar approach Brownstein et al. tested as many as 246 genes that have been linked with HL in humans or their orthologous genes that have been associated with HL in mice [15]. In contrast, owing to technical constraints in the PCR-based NGS HL panels, only a smaller number of genes is usually included, for example, 15 [16], 24 (OtoSeq) [13], or 34 [14]. A compensation for the reduced number of genes tested in PCR-based panels is their more uniform coverage compared to the hybridization panels (for details see Chapter 1). Capture-based methods are limited by the filtering of repetitive regions of the original target area [14].

Comparison of WES with the target-enrichment approach revealed that the targeted capture of deafness genes followed by NGS better fulfills the requirements of a complete molecular screening of the known genes. In an experiment by Schrauwen et al. only about 60% of exonic bases of deafness genes were covered at 20× depth with an overall performance of 86.5% exome coverage at 20× depth [14]. Shearer et al. have demonstrated that regions included in the OtoSCOPE panel were covered only at an average of 70.8% at 10× depth, while overall exome coverage was good, with an average of 92.3% at 10× depth. However, when they compared only coding base pairs of the genes included in the OtoSCOPE panel, exome coverage was better but still lower than OtoSCOPE (94.8% vs 98.2% at 10× depth of coverage). It has been speculated that the difference may be due to omission of many organ-specific isoforms in the standard exome target region [17].

A gene-specific approach based on enrichment of only a selected number of genes using a long-PCR approach (nine genes corresponding to 218 PCR amplicons) was found more efficient than WES also in patients with USH. The coverage of USH gene regions was 95% in the targeted sequencing by long-PCR, whereas in WES only 50% of those regions were covered at a similar overall coverage higher than 25× [18]. It is noteworthy that both procedures were performed using two different NGS platforms. The Genome Sequencer FLX (Roche) or Illumina GAII platform (Illumina) was used for the long-PCR approach, while WES was done on the SOLiD system platform (Life Technologies).

The reliability of NGS data depends on the coverage per base pair. As demonstrated in the different comparison settings presented above, the average coverage of the genes of interest using the currently available technologies remains higher in the targeted enrichments panels than in WES. This limitation of WES may strongly affect the diagnostic yield of an NGS test. However, the NGS technology is evolving rapidly and as of this writing there are no clear recommendations as to whether one should first perform an NGS panel and then WES or whole-genome sequencing or vice versa.

An NGS test for a genetic vision and/or hearing disorder can be ordered in a diagnostic laboratory offering WES or an NGS panel. It is required only to provide the laboratory with the patient's blood sample or isolated DNA as well as all necessary information. A list of selected laboratories testing for hereditary HL and genetic eye disorders is provided in Tables 1–3, respectively. The number of genes

Table 1 NGS Tests for HL (Selected Examples)

Company	Test Name	Genes Tested (No.)	Technology
Otogenetics Corp., Norcross, GA, USA	Human deafness panel	129	Illumina sequencing (MiSeq)
Fulgent Diagnostics, Temple City, CA, USA	Hearing loss NGS panel	103	Illumina sequencing
Radboud University Medical Center, Nijmegen, The Netherlands	Hearing impairment gene panel	102	Roche 454 GS FLX, ABI SOLiD sequencing
Great Ormond Street Hospital for Children, NHS Trust, London, UK	Inherited hearing loss panel	95	Illumina sequencing (MiSeq)
Emory Genetics Laboratory, Atlanta, GA, USA	Hearing loss: sequencing panel	92	Illumina sequencing
University of Iowa Hospital and Clinics, Molecular Otolaryngology and Renal Research Laboratories, Iowa City, IA, USA	OtoSCOPE	90	Roche 454 GS FLX; Illumina sequencing (GAII)
Asper Biotech Ltd., Tartu, Estonia	Hearing loss/deafness multigene panel	76	Illumina sequencing (HiSeq, MiSeq)
Partners HealthCare Personalized Medicine, Laboratory for Molecular Medicine, Cambridge, MA, USA	OtoGenome panel	70	Illumina sequencing
Sistemas Genomicos, Medical Genetics Unit, Paterna, Spain	OTOGeneProfile67	67	ABI SOLiD sequencing
CeGaT GmbH, Tübingen, Germany	Syndromic hearing loss	62	Illumina sequencing (HiSeq)
	Hearing loss, nonsyndromic, autosomal recessive, and X-linked	53	
	Hearing loss, nonsyndromic, autosomal dominant, and X-linked	35	
ARUP Laboratories, Molecular Genetics Laboratory, Salt Lake City, UT, USA	Expanded hearing loss panel	56	Illumina sequencing (HiSeq)
Prevention Genetics, Clinical DNA Testing and DNA Banking, Marshfield, WI, USA	Nonsyndromic hearing loss and deafness NGS panel	48	Illumina sequencing
Centogene AG, Rostock, Germany	Deafness, nonsyndromic sensorineural autosomal recessive panel	44	Roche 454 GS Junior; Life Technologies Ion Torrent sequencing
	Deafness, nonsyndromic sensorineural autosomal dominant panel	32	

Table 1 NGS Tests for HL (Selected Examples)—cont'd

Company	Test Name	Genes Tested (No.)	Technology
Center for Human Genetics and Laboratory Medicine Martinsried, Martinsried, Germany	Hearing loss/deafness multigene panel	42	Illumina sequencing (HiSeq, MiSeq, GAII); Life Technologies Ion Torrent sequencing; Roche 454 GS Junior, Roche 454 GS FLX;
Cincinnati Children's Hospital Medical Center, Molecular Genetics Laboratory, Cincinnati, OH, USA	OtoSeq hearing loss panel	23	Illumina sequencing (HiSeq)

Table 2 Multidisease NGS Tests for Genetic Eye Disorders (Selected Examples)

Company	Test Name	Genes Tested (No.)	Technology
Ocular Genomics Institute, Boston, MA, USA	Genetic eye disease panel for retinal genes GEDi-R	226	Illumina sequencing (MiSeq)
Emory Genetics Laboratory, Atlanta, GA, USA	Eye disorders: comprehensive sequencing	208	Illumina sequencing
Radboud University Medical Center, Nijmegen, The Netherlands	Vision disorders	188	Roche 454 GS FLX, ABI SOLiD sequencing
Fulgent Diagnostics, Temple City, CA, USA	Eye disorders NGS panel	138	Illumina sequencing (MiSeq)
Manchester Centre for Genomic Medicine, Manchester, UK	Retinal degeneration conditions (scan for 105 RD genes)	105	ABI SOLiD sequencing
ARUP Laboratories, Molecular Genetics Laboratory, Salt Lake City, UT, USA	Retinitis pigmentosa/Leber congenital amaurosis	53	Illumina sequencing (HiSeq)
Emory Genetics Laboratory, Atlanta, GA, USA	Achromatopsia, cone, and cone–rod dystrophy	36	Illumina sequencing
Asper Biotech Ltd., Tartu, Estonia	Bardet–Biedl syndrome, McKusick–Kaufman syndrome, Borjeson–Forssman–Lehmann syndrome, Alström syndrome, Albright hereditary osteodystrophy	22	Illumina sequencing (HiSeq, MiSeq)
Ocular Genomics Institute, Boston, MA, USA	Genetic eye disease panel for optic atrophy and early onset glaucoma GEDi-O	9	Illumina sequencing (MiSeq)

Table 3 Single-disease NGS Tests for Genetic Eye Disorders (Selected Examples)

Company	Test Name	Genes Tested (No.)	Technology
Retinitis Pigmentosa Panels			
Fulgent Diagnostics, Temple City, CA, USA	Retinitis pigmentosa NGS panel	88	Illumina sequencing (MiSeq)
Baylor College of Medicine, Medical Genetics Laboratories, Houston, TX, USA	Retinitis pigmentosa panel (BCM-MitomeNGS)	66	Illumina sequencing (HiSeq, MiSeq)
Emory Genetics Laboratory, Atlanta, GA, USA	Retinitis pigmentosa	66	Illumina sequencing
Asper Biotech Ltd., Tartu, Estonia	Autosomal recessive retinitis pigmentosa	56	Illumina sequencing (HiSeq, MiSeq)
	Autosomal dominant retinitis pigmentosa	25	
	X-linked retinitis pigmentosa	3	
CeGaT GmbH, Tübingen, Germany	Retinitis pigmentosa, autosomal recessive	44	Illumina sequencing (HiSeq)
	Retinitis pigmentosa, autosomal dominant and X-linked	26	
Centogene AG, Rostock, Germany	Retinitis pigmentosa panel, autosomal recessive	43	Roche 454 GS Junior; Life Technologies Ion Torrent sequencing
	Retinitis pigmentosa panel, autosomal dominant	28	
Sistemas Genomicos, Medical Genetics Unit, Paterna, Spain	Retinitis pigmentosa (autosomal dominant)	20	ABI SOLiD sequencing
GeneDx, Gaithersburg, MD, USA	arRP sequencing panel	7	Illumina sequencing
Leber Congenital Amaurosis Panels			
Emory Genetics Laboratory, Atlanta, GA, USA	Leber congenital amaurosis	23	Illumina sequencing
Asper Biotech Ltd., Tartu, Estonia	Leber congenital amaurosis	19	Illumina sequencing (HiSeq, MiSeq)
Baylor College of Medicine, Medical Genetics Laboratories, Houston, TX, USA	Leber congenital amaurosis (BCM-MitomeNGS)	19	Illumina sequencing (HiSeq, MiSeq)
Centogene AG, Rostock, Germany	Leber congenital amaurosis panel	19	Roche 454 GS Junior; Life Technologies Ion Torrent sequencing
Sistemas Genomicos, Medical Genetics Unit, Paterna, Spain	Leber congenital amaurosis	17	ABI SOLiD sequencing
CeGaT GmbH, Tübingen, Germany	Leber congenital amaurosis	16	Illumina sequencing (HiSeq)

Table 3 Single-disease NGS Tests for Genetic Eye Disorders (Selected Examples)—cont'd

Company	Test Name	Genes Tested (No.)	Technology
Bardet–Biedl Panels			
CeGaT GmbH, Tübingen, Germany	Bardet–Biedl	19	Illumina sequencing (HiSeq)
Centogene AG, Rostock, Germany	Bardet–Biedl panel	19	Roche 454 GS Junior; Life Technologies Ion Torrent sequencing
Emory Genetics Laboratory, Atlanta, GA, USA	Bardet–Biedl syndrome	18	Illumina sequencing
Prevention Genetics, Clinical DNA Testing and DNA Banking, Marshfield, WI, USA	Bardet–Biedl syndrome NGS panel	17	Illumina sequencing
Fulgent Diagnostics, Temple City, CA, USA	Bardet–Biedl NGS panel	15	Illumina sequencing (MiSeq)
Usher Syndrome Panels			
Asper Biotech Ltd., Tartu, Estonia	Usher syndrome	19	Illumina sequencing (HiSeq, MiSeq)
CeGaT GmbH, Tübingen, Germany	Usher syndrome	13	Illumina sequencing (HiSeq)
Emory Genetics Laboratory, Atlanta, GA, USA	Usher syndrome	13	Illumina sequencing
Fulgent Diagnostics, Temple City, CA, USA	Usher syndrome NGS panel	11	Illumina sequencing (MiSeq)
Baylor College of Medicine, Medical Genetics Laboratories, Houston, TX, USA	Usher syndrome (BCM-MitomeNGS)	9	Illumina sequencing (HiSeq, MiSeq)
Cincinnati Children's Hospital Medical Center, Molecular Genetics Laboratory, Cincinnati, OH, USA	Usher syndrome panel	9	Roche 454 GS Junior; Life Technologies Ion Torrent sequencing
GeneDx, Gaithersburg, MD, USA	Usher syndrome panel	9	Illumina sequencing
Optic Atrophy Panels			
Prevention Genetics, Clinical DNA Testing and DNA Banking, Marshfield, WI, USA	Optic atrophy NGS panel	16	Illumina sequencing
Centogene AG, Rostock, Germany	Optic atrophy panel	11	Roche 454 GS Junior; Life Technologies Ion Torrent sequencing
CeGaT GmbH, Tübingen, Germany	Optic atrophy	5	Illumina sequencing (HiSeq)
Emory Genetics Laboratory, Atlanta, GA, USA	Optic atrophy	5	Illumina sequencing
Asper Biotech Ltd., Tartu, Estonia	Autosomal dominant optic atrophy	3	Illumina sequencing (HiSeq, MiSeq)

Continued

tested in NGS panels for HL varies widely from almost 130 to only 23 (Table 1). NGS panels with a small number of genes are designed usually to detect mutations in the most common genes causing nonsyndromic hearing loss (NSHL), USH, Pendred syndrome, and the branchio-otorenal spectrum disorders. Reducing the number of genes in a panel limits at the same time the number of findings of uncertain clinical significance.

One of the most complex multidisease NGS panels for genetic eye disorders is the genetic eye disease panel for retinal genes (GEDi-R) offered by the Ocular Genomics Institute (Boston, MA, USA). It enables simultaneous testing of 226 genes involved in genetic retinal disorders. In general, the number of genes tested in the panels for eye disorders ranges from 100 to 200. They are mostly focused on retinal dystrophies, which represent a highly heterogeneous group of ocular diseases. However, there are also multidisease panels aimed at testing a more select group of genes involved in RP and LCA or achromatopsia; cone and cone–rod dystrophies; the ciliopathy spectrum, for example, BBS and Alström syndrome; or optic atrophy and glaucoma (Table 2).

Among the single-disease NGS panels for eye disorders, the widest range of diagnostic tests can be found for RP (88–7 genes tested), followed by LCA (23–16 genes tested), BBS (19–15 genes tested), USH (19–9 genes tested), and optic atrophy (16–3 genes tested). Some of the diagnostic tests differentiate between the modes of RP inheritance, and genes for arRP, adRP, or X-linked RP can be tested separately. In other NGS panels most of the RP genes, regardless of the inheritance mode, are tested (Table 3).

All pathogenic, novel variants as well as variants of currently unknown significance, detected using diagnostic NGS panels, are usually confirmed by Sanger sequencing, which is still considered the method of choice for validation of DNA nucleotide sequences.

COMPATIBILITY OF STANDARD ENRICHMENT PANELS WITH GENETIC VISION AND HEARING DISORDERS

An alternative for those having access to an NGS facility is to order a standard enrichment panel and perform target enrichment, sequencing, and data analysis on-site. Standard panels include the TruSight One panel (Illumina, Inc., San Diego, CA, USA), which covers more than 4800 genes associated with known clinical phenotypes; the TruSight inherited diseases panel (Illumina), which targets genes mutated in severe, recessive pediatric-onset disease; or the Ion AmpliSeq inherited disease panel (Life Technologies Ltd., Paisley, UK), which contains a selection of genes that are mutated in over 700 unique inherited diseases, mostly neuromuscular, cardiovascular, developmental, and metabolic diseases. None of the panels is solely dedicated to genetic disorders of vision and hearing.

Using our expanded list of genes containing 356 and 196 genes involved in hereditary disorders of the eye or nonsyndromic and syndromic HL, respectively, only the TruSight One panel was found to cover more than three-fourths of the genes present on both lists. The highest percentage of genes involved in vision and hearing disorders in one panel was found for the Ion AmpliSeq inherited disease panel, in which 20.3% of all genes tested are involved in vision and 9.8% in hearing (Table 4).

As the standard NGS panels may not contain all genes of interest, another, but also the most challenging, possibility is to design a custom panel. The solution is more expensive and laborious as it requires experimental testing of, for example, coverage distribution or optimization of the composition of capture solution or primer mix.

				Eye Disorder-	Ear Disorder-
Product	Vendor	NGS Technology	No. of Genes Tested in the Panel	Compatible Genes (Total = 356 Genes)[a]	Compatible Genes (Total = 196 Genes)[a]
TruSight One	Illumina, Inc., San Diego, CA, USA	Illumina platform (MiSeq, HiSeq)	4813	269 genes (75.6%)	162 genes (82.7%)
TruSight inherited diseases	Illumina, Inc., San Diego, CA, USA	Illumina platform (MiSeq, HiSeq)	552	54 genes (15.2%)	29 genes (14.8%)
Ion AmpliSeq inherited disease	Life Technologies Ltd., Paisley, UK	Ion PGM	325	66 genes (18.5%)	32 genes (16.3%)

Table 4 Standard Enrichment Panels Targeting Genes in Vision and Hearing Disorders

[a]*Number of eye or ear disorder-compatible genes used for calculations.*

UTILITY OF NGS TESTING FOR DIAGNOSTIC PURPOSES OF VISION AND HEARING DISORDERS

The feasibility of applying NGS to patients with genetically heterogeneous diseases has been examined in several studies. Using NGS for 254 known and candidate retinal genes, Audo et al. were able to detect mutations in 57% of the test cases [19]. Targeting 111 RP genes, Neveling et al. have estimated that for a previously unscreened RP population the mutation detection rate with the approach would be about 50% [20]. In an NGS approach that targets 105 retinal disease-associated genes, Gloeckle et al. have identified the causative mutations in 55% of RP and 80% of BBS or USH cases [21]. Bowne et al. targeted in a PCR-based approach 46 genes known to cause adRP and identified the pathogenic mutations in 64% of the cohort [22].

A high diagnostic yield of a 70% mutation detection rate in retinal dystrophies, despite testing of only a small number of 55 disease genes, was obtained by Eisenberger et al. High (250-fold) and complete coverage of the targeted sequences corresponding to more than 99% of the sequences of interest being covered at 15× depth reduced the risk of mutations escaping detection and allowed for systematic analysis of copy number variations (CNVs). Based on their results, Eisenberger et al. pointed out the need for analyzing CNVs and systematic inclusion of the 5′-UTR in disease gene and exome panels [23].

The utility of NGS testing for retinal dystrophy in a health care setting was validated by O'Sullivan et al. When an NGS panel test of 105 genes known to be mutated in all forms of inherited retinal disease was performed on patients with retinal dystrophies, the mutation detection rate increased two times (52% vs 24%) compared to conventional Sanger sequencing [24].

In a post hoc comparison of the utility of Sanger sequencing and exome sequencing for the diagnosis of deafness and blindness, Neveling et al. provided data confirming that NGS has a high diagnostic yield. In this study the total number of genes offered for testing in exome sequencing was 104 for deafness and 164 for blindness. With the introduction of exome sequencing the detection of mutation for blindness increased two times compared to the yield obtained with Sanger sequencing (52% vs 25%), while for deafness the increase was four times (44% vs 10%) [11].

Considering a significant increase in the effectiveness of genetic testing for deafness using NGS as well as a lack of other common genetic causes after exclusion of mutations in the *GJB2* and *GJB6* genes, the American College of Medical Genetics and Genomics, in guidelines published in 2014 for the clinical evaluation and etiologic diagnosis of HL, recommends panel tests as good choices. Regardless of the NGS technology used close attention should always be paid to the performance characteristics of tests, including coverage, analytic sensitivity, the genes that are and are not analyzed, and the types of mutation that are and are not detected. The guideline also advises having tests performed in laboratories that focus on genetic causes of HL. The laboratories may be more likely to report test performance with respect to hearing-related genes. They may have also developed approaches to specifically analyze relevant regions of the genome that may be refractory to more general NGS approaches [25].

CUMULATIVE MUTATION LOAD

NGS tests not only increase the effectiveness of genetic testing but also allow detection of additional mutations acting through various modes of inheritance that may potentially contribute to the phenotype, that is, the cumulative mutation load. After 80 deafness genes were tested using NGS, patients with NSHL were found to have a significantly higher number of genetic variants compared to control subjects. While the median number of variants in controls was 1.0, the median number of variants in the dominant deafness group was 4.0 and 3.0 each in the recessive and the undiagnosed deafness group. Particularly interesting is the significant enrichment of potentially pathogenic variants in the undiagnosed individuals compared to controls. This may suggest a polygenic or multifactorial form of HL inheritance in which rare, possibly pathogenic variants scattered across the genome together with other adverse genetic and/or environmental factors may exceed a critical threshold for phenotypic manifestation [26,27].

DIGENIC AND OLIGOGENIC INHERITANCE

Analyzing the data from NGS testing, which identify a wide range of genetic variants across the genome, it becomes clear that digenic and oligogenic causes of inherited disorders of the eye and HL might be more common than previously thought. Digenic inheritance has been described previously for several deafness genes, for example, cadherin 23 (*CDH23*) and protocadherin 15 (*PCDH15*). Another example is the solute carrier family 26 (anion exchanger), member 4 (*SLC26A4*), and potassium inwardly-rectifying channel J10 (*KCNJ10*) or *SLC26A4* and the transcription factor *FOXI1* [14]. In patients with USH digenic inheritance was reported for the *GPR98* gene and the PDZ domain-containing 7 gene (*PDZD7*) [28]. Using NGS another possibly digenic inheritance of USH was detected in a patient carrying two heterozygous mutations in the *MYO7A* and *PCDH15* genes, respectively [29].

There is a multitude of genes involved in hereditary retinal diseases and NGS makes it possible to identify most of the various genetic changes affecting these genes. However, it should be stressed that according to a 2012 study one in four or five (~20%) individuals from the general population may be a carrier of null mutations in a gene for inherited retinal degeneration. This seems to represent the highest mutation carrier frequency so far measured for a class of Mendelian disorders [30]. Therefore, in patients with possibly multigene contributions to hereditary retinal diseases, to assess if a mutation is causative or has been detected accidentally, an overall variant load in each patient has to be carefully considered and interpreted [23]. NGS has the potential to yield new insights into the role of digenic or oligogenic contributions, such as the interindividual differences in disease severity of otherwise monogenic disorders.

DE NOVO MUTATIONS

An increasingly important aspect of genetic counseling is the identification of apparent *de novo* mutations [31,32]. Data from NGS studies revealed that dominant *de novo* mutations play a significant role, for example, in RP and HL [20,27]. Patients with a negative family history for a genetic disorder are most likely to have a recessive disease and genetic counseling of patients with an isolated disease is strongly biased toward a recessive mode of inheritance. However, identification of a *de novo* dominant mutation significantly shifts the risk of passing the mutated gene to the offspring. Considering a transmission risk of 50% and recurrence risk of <1% for autosomal dominant *de novo* mutations and a transmission risk of <1% and recurrence risk of 25% for an autosomal recessive mode of inheritance, detection of a *de novo* dominant mutation radically changes the genetic counseling of the affected individual and the family.

The increased rate of *de novo* mutations among patients with genetic eye diseases and HL is intriguing as *de novo* mutations have been regarded as occurring more frequently in autosomal dominant disorders with a reduced reproductive fitness [33].

COPY NUMBER VARIATIONS

NGS data can be analyzed qualitatively but also quantitatively, enabling detection of CNVs and thus complementing multiplex-dependent probe amplification or microarray tests. Integrating a customized read-depth approach for variant detection into targeted genomic enrichment coupled with massively parallel sequencing Shearer et al. have identified CNVs as a common contributor to HL. Nearly one in three deafness genes carried a CNV and they were involved in one in five genetic diagnoses of nonsyndromic HL. From all CNVs detected, the greatest number of CNVs was found in the *STRC* and *OTOA* genes comprising 73 and 13% of all identified CNVs, respectively. The carrier frequency of the *STRC* region deletion, the most common CNV, was equal to 4.7% in the study. The data imply that the carrier frequency of deletions of the *STRC* region may in some populations be more common than the carrier frequency of the most common mutations in *GJB2* [34].

In retinal dystrophies several causative CNVs have also been revealed by NGS. CNVs were commonly detected in *PRPF31*, an autosomal dominant RP gene. This was in line with a previous study suggesting that the genetic locus containing *PRPF31* is prone to genomic rearrangements. CNVs were found in the 5′-UTR region of the *EYS* gene and within the coding sequence of the *CRX* gene. In the presented cases identification of CNVs was critical to the diagnosis in previously unsolved genetic constellations. Considering the above, diagnostic NGS pipelines should routinely include detection of CNVs [23].

NOVEL GENES AND "ERRONEOUS" DISEASE GENES

The advent of NGS has significantly accelerated the discovery rate of novel genes for all Mendelian diseases, including HL and genetic eye disorders. Some of the causative genes have been discovered in previously reported loci, others are found in previously unsuspected regions of the genome. Genes identified for HL are involved in several different mechanisms ensuring proper functioning of the organ of Corti. Novel genes for recessive NSHL discovered by WES or targeted enrichment of genomic locus include *ADCY1*, encoding adenylate cyclase 1 [35]; *BDP1*, encoding B double prime 1, a subunit of the RNA polymerase III transcription initiation factor IIIB, which functions as a transcription factor-like nuclear regulator [25]; *SYNE4*, encoding spectrin repeat-containing nuclear envelope protein 4 [36]; *ELMOD3*, encoding elmo/ced12 domain-containing protein 3, which is predicted to function as a GTPase-activating protein [37]; *CABP2*, encoding calcium-binding protein 2 [38]; *GRXCR2*, encoding glutaredoxin, cysteine-rich, 2, which is predicted to play a role in maintaining cochlear stereocilia bundles [39]; *OTOGL*, encoding otogelin-like protein [40]; *TPRN*, encoding taperin, a protein that localizes predominantly in the taper region of hair cell stereocilia [41]; and *TSPEAR*, encoding a protein expressed in the cochlea that is assumed to function as part of a ligand-binding domain [42].

Several phenotypes have been reported for the *TBC1D24* gene, which encodes a protein coordinating Rab proteins and other GTPases for the proper transport of intracellular vesicles. Recessive mutations in the *TBC1D24* gene were previously found in various epileptic disorders [43,44], but in 2014 recessive mutations in this gene were detected in DOORS (deafness, onychodystrophy, osteodystrophy, mental retardation, and seizures) syndrome [45], but also in nonsyndromic autosomal recessive (DFNB86) [46] as well as autosomal dominant deafness, which was simultaneously reported by two independent groups [47,48].

Among other novel genes for dominant NSHL are *CEACAM16*, encoding carcinoembryonic antigen-related cell adhesion molecule 16, an immunoglobulin-related glycoprotein that is expressed in mammalian cochlear outer hair cells and interacts with α-tectorin [49]; *P2RX2*, encoding ion channel gated by extracellular ATP, which is required for lifelong normal hearing and for protection from exposure to noise [50]; and *OSBPL2*, encoding oxysterol binding protein-like 2, which belongs to a family of intracellular lipid receptors [51]. So far, only one novel gene has been discovered for X-linked deafness. The *SMPX* gene was identified in the DFNX4 locus at Xp22.2–p22.1. It encodes the small muscle protein that is suspected to function in the development and/or maintenance of sensory hair cells [52].

The majority of novel genes for inherited eye diseases identified in the genomic studies using NGS were found in patients with arRP. One of the first genes discovered using WES in RP patients was *DHDDS*, which encodes dehydrodolichyl diphosphate synthase and was found in an Ashkenazi Jewish family [53]. Mutations in *MAK*, encoding male germ cell-associated kinase, were reported simultaneously by two independent groups [54,55]. Mutations in *ARL2BP*, encoding ADP-ribosylation factor-like 2-binding protein, were found in a European family with arRP [56]. Other arRP genes include *KIZ*, encoding a centrosomal substrate of Polo-like kinase-1 [57]; *SLC7A14*, encoding a member of the solute carrier family 7, which is predicted to mediate lysosomal uptake of cationic amino acids [58]; and *NEK2*, encoding ciliary serine/threonine-protein kinase [59]. A ciliary gene, *C8orf37*, was mutated in patients with arRP and cone–rod dystrophy [60]. Another novel gene for cone–rod dystrophy is *RAB28* [61].

Well-documented mutations are in *NMNAT1*, encoding a protein involved in metabolic redox reactions that were found in patients with LCA and reported simultaneously by four independent groups [62–65]. Mutations in *KCNJ13*, encoding an inwardly rectifying potassium channel, previously found in snowflake vitreoretinal degeneration, were detected using WES in patients with LCA [66]. Two genes, *LRIT3* and *GPR179*, were discovered for CSNB [67–69]. A novel gene, *LZTFL1*, was also found for BBS [70,71].

Rare genetic variants in *LOXHD1*, a gene for autosomal recessive NSHL, identified by high-throughput sequencing, were implicated in the pathogenesis of Fuchs corneal dystrophy [72]. Common variants in multiple complement pathway genes were previously associated with age-related macular degeneration (AMD) susceptibility. The use of NGS allowed the identification of rare but highly penetrant variants in *CFH* [73] and *CFI* [74] that contribute to the genetic burden of AMD.

Analyzing the data obtained from NGS studies makes it possible not only to discover novel disease genes but also to question the pathogenic role of already known genes. One of the genes disqualified from being a cause of autosomal dominant NSHL is *MYO1A*. Mutations reported previously in the *MYO1A* gene [75] were found to have a population frequency of above 0.1%, which contradicts their pathogenicity under a dominant model. Furthermore, genetic variants in the *MYO1A* gene were detected in families with another clearly defined genetic cause of HL and were present in both affected and unaffected family members [76]. Application of NGS in the study allowed a detailed analysis of the contribution of different genetic factors to HL in these families.

Another example is the gene *RAB40AL*, which was identified as the locus for Martin–Probst syndrome (MPS), an X-linked deafness–intellectual disability syndrome [77], although the sole evidence for this was the segregation of a missense change, p.D59G, with the disease in a single family and in vitro studies. The same genetic variant was identified by WES in two patients; however, further extensive investigations in the patients and their family members unequivocally excluded the diagnosis of MPS. Furthermore, the p.D59G variant was detected with a high prevalence of 2.47% in the general population, which is typical for a common genetic variation observed in asymptomatic individuals [78]. One can speculate that the identification of "erroneous" disease genes will expand with increasing availability of large-scale sequencing data.

CONCLUSIONS

NGS is being successfully adopted by clinical laboratories for the diagnosis of hereditary vision and hearing disorders, and the number of laboratories offering NGS diagnostic tests is constantly increasing. There are differences in the diagnostic yield of the NGS panels, but in approximately one-half of the patients the genetic background can be identified using this technology. It is a major improvement compared to other diagnostic tests currently used in molecular genetics. One of the key challenges now is to distinguish the pathogenic from the nonpathogenic variants that are found in each individual. Classification of variants includes now mainly their frequency in cases and controls with respect to specific ethnic groups and populations, cosegregation with the disease in families, in silico predictions, and, if possible, also the results from in vitro or in vivo studies. It seems plausible that in the future patient-specific genetic information will be incorporated in the interaction networks and cellular pathways to better predict the outcome of a complex genetic study.

ABBREVIATIONS

ad Autosomal dominant
ar Autosomal recessive
CSNB Congenital stationary night blindness
HL Hearing loss
LCA Leber congenital amaurosis
NGS Next generation sequencing
NSHL Nonsyndromic hearing loss
RP Retinitis pigmentosa
USH Usher syndrome
WES Whole-exome sequencing

ACKNOWLEDGMENTS

This work was supported by the Polish National Science Center Grants (5915/B/P01/2011/40 and 5916/B/P01/2011/40) and the project entitled "Integrated System of Tools Designed for Diagnostics and Telerehabilitation of the Sense Organs Disorders (Hearing, Vision, Speech, Balance, Taste, Smell)" INNOSENSE, cofinanced by the National Centre for Research and Development within the STARTEGMED Program.

REFERENCES

[1] Swaroop A, Sieving PA. The golden era of ocular disease gene discovery: race to the finish. Clin Genet August 2013;84(2):99–101.

[2] Bovo R, Ciorba A, Martini A. Environmental and genetic factors in age-related hearing impairment. Aging Clin Exp Res February 2013;23(1):3–10.

[3] Sliwinska-Kowalska M, Pawelczyk M. Contribution of genetic factors to noise-induced hearing loss: a human studies review. Mutat Res January–March 2013;752(1):61–5.

[4] Uchida Y, Sugiura S, Ando F, Nakashima T, Shimokata H. Molecular genetic epidemiology of age-related hearing impairment. Auris Nasus Larynx December 2011;38(6):657–65.

[5] Daiger SP, Sullivan LS, Bowne SJ. Genes and mutations causing retinitis pigmentosa. Clin Genet August 2013;84(2):132–41.

[6] Ratnapriya R, Swaroop A. Genetic architecture of retinal and macular degenerative diseases: the promise and challenges of next-generation sequencing. Genome Med 2013;5(10):84.

[7] Sorusch N, Wunderlich K, Bauss K, Nagel-Wolfrum K, Wolfrum U. Usher syndrome protein network functions in the retina and their relation to other retinal ciliopathies. Adv Exp Med Biol 2014;801:527–33.

[8] Estrada-Cuzcano A, Roepman R, Cremers FP, den Hollander AI, Mans DA. Non-syndromic retinal ciliopathies: translating gene discovery into therapy. Hum Mol Genet October 15, 2012;21(R1):R111–24.

[9] Tonisson N, Kurg A, Kaasik K, Lohmussaar E, Metspalu A. Unravelling genetic data by arrayed primer extension. Clin Chem Lab Med February 2000;38(2):165–70.

[10] Shumaker JM, Tollet JJ, Filbin KJ, Montague-Smith MP, Pirrung MC. APEX disease gene resequencing: mutations in exon 7 of the p53 tumor suppressor gene. Bioorg Med Chem September 2001;9(9):2269–78.

[11] Neveling K, Feenstra I, Gilissen C, Hoefsloot LH, Kamsteeg EJ, Mensenkamp AR, et al. A post-hoc comparison of the utility of sanger sequencing and exome sequencing for the diagnosis of heterogeneous diseases. Hum Mutat December 2013;34(12):1721–6.

[12] Shearer AE, DeLuca AP, Hildebrand MS, Taylor KR, Gurrola 2nd J, Scherer S, et al. Comprehensive genetic testing for hereditary hearing loss using massively parallel sequencing. Proc Natl Acad Sci USA December 7, 2010;107(49):21104–9.

[13] Sivakumaran TA, Husami A, Kissell D, Zhang W, Keddache M, Black AP, et al. Performance evaluation of the next-generation sequencing approach for molecular diagnosis of hereditary hearing loss. Otolaryngol Head Neck Surg June 2013;148(6):1007–16.

[14] Schrauwen I, Sommen M, Corneveaux JJ, Reiman RA, Hackett NJ, Claes C, et al. A sensitive and specific diagnostic test for hearing loss using a microdroplet PCR-based approach and next generation sequencing. Am J Med Genet A January 2013;161A(1):145–52.

[15] Brownstein Z, Friedman LM, Shahin H, Oron-Karni V, Kol N, Abu Rayyan A, et al. Targeted genomic capture and massively parallel sequencing to identify genes for hereditary hearing loss in Middle Eastern families. Genome Biol 2011;12(9):R89.

[16] De Keulenaer S, Hellemans J, Lefever S, Renard JP, De Schrijver J, Van de Voorde H, et al. Molecular diagnostics for congenital hearing loss including 15 deafness genes using a next generation sequencing platform. BMC Med Genomics 2012;5:17.

[17] Shearer AE, Black-Ziegelbein EA, Hildebrand MS, Eppsteiner RW, Ravi H, Joshi S, et al. Advancing genetic testing for deafness with genomic technology. J Med Genet September 2013;50(9):627–34.

[18] Licastro D, Mutarelli M, Peluso I, Neveling K, Wieskamp N, Rispoli R, et al. Molecular diagnosis of Usher syndrome: application of two different next generation sequencing-based procedures. PLoS One 2012;7(8):e43799.

[19] Audo I, Bujakowska KM, Leveillard T, Mohand-Said S, Lancelot ME, Germain A, et al. Development and application of a next-generation-sequencing (NGS) approach to detect known and novel gene defects underlying retinal diseases. Orphanet J Rare Dis 2012;7:8.

[20] Neveling K, Collin RW, Gilissen C, van Huet RA, Visser L, Kwint MP, et al. Next-generation genetic testing for retinitis pigmentosa. Hum Mutat June 2012;33(6):963–72.

[21] Glockle N, Kohl S, Mohr J, Scheurenbrand T, Sprecher A, Weisschuh N, et al. Panel-based next generation sequencing as a reliable and efficient technique to detect mutations in unselected patients with retinal dystrophies. Eur J Hum Genet January 2014;22(1):99–104.

[22] Bowne SJ, Sullivan LS, Koboldt DC, Ding L, Fulton R, Abbott RM, et al. Identification of disease-causing mutations in autosomal dominant retinitis pigmentosa (adRP) using next-generation DNA sequencing. Invest Ophthalmol Vis Sci January 2011;52(1):494–503.

[23] Eisenberger T, Neuhaus C, Khan AO, Decker C, Preising MN, Friedburg C, et al. Increasing the yield in targeted next-generation sequencing by implicating CNV analysis, non-coding exons and the overall variant load: the example of retinal dystrophies. PLoS One 2013;8(11):e78496.

[24] O'Sullivan J, Mullaney BG, Bhaskar SS, Dickerson JE, Hall G, O'Grady A, et al. A paradigm shift in the delivery of services for diagnosis of inherited retinal disease. J Med Genet May 2012;49(5):322–6.

[25] Alford RL, Arnos KS, Fox M, Lin JW, Palmer CG, Pandya A, et al. American College of Medical Genetics and Genomics guideline for the clinical evaluation and etiologic diagnosis of hearing loss. Genet Med April 2014;16(4):347–55.

[26] Riazuddin S, Castelein CM, Ahmed ZM, Lalwani AK, Mastroianni MA, Naz S, et al. Dominant modifier DFNM1 suppresses recessive deafness DFNB26. Nat Genet December 2000;26(4):431–4.

[27] Vona B, Muller T, Nanda I, Neuner C, Hofrichter MA, Schroder J, et al. Targeted next-generation sequencing of deafness genes in hearing-impaired individuals uncovers informative mutations. Genet Med December 2014;16(12):945–53.

[28] Ebermann I, Phillips JB, Liebau MC, Koenekoop RK, Schermer B, Lopez I, et al. PDZD7 is a modifier of retinal disease and a contributor to digenic Usher syndrome. J Clin Invest June 2010;120(6):1812–23.

[29] Yoshimura H, Iwasaki S, Nishio SY, Kumakawa K, Tono T, Kobayashi Y, et al. Massively parallel DNA sequencing facilitates diagnosis of patients with Usher syndrome type 1. PLoS One 2014;9(3):e90688.

[30] Nishiguchi KM, Rivolta C. Genes associated with retinitis pigmentosa and allied diseases are frequently mutated in the general population. PLoS One 2012;7(7):e41902.

[31] Siemiatkowska AM, Collin RW, den Hollander AI, Cremers FP. Genomic approaches for the discovery of genes mutated in inherited retinal degeneration. Cold Spring Harb Perspect Med 2014;4(8).

[32] de Ligt J, Veltman JA, Vissers LE. Point mutations as a source of de novo genetic disease. Curr Opin Genet Dev June 2013;23(3):257–63.

[33] Veltman JA, Brunner HG. De novo mutations in human genetic disease. Nat Rev Genet August 2012;13(8):565–75.

[34] Shearer AE, Kolbe DL, Azaiez H, Sloan CM, Frees KL, Weaver AE, et al. Copy number variants are a common cause of non-syndromic hearing loss. Genome Med 2014;6(5):37.

[35] Santos-Cortez RL, Lee K, Giese AP, Ansar M, Amin-Ud-Din M, Rehn K, et al. Adenylate cyclase 1 (ADCY1) mutations cause recessive hearing impairment in humans and defects in hair cell function and hearing in zebrafish. Hum Mol Genet June 15, 2014;23(12):3289–98.

[36] Horn HF, Brownstein Z, Lenz DR, Shivatzki S, Dror AA, Dagan-Rosenfeld O, et al. The LINC complex is essential for hearing. J Clin Invest February 1, 2013;123(2):740–50.

[37] Jaworek TJ, Richard EM, Ivanova AA, Giese AP, Choo DI, Khan SN, et al. An alteration in ELMOD3, an Arl2 GTPase-activating protein, is associated with hearing impairment in humans. PLoS Genet 2013;9(9): e1003774.

[38] Schrauwen I, Helfmann S, Inagaki A, Predoehl F, Tabatabaiefar MA, Picher MM, et al. A mutation in CABP2, expressed in cochlear hair cells, causes autosomal-recessive hearing impairment. Am J Hum Genet October 5, 2012;91(4):636–45.

[39] Imtiaz A, Kohrman DC, Naz S. A frameshift mutation in GRXCR2 causes recessively inherited hearing loss. Hum Mutat May 2014;35(5):618–24.

[40] Yariz KO, Duman D, Seco CZ, Dallman J, Huang M, Peters TA, et al. Mutations in OTOGL, encoding the inner ear protein otogelin-like, cause moderate sensorineural hearing loss. Am J Hum Genet November 2, 2012;91(5):872–82.

[41] Rehman AU, Morell RJ, Belyantseva IA, Khan SY, Boger ET, Shahzad M, et al. Targeted capture and next-generation sequencing identifies C9orf75, encoding taperin, as the mutated gene in nonsyndromic deafness DFNB79. Am J Hum Genet March 12, 2010;86(3):378–88.

[42] Delmaghani S, Aghaie A, Michalski N, Bonnet C, Weil D, Petit C. Defect in the gene encoding the EAR/EPTP domain-containing protein TSPEAR causes DFNB98 profound deafness. Hum Mol Genet September 1, 2012;21(17):3835–44.

[43] Falace A, Filipello F, La Padula V, Vanni N, Madia F, De Pietri Tonelli D, et al. TBC1D24, an ARF6-interacting protein, is mutated in familial infantile myoclonic epilepsy. Am J Hum Genet September 10, 2010;87(3):365–70.

[44] Guven A, Tolun A. TBC1D24 truncating mutation resulting in severe neurodegeneration. J Med Genet March 2013;50(3):199–202.

[45] Campeau PM, Kasperaviciute D, Lu JT, Burrage LC, Kim C, Hori M, et al. The genetic basis of DOORS syndrome: an exome-sequencing study. Lancet Neurol January 2014;13(1):44–58.

[46] Rehman AU, Santos-Cortez RL, Morell RJ, Drummond MC, Ito T, Lee K, et al. Mutations in TBC1D24, a gene associated with epilepsy, also cause nonsyndromic deafness DFNB86. Am J Hum Genet January 2, 2014;94(1):144–52.

[47] Azaiez H, Booth KT, Bu F, Huygen P, Shibata SB, Shearer AE, et al. TBC1D24 mutation causes autosomal-dominant nonsyndromic hearing loss. Hum Mutat July 2014;35(7):819–23.

[48] Zhang L, Hu L, Chai Y, Pang X, Yang T, Wu H. A dominant mutation in the stereocilia-expressing gene TBC1D24 is a probable cause for nonsyndromic hearing impairment. Hum Mutat July 2014;35(7):814–8.

[49] Zheng J, Miller KK, Yang T, Hildebrand MS, Shearer AE, DeLuca AP, et al. Carcinoembryonic antigen-related cell adhesion molecule 16 interacts with alpha-tectorin and is mutated in autosomal dominant hearing loss (DFNA4). Proc Natl Acad Sci USA March 8, 2011;108(10):4218–23.

[50] Yan D, Zhu Y, Walsh T, Xie D, Yuan H, Sirmaci A, et al. Mutation of the ATP-gated P2X(2) receptor leads to progressive hearing loss and increased susceptibility to noise. Proc Natl Acad Sci USA February 5, 2013;110(6):2228–33.

[51] Xing G, Yao J, Wu B, Liu T, Wei Q, Liu C, et al. Identification of OSBPL2 as a novel candidate gene for progressive nonsyndromic hearing loss by whole-exome sequencing. Genet Med March 2015;17(3):210–8.

[52] Schraders M, Haas SA, Weegerink NJ, Oostrik J, Hu H, Hoefsloot LH, et al. Next-generation sequencing identifies mutations of SMPX, which encodes the small muscle protein, X-linked, as a cause of progressive hearing impairment. Am J Hum Genet May 13, 2011;88(5):628–34.

[53] Zuchner S, Dallman J, Wen R, Beecham G, Naj A, Farooq A, et al. Whole-exome sequencing links a variant in DHDDS to retinitis pigmentosa. Am J Hum Genet February 11, 2011;88(2):201–6.

[54] Tucker BA, Scheetz TE, Mullins RF, DeLuca AP, Hoffmann JM, Johnston RM, et al. Exome sequencing and analysis of induced pluripotent stem cells identify the cilia-related gene male germ cell-associated kinase (MAK) as a cause of retinitis pigmentosa. Proc Natl Acad Sci USA August 23, 2011;108(34):E569–76.

[55] Ozgul RK, Siemiatkowska AM, Yucel D, Myers CA, Collin RW, Zonneveld MN, et al. Exome sequencing and cis-regulatory mapping identify mutations in MAK, a gene encoding a regulator of ciliary length, as a cause of retinitis pigmentosa. Am J Hum Genet August 12, 2011;89(2):253–64.

[56] Davidson AE, Schwarz N, Zelinger L, Stern-Schneider G, Shoemark A, Spitzbarth B, et al. Mutations in ARL2BP, encoding ADP-ribosylation-factor-like 2 binding protein, cause autosomal-recessive retinitis pigmentosa. Am J Hum Genet August 8, 2013;93(2):321–9.

[57] El Shamieh S, Neuille M, Terray A, Orhan E, Condroyer C, Demontant V, et al. Whole-exome sequencing identifies KIZ as a ciliary gene associated with autosomal-recessive rod-cone dystrophy. Am J Hum Genet April 3, 2014;94(4):625–33.

[58] Jin ZB, Huang XF, Lv JN, Xiang L, Li DQ, Chen J, et al. SLC7A14 linked to autosomal recessive retinitis pigmentosa. Nat Commun 2014;5:3517.

[59] Nishiguchi KM, Tearle RG, Liu YP, Oh EC, Miyake N, Benaglio P, et al. Whole genome sequencing in patients with retinitis pigmentosa reveals pathogenic DNA structural changes and NEK2 as a new disease gene. Proc Natl Acad Sci USA October 1, 2013;110(40):16139–44.

[60] Estrada-Cuzcano A, Neveling K, Kohl S, Banin E, Rotenstreich Y, Sharon D, et al. Mutations in C8orf37, encoding a ciliary protein, are associated with autosomal-recessive retinal dystrophies with early macular involvement. Am J Hum Genet January 13, 2012;90(1):102–9.

[61] Roosing S, Rohrschneider K, Beryozkin A, Sharon D, Weisschuh N, Staller J, et al. Mutations in RAB28, encoding a farnesylated small GTPase, are associated with autosomal-recessive cone-rod dystrophy. Am J Hum Genet July 11, 2013;93(1):110–7.

[62] Koenekoop RK, Wang H, Majewski J, Wang X, Lopez I, Ren H, et al. Mutations in NMNAT1 cause Leber congenital amaurosis and identify a new disease pathway for retinal degeneration. Nat Genet September 2012;44(9):1035–9.

[63] Perrault I, Hanein S, Zanlonghi X, Serre V, Nicouleau M, Defoort-Delhemmes S, et al. Mutations in NMNAT1 cause Leber congenital amaurosis with early-onset severe macular and optic atrophy. Nat Genet September 2012;44(9):975–7.

[64] Falk MJ, Zhang Q, Nakamaru-Ogiso E, Kannabiran C, Fonseca-Kelly Z, Chakarova C, et al. NMNAT1 mutations cause Leber congenital amaurosis. Nat Genet September 2012;44(9):1040–5.

[65] Chiang PW, Wang J, Chen Y, Fu Q, Zhong J, Yi X, et al. Exome sequencing identifies NMNAT1 mutations as a cause of Leber congenital amaurosis. Nat Genet September 2012;44(9):972–4.

[66] Sergouniotis PI, Davidson AE, Mackay DS, Li Z, Yang X, Plagnol V, et al. Recessive mutations in KCNJ13, encoding an inwardly rectifying potassium channel subunit, cause leber congenital amaurosis. Am J Hum Genet July 15, 2011;89(1):183–90.

[67] Zeitz C, Jacobson SG, Hamel CP, Bujakowska K, Neuille M, Orhan E, et al. Whole-exome sequencing identifies LRIT3 mutations as a cause of autosomal-recessive complete congenital stationary night blindness. Am J Hum Genet January 10, 2013;92(1):67–75.

[68] Audo I, Bujakowska K, Orhan E, Poloschek CM, Defoort-Dhellemmes S, Drumare I, et al. Whole-exome sequencing identifies mutations in GPR179 leading to autosomal-recessive complete congenital stationary night blindness. Am J Hum Genet February 10, 2012;90(2):321–30.

[69] Peachey NS, Ray TA, Florijn R, Rowe LB, Sjoerdsma T, Contreras-Alcantara S, et al. GPR179 is required for depolarizing bipolar cell function and is mutated in autosomal-recessive complete congenital stationary night blindness. Am J Hum Genet February 10, 2012;90(2):331–9.

[70] Marion V, Stutzmann F, Gerard M, De Melo C, Schaefer E, Claussmann A, et al. Exome sequencing identifies mutations in LZTFL1, a BBSome and smoothened trafficking regulator, in a family with Bardet–Biedl syndrome with situs inversus and insertional polydactyly. J Med Genet May 2012;49(5):317–21.

[71] Schaefer E, Lauer J, Durand M, Pelletier V, Obringer C, Claussmann A, et al. Mesoaxial polydactyly is a major feature in Bardet-Biedl syndrome patients with LZTFL1 (BBS17) mutations. Clin Genet May 2014;85(5):476–81.

[72] Riazuddin SA, Parker DS, McGlumphy EJ, Oh EC, Iliff BW, Schmedt T, et al. Mutations in LOXHD1, a recessive-deafness locus, cause dominant late-onset Fuchs corneal dystrophy. Am J Hum Genet March 9, 2012;90(3):533–9.

[73] Raychaudhuri S, Iartchouk O, Chin K, Tan PL, Tai AK, Ripke S, et al. A rare penetrant mutation in CFH confers high risk of age-related macular degeneration. Nat Genet December 2011;43(12):1232–6.

[74] van de Ven JP, Nilsson SC, Tan PL, Buitendijk GH, Ristau T, Mohlin FC, et al. A functional variant in the CFI gene confers a high risk of age-related macular degeneration. Nat Genet July 2013;45(7):813–7.

[75] Donaudy F, Ferrara A, Esposito L, Hertzano R, Ben-David O, Bell RE, et al. Multiple mutations of MYO1A, a cochlear-expressed gene, in sensorineural hearing loss. Am J Hum Genet June 2003;72(6):1571–7.

[76] Eisenberger T, Di Donato N, Baig SM, Neuhaus C, Beyer A, Decker E, et al. Targeted and genomewide NGS data disqualify mutations in MYO1A, the "DFNA48 gene", as a cause of deafness. Hum Mutat May 2014;35(5):565–70.

[77] Bedoyan JK, Schaibley VM, Peng W, Bai Y, Mondal K, Shetty AC, et al. Disruption of RAB40AL function leads to Martin–Probst syndrome, a rare X-linked multisystem neurodevelopmental human disorder. J Med Genet May 2012;49(5):332–40.

[78] Oldak M, Sciezynska A, Mlynarski W, Borowiec M, Ruszkowska E, Szulborski K, et al. Evidence against RAB40AL being the locus for Martin-Probst X-linked deafness-intellectual disability syndrome. Hum Mutat October 2014;35(10):1171–4.

NEXT GENERATION SEQUENCING AS A TOOL FOR NONINVASIVE PRENATAL TESTS

Ozgur Cogulu[1,2]

[1]*Department of Pediatric Genetics, Faculty of Medicine, Ege University, Izmir, Turkey;* [2]*Department of Medical Genetics, Faculty of Medicine, Ege University, Izmir, Turkey*

CHAPTER OUTLINE

INTRODUCTION

Since the dawn of time, the basic quest of human beings has always been the discovery and exploration of novel things, to answer the questions of "How?" and "Why?", and to be the first. The results of this thirst are unpredictable! Many new discoveries do benefit humankind; unfortunately, however, some are to its detriment.

In accordance with this trend, developments in genetic and computer technologies have been gaining great interest, especially following the completion of the Human Genome Project, which is on the cutting edge of both fields. This project has opened up new horizons in terms of human health and genetic science.

The dream and wish of pregnant couples has always been the arrival of a healthy child. A very important, and often overlooked, aspect of this is the service of genetic counseling. In parallel with technological developments in the field of genetics, the genetic basis of an increasingly large number of inheritable disorders has been gradually identified. Unfortunately, genetic treatment modalities have not kept pace with discoveries, meaning very few treatments are currently available for genetic disease.

Nevertheless, when genetic concerns about a pregnancy outcome exist, and genetic counseling services become involved, the diagnosis of disorders in utero means that parents can now be informed with regard to the genetic basis of the known disorders and the likely outcomes and then have the opportunity to make informed decisions concerning the maintenance of the pregnancy.

The incidence of genetic-related diseases has been reported in about 4% of all neonates [1]. If we focus on congenital anomalies, specifically those that accompany known genetic disorders, they affect approximately 1 in 33 infants; with nearly 300,000 every year not surviving the newborn period [2]. Genetically based diseases are caused by either single-gene or gene-group defects. When all genetic mechanisms affecting gene functions are taken into consideration, several groups of diseases can be observed, as summarized in Table 1.

A considerable number of congenital anomalies are associated with morbidity, mortality, and mental retardation [3], having significantly unfavorable impacts on individuals, families, and the general health care system [2]. As the genetic basis of the above-mentioned diseases is well understood, identifying them in the prenatal period becomes very important. Prenatal testing and diagnosis have therefore become a crucial component of both obstetric and genetic science. Factually relevant information provides parents with real options; helps to prepare them for all possibilities including psychological, social, financial, and medical problems; and, where serious defects are detected, also provides the option of early termination of pregnancy [4]. Chromosomal anomalies, occurring in 1 in 160 live births, are the most common reason for prenatal testing [5–7]. The most frequent, trisomy 21, currently affecting 1 in 700–1000 live births worldwide, has become the most common referral reason for reproductive genetic counseling [8]. However, because the American College of Obstetricians and Gynecologists (ACOG) recommends prenatal screening for aneuploidies in all pregnancies [9], methodical approaches to reproductive genetic counseling and selection criteria for appropriate prenatal genetic testing are crucial.

Table 1 Genetic Diseases and Underlying Mechanisms	
Genetic diseases	Chromosomal diseases
	Single-gene diseases
	Multifactorial–polygenic diseases
	Mitochondrial diseases
Underlying mechanisms	
Quantitative genomic disorders	Deletions
	Amplifications
Qualitative genomic disorders	Substitutions
	Inversions
	Imprinting defects
	Epigenetic changes

CONVENTIONAL PRENATAL DIAGNOSTICS AND METHODS

The principle of information gathering regarding the condition of the embryo or fetus has always been the application of diagnostic or screening modalities in utero without harming the fetus or the mother. This is achieved by both noninvasive and invasive methods. The critical difference between noninvasive and invasive methods concerns the level of risk to the unborn child. Unfortunately, constitutional non-invasive tests do not directly evaluate the fetal condition; invasive methods, however, while tradition-ally carrying far greater risk, do provide the fetal material necessary for direct genetic analysis.

Noninvasive prenatal screening tests generally include the measurement of multiple analytes in the maternal serum, as well as ultrasonographic evaluation of the fetus including such features as nuchal translucency. These tests describe epiphenomena and allow for a statistical assessment regarding the probability of the occurrence of a genetic condition, particularly Down syndrome, Edwards syndrome, and neural tube defects. Biochemical and ultrasound-based noninvasive prenatal screening tests have poor accuracy, with false positive and false negative rates between 3 and 5% and suboptimal overall sensitivity and specificity of around 85–90% [10–15]. On the other hand, for more conclusive diagnos-tic results, testing has traditionally required invasive fetal material sampling through amniocentesis, chorionic villus sampling, or fetal blood sampling followed by cell culture for direct analysis reflecting the nature of the suspected genetic condition. Such techniques include cytogenetics to screen for fetal chromosomal anomalies and molecular genetic techniques generally applied to single-gene disorders. The diagnostic yield of cytogenetic methods is estimated to be 98–99%; however, the risk of fetal loss postprocedure is reported to be 0.5–1% and has always been the major limitation or concern for the families and doctors [16–18]. Owing to the high risk:result ratios of the aforementioned testing proce-dures, there is a constant need for safer, less expensive, and more rapid methods.

USE OF FETAL BIOLOGICAL MATERIAL IN THE MATERNAL CIRCULATION FOR PRENATAL DIAGNOSIS

For years, both conventional invasive and noninvasive procedures were the only available tools provid-ing information regarding fetus health. As the first prenatal diagnostic tool, chromosome analysis of human amniotic fluid cells was introduced in the mid-1960s [19]; and sonographic and serum markers have been widely implemented since the 1970s for noninvasive screening of fetal anomalies, although the diagnostic accuracy of these methods was limited. In the light of current recommendations, these methods are still in use, while invasive prenatal diagnostic testing is applied only in high-risk pregnan-cies [20]. Further studies have focused on the combination of noninvasive isolation techniques followed by direct genetic analysis of the fetal material. Noninvasive prenatal screening in its current form involves the genetic testing of fetal biological material obtained from the maternal blood by means of advanced genomic technologies, predominantly next generation sequencing (NGS). Two modalities of noninvasive prenatal testing (NIPT) exist: cell-based and cell-free assay [4,21] (Table 2).

In 1969, Walknowska et al. first reported the fetomaternal transfer of lymphocytes, and conse-quently attempts to isolate and to examine those cells were undertaken [22–26]. The discovery of bidirectional cell traffic between the fetus and the mother continuously inspires the scientific commu-nity, providing new insight into the field. The average number of fetal cells carried in the maternal

Table 2 Fetal Biological Materials in the Maternal Blood [21]	
Cells	Trophoblasts
	Lymphocytes
	Granulocytes
	Erythroblasts (the most commonly used cell type)
	Platelets
DNA	
RNA	

Table 3 The limitations of using fetal cells for NIPT [11,21,27–29]

- Low number of fetal cells in maternal circulation (1 cell/ml maternal blood)
- Low efficiency of enrichment protocols
- High density of fetal cells nuclei limits feasibility of fluorescence in-situ hybridization (FISH)
- Long life span of fetal cells in the maternal blood, particularly fetal hematopoietic progenitor cells, has been shown to persist for years after delivery

blood, in the second trimester of an uneventful pregnancy, has been reported at around 1.2 cells/ml [26]. The fetal cell types that circulate in the maternal blood and could be used for prenatal genetic testing are listed in Table 2. Unfortunately, a number of factors prevent this technique from being widely used in routine clinical practice [11,21,27–29]. Those limitations are summarized in Table 3. The approach of using isolated fetal cells in NIPT is more difficult than using free fetal DNA, as the amounts of fetal genomic material available from fetal nucleated blood cells are much lower, and the procedures are more complicated, more expensive, and time-consuming.

FETAL DNA IN PRENATAL DIAGNOSIS

In 1997, Lo et al. showed that fetal DNA present in maternal blood allows for the development of noninvasive genetic prenatal diagnosis [30]. DNA isolated from maternal blood is a mixture of short fragments of both maternal and fetal genomic material. Fetal DNA is a minor portion in the large background of nucleic acids in maternal blood, which can be detected around 3 weeks after conception [30].

PROPERTIES OF CELL-FREE FETAL DNA

The major sources of cell-free fetal DNA (cffDNA) in maternal blood are the apoptotic trophoblastic placental cells, fetal hematopoietic system, and lysis of other fetal cells within the maternal circulation [31–33]. Fetal DNA is present in both maternal serum and plasma in similar concentrations. It is preferable, however, to use maternal plasma instead of serum because serum contains a larger quantity of

Table 4 Challenges Associated with cffDNA in NIPT [45]
• Low concentration of fetal DNA in total maternal DNA • Fetal DNA concentration variation between individuals • Half of the fetal genome is identical with that of the mother • Inability to separate fetal DNA from maternal DNA

background maternal DNA, owing to the liberation of maternal DNA from blood cells during the clotting process [34].

cffDNA can be detected from the 18th day of gestation and gradually increases in the course of the pregnancy [35]. cffDNA presents as fragmented short strands, which are known as the fetal DNA fraction in the maternal circulation. The length of circulating DNA fragments in nonpregnant women is shorter than in pregnant women. Chan et al. investigated 34 nonpregnant and 31 pregnant women with regard to the size of plasma DNA fragments and reported that the median percentage of DNA shorter than 201 bp was lower in nonpregnant compared to pregnant women [36]. Fetal-derived DNA fragments in the maternal blood are much shorter than maternal-derived DNA, being less than 313 bp with >99% of all fragments falling between 160 and 180 bp [36–39].

The complete fetal genome is available within the maternal blood in the form of fetal cells or fetal DNA [39,40]. Real-time quantitative polymerase chain reaction (PCR) studies have shown that fetal DNA is present in maternal blood in the amount of 25.4 genome equivalents/ml in early pregnancy and 292.2 genome equivalents/ml in late pregnancy [34]. According to other authors, fetal DNA comprises around 3.4 and 6.2–10% of the total plasma DNA in early and late gestation, respectively [34,41]. The amount of cffDNA is 25 times higher than the amount of DNA extracted from nucleated blood cells that isolated a similar volume of whole maternal blood [32]. Further studies, using the digital PCR assay, revealed median fractional fetal DNA concentrations of 9.7, 9.0, and 20.4% for the first, second, and third trimesters, respectively [42]. This means that while 16 fetal genomes per milliliter are found in the first trimester, this number increases up to 80 by the third trimester [26,34]. The two likely explanations for this are: (1) the gradual increase in size of the fetomaternal interface and (2) the decrease in DNA clearance [34]. It has been reported that there exists a significant variation in DNA concentrations between ethnic groups and this information should always be considered in cffDNA studies [43].

Full separation of fetal DNA from maternal DNA is not technically feasible; genetic tests, therefore, are routinely performed using whole cffDNA, which includes both maternal and fetal components [32]. Immediately following the termination of the pregnancy, cffDNA disappears from maternal blood, becoming undetectable by 1 day postdelivery. The clearance rate among individuals also varies [44]; and as mentioned above, fetal cells may also persist in maternal blood for years, having implications for future genetic testing. Accordingly, fetal cells may persist from previous pregnancies, and isolation strategies are limited by the lack of availability of unique cell markers to differentiate leukocytes of different origin. The developments associated with the direct study of fetal DNA from maternal blood are still in their infancy, and many problems associated with the application of NIPT procedures still need to be solved [45] (Table 4); however, compared to fetal cells, the study of cffDNA currently presents as a safer, less labor-intensive, less time-consuming, and more cost-effective option.

APPLICATIONS OF NONINVASIVE PRENATAL TESTS

NIPT, sometimes also called noninvasive prenatal diagnosis (NIPD), uses fetal genetic material from maternal blood to detect selected inherited disorders. Sometimes this terminology may be misleading, given that NIPT/NIPD are not diagnostic tests, but can be used only as screening tools, which require confirmation with further invasive diagnostic testing.

Currently, **noninvasive genetic prenatal tests** include several options:

- **NGS of cffDNA**—the most widely applied NIPT (see below) is most frequently used for aneuploidy detection; however, there are attempts to apply prenatal NGS testing for single-gene disorders or even to reconstruct the whole genome sequence of a human fetus. An attempt to perform whole-genome sequencing (WGS) of the fetal genome, using paternal (buccal) DNA, maternal DNA, and cffDNA from the pregnant mother's plasma, has been undertaken by Kitzman et al. [46]. The mother's haplotypes, in combination with genome sequencing results of the father's DNA, were used to predict which genetic variants passed to the fetus [19]. Using cord blood taken at delivery, it was found that such approach allowed the prediction of inheritance with 98.1% accuracy. The study sequenced DNA from only two fetuses and the cost of such analysis was of $50,000 each [46].
- **PCR-based methods**—real-time RT-PCR-, digital PCR-, and methylation-specific PCR-based assays are currently in use but their robustness is limited by PCR primer specificity and assay sensitivity. Also, given that most cffDNA in the circulation is of maternal origin, and there is over 99% homology between maternal and fetal DNA, it is difficult to use the traditional PCR technology for fetal aneuploidy detection.
- **Microarrays** provide efficient genetic analysis of *invasively* collected samples (amniotic samples, chorionic villus sampling, and preimplantation embryos); array comparative genome hybridization (aCGH) has largely replaced karyotyping in this regard [47–49]. As for noninvasive prenatal diagnostics, as of this writing, almost all commercially available NIPTs utilize NGS; however, microarray analysis can also identify aneuploidy as well as other genetic changes. As an illustration, Ariosa provides the Harmony™ noninvasive prenatal test, combining targeted DNA analysis with microarray technology, as a first application of microarray technology to NIPT with cffDNA. The test offers significant advantages over traditional sequencing-based NIPT, including very good performance and shorter turnaround time [50]. Together with Ariosa's proprietary FORTE™ algorithm, the new custom microarrays improve fetal fraction quantification and significantly lower the variance of chromosome concentration, compared to DNA sequencing. Completely concordant test results were achieved with a significant reduction in turnaround time [50]. Designed microarrays also can target copy number variants (CNVs) of known pathogenicity instead of testing the entire genome.
- **Single fetal cell genome analysis** can be done by quantitative fluorescence PCR, FISH, and even karyotyping on cultured fetal cells. Advances in single-cell isolation by a high-performance cell sorter and further single-cell genomic analysis by aCGH and NGS further enabled single fetal cell NIPD [51]. However, the concentration of these cells in the maternal circulation is low, meaning the tests have low sensitivity and specificity [51].

The clinical applications of NIPT as of this writing are limited to defined conditions as presented in Table 5. *The critical underlying principle of the analysis for single-gene disorders is to investigate*

Table 5 Possible Clinical Applications of Qualitative and Quantitative NIPT

Qualitative
1. Prevention of hemolytic disease of newborn; determination of fetal RhD genotype in an RhD(−) woman
2. X-linked diseases; sex determination, as Y chromosome sequences in maternal blood indicate a male fetus
3. Autosomal dominant diseases for which the father carries a mutation
4. Autosomal recessive diseases for which a compound heterozygosity is present

Quantitative
1. Autosomal dominant diseases for which the mother carries a mutation
2. Autosomal recessive diseases for which both father and mother are carrying the same mutation
3. Aneuploidy detection

---

the presence of targeted alleles, which are not carried by the mother, and to make a dosage comparison between the targeted and the reference alleles. The dosage comparison emerges from the difficulties associated with the separation of fetal DNA from maternal DNA. In general, there are two approaches used in NIPT: qualitative and quantitative analysis. Qualitative analysis detects alleles that are absent from the maternal genome. Quantitative analysis is based on quantification of DNA molecules or chromosomes in the maternal blood. The dosage comparison of mutant and wild-type alleles is known as relative mutation dosage (RMD) and is commonly used for single-gene disorders. The dosage comparison of chromosomes, known as relative chromosome dosage (RCD), is also possible, being used where aneuploidies are suspected [52]. The procedure is based on the principle that in fetal trisomies alleles carried by the fetus are detected in higher relative amounts compared to the alleles of only maternal origin [53,54]. In RCD, the dosage of the targeted fetal chromosome is compared to a reference chromosome. The indications for both qualitative and quantitative analysis are summarized in Table 5.

1. **Fetal Rhesus D (Blood group antigen D) (RhD) genotyping** [55,56]: NIPT can be used for fetal RhD genotyping. Such testing may prevent unnecessary intrauterine blood transfusion to correct fetal anemia [57] and is routinely used in clinical practice. Furthermore, NIPT can support clinical decision-making on antenatal Rh prophylaxis.
2. **Fetal sex determination for X-linked diseases**: This procedure can be performed after 7 weeks of gestation [58–60]. The presence of Y chromosome sequences in the maternal blood is an indication of a male fetus. Early sex determination is very important for women at high risk of serious genetic disorders affecting male sex, such as hemophilia, Duchenne muscular dystrophy, and congenital adrenal hypoplasia [61]. Bustamante–Aragones et al., in a group of pregnant women carrying hemophilia, achieved 100% accuracy in early fetal sex assessment based on the detection of specific Y chromosome sequences [62]. Prenatal sex determination is currently adopted for routine clinical practice [63].
3. **Single-gene disorders**: Although there have been several studies regarding the use of NIPT for detection of single-gene disorders, as of this writing there are no elaborated guidelines with reference to such testing in routine clinical practice [46,59,61,63] (Table 6).

NIPT is available for selected single-gene disorders that are inherited in a dominant fashion from the father or *de novo* mutations such as Huntington disease, achondroplasia, myotonic dystrophy,

Table 6 The Most Common Single-Gene Disorders Studied in the Prenatal Period Using NGS-Based NIPT [61,63]

Autosomal Dominant	Autosomal Recessive	X-Linked
Huntington disease	Thalassemia	Hemophilia
Achondroplasia	Sickle cell anemia	
Myotonic dystrophy	Propionic acidemia	
	Leber congenital amaurosis	
	Cystic fibrosis	

thanatophoric dysplasia, and Apert syndrome (i.e., *FGFR2* and *FGFR3* alterations) [26,61,64–70]. In these situations, cffDNA contains deleterious alleles that are not present in the maternal genome [61].

In autosomal recessive conditions, there are two distinct possibilities: either maternal and paternal pathogenic alleles are identical (homozygous condition) or they are different (compound heterozygosity). In the heterozygous form, the lack of a paternal mutated allele in the maternal blood indicates that the risk the fetus is affected is very low and obviates the need for invasive diagnostic procedures. If both parents are carrying the same mutation, it can be difficult to reveal the genotype of the fetus and an invasive test could be offered to determine if the fetus is affected by the condition. As an illustration, NIPT for cystic fibrosis to exclude the presence of the pathogenic variant of parental origin is available as a research tool.

A quantitative approach known as RMD has been developed using digital PCR. This technique is specifically useful in homozygous conditions and is based on the overrepresentation of wild-type or mutant fetal alleles [59]. Thalassemia, sickle cell anemia, propionic acidemia, Leber congenital amaurosis, and cystic fibrosis are among the most common autosomal recessive diseases diagnosed prenatally [26,52,61,67–70]. In addition, using a sex determination approach for hemophilia, Tsui et al. screened 12 maternal plasma samples and detected fetal genotypes responsible for hemophilia with a 100% accuracy [71].

4. **Fetal aneuploidy detection**: NIPT is frequently used to detect most common aneuploidies in live births such as Down syndrome, trisomy 18, and trisomy 13 [39,72–74]. Several studies have assessed the usefulness of NIPT in aneuploidies and in a number of other conditions with chromosomal abnormalities such as mosaicism, translocation, and polyploidies [39]. The detection rates of the most commonly studied chromosomal anomalies are summarized in Table 7 [39,74–76].

Fetal aneuploidy can be detected by either PCR-based methods or, preferably, NGS. Fetal-specific DNA fragments in maternal plasma, such as *PLAC4* or *SERPINB5*, can be recognized using PCR-based techniques followed by calculation of the allelic ratio. *SERPINB5* is differentially methylated in the mother and the fetus (hypomethylated in fetal DNA and hypermethylated in maternal DNA). Maternal plasma mRNA encoded by the *PLAC4* (placenta-specific 4), which is transcribed from chromosome 21 in placental cells, is a potential marker for the noninvasive assessment of chromosome 21 dosage in the fetus and can be used to identify fetal aneuploidies. This is, however, applicable only in heterozygous conditions. In the NGS-based approach, chromosome dosage is calculated and the amount of the chromosome of interest is compared with a reference chromosome. Normally, the ratio of the reads of chromosomes in a pregnant woman carrying a euploid fetus will be 1:1; in contrast, the sample is

Table 7 Detection Rates (DR) and False Positive Rates (FPR) of the Most Common Chromosomal Anomalies by Using Massively Parallel Sequencing [39,74–76]

Aneuploidies	Incidence	DR (%)	FPR (%)
+,21	1/700	99.3–100	0.03–0.16
+,18	1/3000	97.4–100	0.03–0.28
+,13	1/10000	78.9–91.7	0.41–0.97
Turner syndrome	1/2500	83	0.2
Klinefelter syndrome	1/1000	85	–
47,XXX	1/1000	83	–
47,XYY	1/1000	75	–
Sex chromosomal abnormality (SCA)	1/400	96.2	0.3

identified as aneuploid when the calculated ratio falls beyond an established threshold. Unbalanced translocations, partial duplication of a chromosome, and mosaic conditions are also detectable using NGS [77]. Further efforts are focused on expanding the diagnostic scope of NIPT to microdeletions.

NGS IN THE DETERMINATION OF GENOMIC DISORDERS BY USING cffDNA IN MATERNAL PLASMA

NGS is the term to describe a series of newer sequencing technologies [78], including massively parallel sequencing, deep sequencing, and second generation sequencing [79–81]. This technology is currently the basis of commercially available sequencing platforms such as those described earlier in this book (Chapter 1).

Nowadays, there are two primary NGS-based approaches for cffDNA testing: massively parallel shotgun sequencing, which sequences DNA fragments from the whole genome, and targeted sequencing, selectively testing targeted genomic regions [10,21,82,83]. Both techniques are based on counting sequenced DNA fragments obtained from maternal blood [21,83]. In both approaches GC base content of the chromosome directly affects the accuracy of the test, and bioinformatic algorithms have been introduced with the purpose of eliminating the GC bias. The targeted approach, therefore, holds advantages over WGS by means of reducing the workload for the assessment of the procedure and selection of the regions reflecting the GC content of the chromosome [72]. As first reports, in 2008, two groups used massively parallel genomic sequencing for the noninvasive prenatal screening of fetal trisomy 21. Encouraging results of these studies gave rise to a new noninvasive approach for identifying fetal trisomy 21 as well as other fetal chromosomal aneuploidies [77,84]. The minimum required fetal fraction of total DNA to detect fetal aneuploidy is 4% [39]. As an example, there should be a slight increase in the amount of DNA from chromosome 21 in plasma of a pregnant woman carrying a fetus with trisomy 21. If the fetal DNA fraction is 20%, the relative dosage of chromosome 21 DNA fragments will be $(0.8 \times 2) + (0.2 \times 3) = 2.2$. This value is then compared with the DNA count of a pregnant woman carrying a euploid fetus $(0.8 \times 2) + (0.2 \times 2) = 2$. The comparison could be done in two ways: either the counts of a targeted chromosome are normalized against another chromosome expected to be disomic within the same test run or the counts of the targeted chromosome are compared with the results of other euploid cases. The results are given as positive or negative according to a statistical value known as the Z-score or modified Z-score [83]. The Z-score determines how many times the standard deviation of the proportion of reads is above or below the mean; in other words, how many times the percentage of targeted chromosome sequences differs from the mean levels in a reference sample. Consistently, if a Z-score of >2.5 is reported, fetal trisomy is inferred and is reported as high risk for trisomy for that chromosome. On the other hand, a modified Z-score has been recommended in these calculations, as presented below;

$$Z(Cn)S = (\%(Cn)S - Median(Cn)R)/MAD(Cn)R$$

Cn is the targeted chromosome, R is the reference value, S is the sample, and MAD is the median absolute deviation. In pregnancies without fetal trisomy the Z-score is expected to be lower than 3; and any score exceeding 3 is an indication that the targeted chromosomes are present in statistically abnormal quantities, that is, the presence of trisomy 21. Such approach is very efficient in the detection of trisomy 21; however, owing to technical reasons and individual sequencing variations, it does not detect other aneuploidies with the same level of accuracy [33]. Another counting algorithm, therefore, has been developed; and by using normalized chromosome values, the difficulties associated with using the Z-score algorithm have been corrected [33,85,86].

The procedure can be summarized as follows [21,87]:

1. Isolation of DNA molecules (mixed maternal and fetal DNA) from the maternal plasma; no further fragmentation is necessary because the DNA is already fragmented in the maternal plasma [6,46,77,83].
2. Library preparation with special sequence tags.
3. Clonal amplification.
4. Random parallel sequencing on a selected platform in a single run.
5. Comparison of the sequenced data with a reference genome.
6. Categorization of fragments according to chromosome. Tens of thousands of sequence reads per chromosome are produced, with sequence tags mapped to each chromosome and then counted. High numbers, such as the 60,000–70,000 reads recorded for chromosome 21, are not uncommon.
7. The number of reads mapping to the chromosome of interest is compared with the number of reads mapping to the matched normal reference chromosome. In the case of trisomy 21, the proportion of the DNA fragments of chromosome 21 will be increased compared to the reference chromosomes [88]. The accuracy of such NGS-based approach has been validated in several studies and has been found to be very high [74,85,86,89,90].

The procedure applied by Illumina is summarized in Figure 1.

SINGLE-NUCLEOTIDE POLYMORPHISM SEQUENCING OF CELL-FREE FETAL DNA

Single-nucleotide polymorphism (SNP) sequencing enables for a more accurate cffDNA analysis, yielding much more information than methods considering only the number of reads to identify numeric abnormalities of fetal chromosomes. SNP-based sequencing allows for a more qualitative

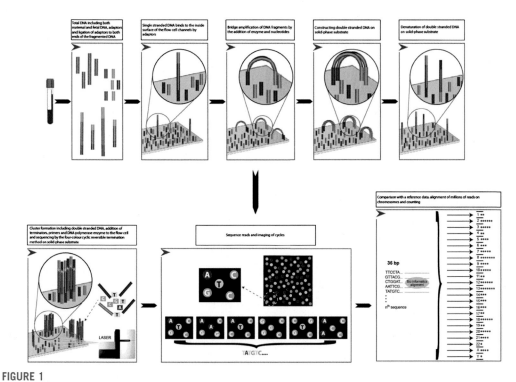

FIGURE 1

Summary of the chromosome counting by NGS-based NIPT.

approach differentiating between maternal and fetal input. Such method may also reveal other abnormalities such as uniparental disomy. Furthermore, this method is able to detect triploidy [21]. SNP sequencing can be applied as an allele ratio (see above) or specifically target amplification of polymorphic loci, followed by NGS and bioinformatics analysis to identify fetal chromosomal copy number and to calculate a risk score for each chromosome. In such approach, allelic information from both parents is included in the analysis, taking into consideration different genetic inheritance patterns. In contrast to the allele ratio, a reference chromosome is not used, as only the relative amount of polymorphic allele is analyzed [21]. The overall sensitivity for trisomy 21, trisomy 18, and trisomy 13 detection is >99%. The commercially available NIPT based on this methodology reports copy numbers for all five chromosomes implicated in the most common abnormalities in newborns (13, 18, 21, X, and Y) [21]. As an example, the Panorama test from Natera specifically sequences SNPs and thus can differentiate between maternal and fetal genotypes. Panorama uses its own algorithm called Next Generation Aneuploidy Test Using SNPs (NATUS) [91]. The test utilizes the mother's white blood cells to isolate and identify her DNA and further uses this information to subtract out the maternal genotype to increase the accuracy of the test even when the fetal fraction is very low [91].

LIMITATIONS AND CHALLENGES OF NGS-BASED NONINVASIVE PRENATAL TESTING

Despite potential benefits, there are a number of constraints and limitations of NGS-based NIPT [53,72]. The major shortcomings of currently available tests are listed below:

- The accuracy of NIPT methods depends on the cffDNA fraction. Increased fetal DNA fraction in total DNA of maternal plasma, such as multiple pregnancies or vanishing twins, yields a higher detection rate in the counting method for aneuploidy studies [92,93]. Furthermore, the relative proportion of cffDNA is decreased in obese women, as they have more of their own free DNA in their blood [90,94,95]. Furthermore, a decrease in the relative amount of fetal DNA is possibly caused by a dilution of cffDNA owing to a larger total plasma volume in women with higher body mass index (BMI) [96]. Such hypothesis was confirmed by Vora et al., who showed that total cffDNA levels correlate with maternal BMI [94].

- The NIPT presumes a normal maternal/placental karyotype. Maternal/placental mosaicism will result in various maternal or trophoblastic contributions to the circulating DNA and can decrease the diagnostic accuracy of NIPT. The potential for discordance among maternal, fetal, and placental chromosomes requires additional studies assessing the correlation between NIPT and actual maternal and fetal karyotype. Placental trisomy 21 accompanied by normal fetal karyotype has been previously reported [39,97].

- False positive and false negative test results may occur. Large-scale clinical validation studies of NIPT as of this writing are too limited to draw far-reaching conclusions. As an example, Zhang et al. analyzed NIPT results from 112,669 samples. The overall sensitivity of the test was 99.17% for T21, 98.24% for T18, and 100% for T13, and the specificity was 99.95% for T21, 99.95% for T18, and 99.96% for T13. Aneuploidy was confirmed in 720 of 781 T21-positive cases, 167 of 218 T18-positive cases, and 22 of 67 T13-positive cases. There were 9 false negatives identified, including 6 T21 and 3 T18 cases. There was no significant difference in test performance between high-risk and low-risk subjects (sensitivity 99.21% vs 98.97%, $p=0.82$; specificity 99.95% vs 99.95%, $p=0.98$). The major factors contributing to NIPT false positive and false negative results were maternal CNV and fetal/placental mosaicism, but not fetal fraction [98].

- GC content of the chromosome also affects the accuracy of the test, as the GC reach sequences are often underrepresented in NGS libraries. Several studies have proposed alternative algorithms to solve this problem [33,74,85,86,99]. Bianchi et al., for example, used normalized chromosome values, and the normalization procedure was performed referencing unaffected samples [86]. In this algorithm a value of more than 4 classified the chromosome as affected, less than 2.5 as unaffected, and between 2.5 and 4.0 as "unclassified" [86].

- Certain birth defects such as neural tube defects cannot be detected by NIPT [72].

- There are significant novel ethical issues generated by NIPT. The procedure creates the possibility of discovering maternal chromosomal variations, not previously known, with a potential impact on the mother's health. Furthermore, NIPT can raise a suspicion of cancer in the mother [100]. There are also the obvious ethical issues with regard to adult-onset conditions.

- The cost of NGS-based NIPT is high, and such testing cannot be readily afforded by social security systems.

- Complex characteristics of technique and long test turnaround times all limit its use. There is also a need for specific expertise, especially in bioinformatics [58].

CLINICAL IMPLEMENTATION OF NONINVASIVE PRENATAL TESTING FOR ANEUPLOIDIES

Although NIPT can be performed in all pregnancies without specific indications, the highest confidence in this test is in screening for the most common fetal chromosome anomalies, especially trisomy 21 (Down syndrome) [4]. NIPT should be regarded as a screening test for aneuploidies, but cannot be considered as diagnostic, and abnormal results should always be confirmed by a definitive test such as trophoblast biopsy or amniocentesis. On the other hand, it has been pointed out that NIPT would reduce the need for invasive prenatal tests by 99% [72,87,96]. NIPT could, however, serve either as a primary or an advanced screening test, in conjunction with conventional existing screening procedures. This means that it should be available to pregnant women, especially to groups at high risk for fetal abnormalities, as a second tier after conventional screening tests [10,21].

The California Technology Assessment Forum (CTAF) is a community platform providing information regarding the safety and effectiveness of medical technologies, treatments, and models of care. According to the CTAF, screening for trisomy 18 and trisomy 21 using NGS-based NIPT meets the criteria of safety and efficacy and improvement in health outcomes and is as beneficial as any of the established alternatives for noninvasive prenatal testing [10].

The ACOG recommends that pregnant women are offered screening for chromosomal abnormalities, regardless of maternal age; however, more research is needed to determine how these tests perform in low-risk populations. As of this writing, universal screening with NIPT seems not to be feasible. As evidenced, NIPT is especially dedicated to pregnant women who fall into the category of "high risk" for fetal abnormalities (Table 8).

The ideal time for performing NGS-based NIPT is between 10 and 15 weeks of gestation [7]. There are currently a number of companies providing NIPT for the most common aneuploidies worldwide; these include the Sequenom Center for Molecular Medicine, Verinata Health, Ariosa Diagnostics, and Natera, as well as several clinical diagnostics laboratories; and they are all using sequencing-based techniques [21]. The technical approach of Natera and Ariosa Diagnostics is targeted sequencing and of the Sequenom Center for Molecular Medicine and Verinata Health is massively parallel sequencing.

Table 8 Clinical Indications for NIPT [9]
1. Maternal age of 35 years or older at delivery
2. Fetal ultrasonographic findings consistent with an increased risk of aneuploidy
3. History of a prior pregnancy with a trisomy
4. Positive biochemical test result for aneuploidy, including first trimester, sequential, or integrated screen or a quadruple screen
5. Parental balanced Robertsonian translocation with increased risk of fetal trisomy 13 or trisomy 21

CONCLUSION

NIPT for selected fetal trisomies became clinically available in 2011. NGS is the newest and probably the most useful technique applied in prenatal screening approaches and today NIPT is by far the most prevalent clinical application of NGS. Nowadays, NGS-based NIPT is routinely used for a very few diseases; however, in the near future, NIPT is expected to detect both chromosomal aneuploidies and single-gene disorders, thus facilitating broader adoption in the clinic. However, according to ACOG recommendations "whilst NIPT via NGS can provide very statistically accurate information regarding the health of a pregnancy, all positive test results need to be confirmed using invasive prenatal diagnostic methods." Several factors may need to be considered if NIPT is incorporated into the current clinical practice, including ethical and social issues, appropriate reproductive genetic counseling and documentation of informed consent is warranted. Furthermore, genetic information provided by NIPT reassures future parents, but may also cause anxiety or even unwanted fetal selection. Parents need to be counseled about the possibility of false negative and false positive results, and information about the risk should be provided. Genetic counseling and discussion about further invasive testing is recommended to follow up a screen-positive result, and recent information about the conditions being tested should be provided to parents. In addition, systems for evaluating and sharing experiences with NIPT will be helpful to elaborate clinical recommendations. Finally, all these efforts must include eugenics concerns and regulatory and legislative issues.

REFERENCES

[1] Wieacker P, Steinhard J. The prenatal diagnosis of genetic diseases. Dtsch Arztebl Int 2010;107:857–62.
[2] http://www.who.int/mediacentre/factsheets/fs370/en.
[3] Binns V, Hsu N. Prenatal diagnosis, encyclopedia of life sciences. Macmillan Publishers Ltd., Nature Publishing Group; 2002. www.els.net.
[4] Papageorgiou EA, Patsalis PC. Non-invasive prenatal diagnosis of aneuploidies: new technologies and clinical applications. Genome Med 2012;4:46.
[5] Driscoll DA, Gross S. Clinical practice. Prenatal screening for aneuploidy. N Engl J Med 2009;360:2556–62.
[6] Dan S, Chen F, Choy KW, Jiang F, Lin J, Xuan Z, et al. Prenatal detection of aneuploidy and imbalanced chromosomal arrangements by massively parallel sequencing. PLoS One 2012;7:e27835.
[7] Go AT, van Vugt JM, Oudejans CB. Non-invasive aneuploidy detection using free fetal DNA and RNA in maternal plasma: recent progress and future possibilities. Proc Natl Acad Sci USA 2014;111:8583–8.
[8] http://www.who.int/genomics/public/geneticdiseases/en/index1.html.
[9] https://www.oxhp.com/secure/policy/noninvasive prenatal diagnosis fetal aneuploidy.pdf (ACOG Committee).
[10] Walsh JM, Goldberg JD. Fetal aneuploidy detection by maternal plasma DNA sequencing: a technology assessment. Prenat Diagn 2013;33:514–20.
[11] Lo YM. Non-invasive prenatal diagnosis by massively parallel sequencing of maternal plasma DNA. Open Biol 2012;2:120086.
[12] Nicolaides KH. Screening for fetal aneuploidies at 11 to 13 weeks. Prenat Diagn 2011;31:7–15.
[13] Malone FD, Canick JA. First-trimester or second-trimester screening, or both, for Down's syndrome. N Engl J Med 2005;353:2001–11.
[14] Wald NJ. Prenatal screening for open neural tube defects and Down syndrome: three decades of progress. Prenat Diagn 2010;30:619–21.

[15] Canick J. Prenatal screening for trisomy 21: recent advances and guidelines. Clin Chem Lab Med 2012;50:1003–8.

[16] Chachkin CJ. What potent blood: non-invasive prenatal genetic diagnosis and the transformation of modern prenatal care. Am J Law Med 2007;33:9–53.

[17] Caughey AB, Hopkins LM, Norton ME. Chorionic villus sampling compared with amniocentesis and the difference in the rate of pregnancy loss. Obstet Gynecol 2006;108(3 Pt 1):612–6.

[18] Odibo AO, Dicke JM, Gray DL, Oberle B, Stamilio DM, Macones GA, et al. Evaluating the rate and risk factors for fetal loss after chorionic villus sampling. Obstet Gynecol 2008;112:813–9.

[19] Steele MW, Breg WR. Chromosome analysis of human amniotic-fluid cells. Lancet 1966;1(7434):383–5.

[20] http://www.dddmag.com/articles/2013/05/advances-non-invasive-prenatal-diagnostic-testing.

[21] Norwitz ER, Levy B. Noninvasive prenatal testing: the future is now. Rev Obstet Gynecol 2013;6:48–62.

[22] Walknowska J, Conte FA, Grumbach MM. Practical and theoretical implications of fetal-maternal lympho-cyte transfer. Lancet 1969;1(7606):1119–22.

[23] Lo YM, Patel P, Wainscoat JS, Sampietro M, Gillmer MD, Fleming KA. Prenatal sex determination by DNA amplification from maternal peripheral blood. Lancet 1989;2(8676):1363–5.

[24] Bianchi DW, Flint AF, Pizzimenti MF, Knoll JH, Latt SA. Isolation of fetal DNA from nucleated erythro-cytes in maternal blood. Proc Natl Acad Sci USA 1990;87:3279–83.

[25] Lo YM, Lo ES, Watson N, Noakes L, Sargent IL, Thilaganathan B, et al. Two-way cell traffic between mother and fetus: biologic and clinical implications. Blood 1996;88:4390–5.

[26] Tounta G, Kolialexi A, Papantoniou N, Tsangaris GT, Kanavakis E, Mavrou A. Non-invasive prenatal diag-nosis using cell-free fetal nucleic acids in maternal plasma: progress overview beyond predictive and per-sonalized diagnosis. EPMA J 2011;2:163–71.

[27] Bianchi DW, Williams JM. PCR quantitation of fetal cells in maternal blood in normal and aneuploid preg-nancies. Am J Hum Genet 1997;61:822–9.

[28] Babochkina T, Mergenthaler S. Numerous erythroblasts in maternal blood are impervious to fluorescent in situ hybridization analysis, a feature related to a dense compact nucleus with apoptotic character. Haemato-logica 2005;90:740–5.

[29] Bianchi DW, Zickwolf GK, Weil GJ, Sylvester S, DeMaria MA. Male fetal progenitor cells persist in mater-nal blood for as long as 27 years postpartum. Proc Natl Acad Sci USA 1996;93:705–8.

[30] Lo YM, Corbetta N, Chamberlain PF, Rai V, Sargent IL, Redman CW, et al. Presence of fetal DNA in maternal plasma and serum. Lancet 1997;350(9076):485–7.

[31] Alberry M, Maddocks D, Jones M, Abdel Hadi M, Abdel-Fattah S, Avent N, et al. Free fetal DNA in maternal plasma in anembryonic pregnancies: confirmation that the origin is the trophoblast. Prenat Diagn 2007;27:415–8.

[32] Bianchi DW. Fetal DNA in maternal plasma: the plot thickens and the placental barrier thins. Am J Hum Genet 1998;62:763–4.

[33] Swanson A, Sehnert AJ, Bhatt S. Non-invasive prenatal testing: technologies, clinical assays and implemen-tation strategies for women's healthcare practitioners. Curr Genet Med Rep 2013;1:113–21.

[34] Lo YM, Tein MS, Lau TK, Haines CJ, Leung TN, Poon PM, et al. Quantitative analysis of fetal DNA in maternal plasma and serum: implications for noninvasive prenatal diagnosis. Am J Hum Genet 1998;62:768–75.

[35] Guibert J, Benachi A, Grebille AG, Ernault P, Zorn JR, Costa JM. Kinetics of SRY gene appearance in maternal serum: detection by real time PCR in early pregnancy after assisted reproductive technique. Hum Reprod 2003;18:1733–6.

[36] Chan KC, Zhang J, Hui AB, Wong N, Lau TK, Leung TN, et al. Size distributions of maternal and fetal DNA in maternal plasma. Clin Chem 2004;50:88–92.

[37] Fan HC, Blumenfeld YJ, Chitkara U, Hudgins L, Quake SR. Analysis of the size distributions of fetal and maternal cell-free DNA by paired-end sequencing. Clin Chem 2010;56:1279–86.

[38] Li Y, Zimmermann B, Rusterholz C, Kang A, Holzgreve W, Hahn S. Size separation of circulatory DNA in maternal plasma permits ready detection of fetal DNA polymorphisms. Clin Chem 2004;50:1002–11.

[39] Benn P, Cuckle H, Pergament E. Non-invasive prenatal testing for aneuploidy: current status and future prospects. Ultrasound Obstet Gynecol 2013;42:15–33.

[40] Lam KW, Jiang P, Liao GJ, Chan KC, Leung TY, Chiu RW, et al. Noninvasive prenatal diagnosis of monogenic diseases by targeted massively parallel sequencing of maternal plasma: application to β- thalassemia. Clin Chem 2012;58:1467–75.

[41] Greene MF, Phimister EG. Screening for trisomies in circulating DNA. N Engl J Med 2014;370:874–5.

[42] Lun FM, Chiu RW, Chan KC, Leung TY, Lau TK, Lo YM. Microfluidics digital PCR reveals a higher than expected fraction of fetal DNA in maternal plasma. Clin Chem 2008;54:1664–72.

[43] Gerovassili A, Nicolaides KH, Thein SL, Rees DC. Cell-free DNA levels in pregnancies at risk of sickle-cell disease and significant ethnic variation. Br J Haematol 2006;135:738–41.

[44] Lo YM, Zhang J, Leung TN, Lau TK, Chang AM, Hjelm NM. Rapid clearance of fetal DNA from maternal plasma. Am J Hum Genet 1999;64:218–24.

[45] Wright CF, Burton H. The use of cell-free fetal nucleic acids in maternal blood for non-invasive prenatal diagnosis. Hum Reprod Update 2009;15:139–51.

[46] Kitzman JO, Snyder MW, Ventura M, Lewis AP, Qiu R, Simmons LE, et al. Non-invasive whole genome sequencing of a human fetus. Sci Transl Med 2012;4(137):137ra76. http://dx.doi.org/10.1126/scitranslmed.3004323.

[47] Wapner RJ, Martin CL, Levy B, Ballif BC, Eng CM, Zachary JM, et al. Chromosomal microarray versus karyotyping for prenatal diagnosis. N Engl J Med 2012;367:2175–84.

[48] Colls P, Escudero T, Fischer J, Cekleniak NA, Ben-Ozer S, Meyer B, et al. Validation of array comparative genome hybridization for diagnosis of translocations in preimplantation human embryos. Reprod Biomed Online 2012;24:621–9.

[49] Handyside AH, Montag M, Magli MC, Repping S, Harper J, Schmutzler A, et al. Multiple meiotic errors caused by predivision of chromatids in women of advanced maternal age undergoing in vitro fertilisation. Eur J Hum Genet 2012;20:742–7.

[50] Juneau K, Bogard PE, Huang S, Mohseni M, Wang ET, Ryvkin P, et al. Microarray-based cell-free DNA analysis improves noninvasive prenatal testing. Fetal Diagn Ther 2014;36:282–6.

[51] Chan WK, Kwok YK, Choy KW, Leung TY, Wang CC. Single fetal cells for non-invasive prenatal genetic diagnosis: old myths new prospective. Med J Obstet Gynecol 2013;1:1004.

[52] Lun FMF, Tsui NBY, Chan KCA, Leung TY, Lau TK, Charoenkwan P, et al. Noninvasive prenatal diagnosis of monogenic diseases by digital size selection and relative mutation dosage on DNA in maternal plasma. Proc Natl Acad Sci USA 2008;105:19920–5.

[53] Chan KC. Clinical applications of the latest molecular diagnostics in noninvasive prenatal diagnosis. Top Curr Chem 2014;336:47–65.

[54] Chiu RW, Lo YM. Non-invasive prenatal diagnosis by fetal nucleic acid analysis in maternal plasma: the coming of age. Semin Fetal Neonatal Med 2011;16:88–93.

[55] Lo YMD, Hjelm NM, Fidler C, Sargent IL, Murphy MF, Chamberlain PF, et al. Prenatal diagnosis of fetal RhD status by molecular analysis of maternal plasma. N Engl J Med 1998;339:1734–8.

[56] Finning K, Martin P, Daniels G. The use of maternal plasma for prenatal RhD blood group genotyping. Methods Mol Biol 2009;496:143–57.

[57] Daniels G, Finning K, Martin P, Massey E. Noninvasive prenatal diagnosis of fetal blood group phenotypes: current practice and future prospects. Prenat Diagn 2009;29:101–7.

[58] Raymond FL, Whittaker J, Jenkins L, Lench N, Chitty LS. Molecular prenatal diagnosis: the impact of modern technologies. Prenat Diagn 2010;30:674–81.

[59] Hill M, Compton C, Karunaratna M, Lewis C, Chitty L. Client views and attitudes to non-invasive prenatal diagnosis for sickle cell disease, thalassaemia and cystic fibrosis. J Genet Couns 2014. May 3, 2014.

[60] Hill M, Finning K, Martin P, Hogg J, Meaney C, Norbury G, et al. Non-invasive prenatal determination of fetal sex: translating research into clinical practice. Clin Genet 2011;80:68–75.

[61] Lench N, Barrett A, Fielding S, McKay F, Hill M, Jenkins L, et al. The clinical implementation of non-invasive prenatal diagnosis for single-gene disorders: challenges and progress made. Prenat Diagn 2013;33:555–62.

[62] Bustamante-Aragones A, Rodriguez de Alba M, Gonzalez-Gonzalez C, Trujillo-Tiebas MJ, Diego-Alvarez D, Vallespin E, et al. Foetal sex determination in maternal blood from the seventh week of gestation and its role in diagnosing haemophilia in the foetuses of female carriers. Haemophilia 2008;14:593–8.

[63] Bustamante-Aragonés A, Rodríguez de Alba M, Perlado S, Trujillo-Tiebas MJ, Arranz JP, Díaz-Recasens J, et al. Non-invasive prenatal diagnosis of single-gene disorders from maternal blood. Gene 2012;504:144–9.

[64] Amicucci P, Gennarelli M, Novelli G, Dallapiccola B. Prenatal diagnosis of myotonic dystrophy using fetal DNA obtained from maternal plasma. Clin Chem 2000;46:301–2.

[65] Li Y, Page-Christiaens GC, Gille JJ, Holzgreve W, Hahn S. Non-invasive prenatal detection of achondroplasia in size-fractionated cell-free DNA by MALDI-TOF MS assay. Prenat Diagn 2007;27:11–7.

[66] Li Y, Holzgreve W, Page-Christiaens GC, Gille JJ, Hahn S. Improved prenatal detection of a fetal point mutation for achondroplasia by the use of size-fractionated circulatory DNA in maternal plasma–case report. Prenat Diagn 2004;24:896–8.

[67] Chiu RWK, Lau TK, Leung TN, Chow KC, Chui DH, Lo YMD. Prenatal exclusion of β thalassaemia major by examination of maternal plasma. Lancet 2002;360:998–1000.

[68] Bustamante-Aragones A, Vallespin E, Rodriguez de Alba M, Trujillo-Tiebas MJ, Gonzalez-Gonzalez C, Diego-Alvarez D, et al. Early noninvasive prenatal detection of a fetal CRB1 mutation causing Leber congenital amaurosis. Mol Vis 2008;14:1388–94.

[69] Bustamante-Aragones A, Pérez-Cerdá C, Pérez B, de Alba MR, Ugarte M, Ramos C. Prenatal diagnosis in maternal plasma of a fetal mutation causing propionic acidemia. Mol Genet Metab 2008;95:101–3.

[70] González-González MC, García-Hoyos M, Trujillo MJ, Rodríguez de Alba M, Lorda-Sánchez I, Díaz-Recasens J, et al. Prenatal detection of a cystic fibrosis mutation in fetal DNA from maternal plasma. Prenat Diagn 2002;22:946–8.

[71] Tsui NB, Kadir RA, Chan KC, Chi C, Mellars G, Tuddenham EG, et al. Noninvasive prenatal diagnosis of hemophilia by microfluidics digital PCR analysis of maternal plasma DNA. Blood 2011;117:3684–91.

[72] Nepomnyashchaya YN, Artemov AV, Roumiantsev SA, Roumyantsev AG, Zhavoronkov A. Non-invasive prenatal diagnostics of aneuploidy using next-generation DNA sequencing technologies, and clinical considerations. Clin Chem Lab Med 2013;51:1141–54.

[73] Chen CP, Chern SR, Wang W. Fetal DNA in maternal plasma: the prenatal detection of a paternally inherited fetal aneuploidy. Prenat Diagn 2000;20:355–7.

[74] Palomaki GE, Deciu C, Kloza EM, Lambert-Messerlian GM, Haddow JE, Neveux LM, et al. DNA sequencing of maternal plasma reliably identifies trisomy 18 and trisomy 13 as well as Down syndrome: an international collaborative study. Genet Med 2012;14:296–305.

[75] Mazloom AR, Džakula Ž, Oeth P, Wang H, Jensen T, Tynan J, et al. Noninvasive prenatal detection of sex chromosomal aneuploidies by sequencing circulating cell-free DNA from maternal plasma. Prenat Diagn 2013;33:591–7.

[76] Dan S, Wang W, Ren J, Li Y, Hu H, Xu Z, et al. Clinical application of massively parallel sequencing-based prenatal noninvasive fetal trisomy test for trisomies 21 and 18 in 11 105 pregnancies with mixed risk factors. Prenat Diagn 2012;32:1225–32.

[77] Fan HC, Blumenfeld YJ, Chitkara U, Hudgins L, Quake SR. Noninvasive diagnosis of fetal aneuploidy by shotgun sequencing DNA from maternal blood. Proc Natl Acad Sci USA 2008;105:16266–71.

[78] Metzker ML. Sequencing technologies-the next generation. Nat Rev Genet 2010;11:31–46.

[79] Gut IG. New sequencing technologies. Clin Transl Oncol 2013;5:879–81.

[80] Behjati S, Tarpey PS. What is next generation sequencing? Arch Dis Child Educ Pract Ed 2013;98:236–8.

[81] Wang W, Wei Z, Lam TW, Wang J. Next generation sequencing has lower sequence coverage and poorer SNP-detection capability in the regulatory regions. Sci Rep 2011;1:55.

[82] http://www.rcog.org.uk/files/rcog-corp/SIP_15_04032014.pdf.

[83] Stumm M, Entezami M, Trunk N, Beck M, Löcherbach J, Wegner RD, et al. Noninvasive prenatal detection of chromosomal aneuploidies using different next generation sequencing strategies and algorithms. Prenat Diagn 2012;32:569–77.

[84] Chiu RW, Chan KC, Gao Y, Lau VY, Zheng W, Leung TY, et al. Noninvasive prenatal diagnosis of fetal chromosomal aneuploidy by massively parallel genomic sequencing of DNA in maternal plasma. Proc Natl Acad Sci USA 2008;105:20458–63.

[85] Sehnert AJ, Rhees B, Comstock D, de Feo E, Heilek G, Burke J, et al. Optimal detection of fetal chromosomal abnormalities by massively parallel DNA sequencing of cell-free fetal DNA from maternal blood. Clin Chem 2011;57:1042–9.

[86] Bianchi DW, Platt LD, Goldberg JD, Abuhamad AZ, Sehnert AJ, Rava RP, MatErnal BLood IS Source to Accurately diagnose fetal aneuploidy (MELISSA) Study Group. Genome-wide fetal aneuploidy detection by maternal plasma DNA sequencing. Obstet Gynecol 2012;119:890–901.

[87] Hahn S, Lapaire O, Tercanli S, Kolla V, Hösli I. Determination of fetal chromosome aberrations from fetal DNA in maternal blood: has the challenge finally been met? Expert Rev Mol Med 2011;13:e16.

[88] Hyett J. Non-invasive prenatal testing for Down syndrome. Aust Prescriber 2014;37:51–5.

[89] Chiu RW, Akolekar R, Zheng YW, Leung TY, Sun H, Chan KC, et al. Non-invasive prenatal assessment of trisomy 21 by multiplexed maternal plasma DNA sequencing: large scale validity study. BMJ 2011;342:c7401.

[90] Palomaki GE, Kloza EM, Lambert-Messerlian GM, Haddow JE, Neveux LM, Ehrich M, et al. DNA sequencing of maternal plasma to detect Down syndrome: an international clinical validation study. Genet Med 2011;13:913–20.

[91] www.panoramatest.com.

[92] Canick JA, Palomaki GE, Kloza EM, Lambert-Messerlian GM, Haddow JE. The impact of maternal plasma DNA fetal fraction on next generation sequencing tests for common fetal aneuploidies. Prenat Diagn 2013;33:667–74.

[93] Grömminger S, Yagmur E, Erkan S, Sándor Nagy S, Schöck U, Bonnet J, et al. Fetal aneuploidy detection by cell-free DNA sequencing for multiple pregnancies and quality issues with vanishing twins. J Clin Med 2014;2014(3):679–92.

[94] Vora NL, Johnson KL, Basu S, Catalano PM, Hauguel-De Mouzon S, Bianchi DW. A multifactorial relationship exists between total circulating cell-free DNA levels and maternal BMI. Prenat Diagn 2012;32:912–4.

[95] Wataganara T, PetCorbettaer I, Messerlian GM, Borgatta L, Bianchi DW. Inverse correlation between maternal weight and second trimester circulating cell-free fetal DNA levels. Obstet Gynecol 2004;104:545–50.

[96] Chiu RW, Lo YM. Noninvasive prenatal diagnosis empowered by high-throughput sequencing. Prenat Diagn 2012;32:401–6.

[97] Babkina N, Graham Jr JM. New genetic testing in prenatal diagnosis. Semin Fetal Neonatal Med 2014;19:214–9.

[98] Zhang H, Gao Y, Jiang F, Fu M, Yuan Y, Guo Y, et al. Noninvasive prenatal testing for trisomy 21, 18 and 13 — clinical experience from 146,958 pregnancies. Ultrasound Obstet Gynecol 2015. http://dx.doi.org/10.1002/uog.14792.

[99] Fan HC, Quake SR. Sensitivity of noninvasive prenatal detection of fetal aneuploidy from maternal plasma using shotgun sequencing is limited only by counting statistics. PLoS One 2010;5:e10439.

[100] Osborne CM, Hardisty E, Devers P, Kaiser-Rogers K, Hayden MA, et al. Discordant noninvasive prenatal testing results in a patient subsequently diagnosed with metastatic disease. Prenat Diagn 2013;33:609–11.

CLINICAL APPLICATIONS FOR NEXT GENERATION SEQUENCING IN CARDIOLOGY

10

Joanna Ponińska[1], Rafał Płoski[2], Zofia T. Bilińska[3]

[1]Laboratory of Molecular Biology, Institute of Cardiology, Warsaw, Alpejska, Poland; [2]Department of Medical Genetics, Centre of Biostructure, Medical University of Warsaw, Warsaw, Poland; [3]Unit for Screening Studies in Inherited Cardiovascular Diseases, Institute of Cardiology, Alpejska, Warsaw, Poland

CHAPTER OUTLINE

INTRODUCTION

The management of patients with certain cardiovascular diseases requires genetic counseling regarding their highly hereditary character and the necessity to evaluate risk for asymptomatic family members. Diseases that belong to this group include cardiomyopathies, primary arrhythmias, thoracic aortic aneurysms and dissections (TAAD), familial hypercholesterolemia, and congenital heart defects (CHD). The results of genetic testing may also influence the implementation of lifestyle adaptations, protective measures, or therapy. A determination of hereditary background of the disease usually implies earlier onset and worse prognosis.

An important aspect of hereditary cardiac conditions is that their boundaries are often blurred and the phenotypes overlap, coexisting within the same individual or occurring in one family as different manifestations caused by a common genetic defect. Arrhythmia can develop secondary to cardiomyopathy and vice versa. In some systemic disorders cardiomyopathies coincide with congenital malformations, and TAAD may accompany bicuspid aortic valve. However, cardiovascular diseases are greatly heterogeneous, and there are over 100 genes directly involved in the pathogenesis of these conditions, a lot of them underlying different clinical entities. For instance, sarcomeric genes are involved in several kinds of cardiomyopathy as well as CHD, and mutations in the *SCN5A* gene commonly known to cause arrhythmic phenotypes have been reported to result in dilated cardiomyopathy. Apart from strictly monogenic dominant inheritance, a substantial fraction of cases are caused by two or more genetic defects located in the same or different genes, which is often associated with a more severe phenotype. In addition, common polymorphisms can modify symptoms of a monogenic disorder, resulting in poorer prognosis. Taking into account the above-mentioned genetic complexity, molecular diagnostics requires a comprehensive approach with parallel analysis of a large number of genes.

Although molecular genetic testing used to be predominantly a research tool, it has now entered into clinical diagnostics because of its potential to provide more individualized and informative counseling in families [1].

While new causative mutations and disease-associated genes are being discovered, some of them still turn out to be false positives, which need to be eliminated from the literature and databases [2]. Laboratories offering fee-for-service clinical genetic testing are usually up to date with novel scientific findings. Access to mutation analyses for most known genes related to inherited cardiac diseases is provided by the Orphanet, the European database for rare diseases and orphan drugs (www.orpha.net), and GeneTests (www.genetests.org) Web sites [1].

Genetic testing for inherited cardiac disorders is best viewed as a family test rather than a test of an individual since results are most accurately interpreted after integrating genetic and medical test results from multiple family members. The general approach to identifying genetic causes in individuals with a familial disorder includes application of multigene panel testing for the index patient. Identification of likely or uncertain pathogenic variants followed by positive verification by segregation analysis means that the confirmed presence of a mutation in a family member qualifies for further clinical evaluations at recommended intervals and/or implementation of therapeutic and lifestyle modifications. Absence of mutation leads to dismissal from further routine evaluations. Despite the continuous increase in knowledge regarding the causes of inherited cardiac disorders, the yield of genetic testing depending on the disease varies between 20 and 75%, and it is important to keep in mind that a negative genetic test does not exclude any assumed diagnosis and does not dismiss family members from routine evaluations. In the case of a negative genetic test for a patient with a clear clinical phenotype, experts encourage establishing cooperation with a scientific laboratory looking out for novel disease-causing variants [3], but it is also possible to consider commercially available clinical exome or genome sequencing.

An important aspect of next generation sequencing (NGS) in cardiology concerns incidental findings made while performing extensive analyses such as whole-exome sequencing (WES) or whole-genome sequencing (WGS) in subjects tested for reasons not linked with heart disease. The American College of Medical Genetics published a list of clinically important and medically actionable genes that should be screened for pathogenic variants regardless of the specific purpose of genetic testing [4]. This list includes loci associated with cardiovascular diseases such as Marfan syndrome; Loeys–Dietz syndromes; familial thoracic aortic aneurysms and dissections; hypertrophic cardiomyopathy; dilated cardiomyopathy; catecholaminergic polymorphic ventricular tachycardia; arrhythmogenic right-ventricular cardiomyopathy; Romano–Ward long QT syndromes types 1, 2, and 3; Brugada syndrome; and familial hypercholesterolemia. Given the increasing use of WES and WGS it is likely that there will also be a growing demand to interpret and clinically follow up such findings.

This chapter summarizes the current knowledge on the genetics of cardiomyopathies, arrhythmias, thoracic aortic aneurysms and dissections, familial hypercholesterolemia, and congenital heart defects and provides information on the available next generation-based multigene panels.

GENERAL INFORMATION ON MULTIGENE PANELS

Information regarding commercially available diagnostic multigene panels for the purposes of this chapter was obtained through the GeneTests Web site. First, a list of laboratories that offered NGS panels dedicated to at least one of the cardiac phenotypes of interest was made. Offers were subsequently verified via laboratories' Web sites and e-mails. Finally, a list of over 130 up-to-date cardiac panels offered by 19 laboratories in eight countries from Europe and North America was made. Pricing data are based on 17 laboratories that made such information available on their Web site or answered e-mails inquiring for prices.

Generally the preferred specimen is 2–10 ml of blood in EDTA (lavender-top tube) shipped overnight at room temperature or at 4 °C. In several cases it was emphasized that blood should not be frozen. Alternatively, 5–25 µg of isolated DNA or 3 ml of saliva may be accepted. Some laboratories offer prenatal testing and accept chorionic villus samples (CVS), CVS culture, or amniocentesis culture. Prices vary from US$1340 to US$7045 (US$2715 average) and surprisingly depend more on the particular laboratory's pricing policy than on the number of genes in the panel. A similar relationship may be observed considering turnaround time, ranging from 2 to 20 weeks (8 weeks average). Whenever the facility provided information on the applied sequencing apparatus it was one of Illumina platforms. Generally laboratories offer full sequencing services with bioinformatic analysis and clinical report indicating the most probable causative variants and including information about discovered variants of unknown significance, but it is also possible to order the sequencing service alone with results sent back for at-home interpretation. If an institution has its own NGS platform, an optimized kit for a chosen panel can be implemented at home.

Aside from well-established phenotype-associated genes, many panels in the analyzed offers include genes not listed in the Online Mendelian Inheritance in Man database in reference to a given phenotype but selected based on single reports or their function deemed to be plausible with disease causation. This practice is not limited to cardiac panels and has been discussed in the literature [5]. Therefore the current literature should be consulted before choosing a testing panel to avoid confusing results.

The current technical ability of NGS to detect copy number variations (CNVs) is of limited value; therefore many laboratories offer deletion/duplication analysis using other methods (CGH array, Q-PCR, or MLPA) as an extra option or a part of the sequencing test. Deletion/duplication analysis is likely to increase the cost by approximately US$1000, but may raise the yield of genetic testing by up to 10% especially in cases of CHD [6].

Although there are numerous testing panels dedicated to diagnosing particular types of hereditary cardiac disorders, taking into account substantial genetic and phenotypic overlap leading to the existence of borderline cases and varying symptoms within one family, it may be worth considering using one of the comprehensive cardiovascular panels, especially as the pricing is comparable to the smaller ones. The use of a broader panel may not only increase the probability of identifying the causative mutation, but also allow finding additional phenotype-modifying variants that could explain potential diverse expression of the disease in the family.

Some laboratories offer vast comprehensive panels for cardiovascular diseases, called generally "heart panels." Four such panels have been found among the analyzed offers and one combining only cardiomyopathy and arrhythmia genes. These tests cover 39–133 genes and cost between US$1450 and US$3200. The most expensive test included sequencing of 105 genes and deletion/duplication analysis and, as the only panel, covered TAAD genes.

CARDIOMYOPATHIES

Cardiomyopathy is defined as a myocardial disorder in which the heart muscle is structurally and functionally abnormal, in the absence of coronary artery disease, hypertension, valvular disease, or congenital heart disease sufficient to cause the observed myocardial abnormality [7].

According to structural and functional changes to the heart, the majority of cardiomyopathies are classified as follows: hypertrophic cardiomyopathy (HCM), dilated cardiomyopathy (DCM), restrictive cardiomyopathy (RCM), arrhythmogenic right-ventricular cardiomyopathy (ARVC)/arrhythmogenic ventricular cardiomyopathy, and unclassified cardiomyopathies including left-ventricular noncompaction (LVNC) [7]. Although treatment protocols are currently based on the phenotype-based classification, this assignment does not reflect genetic origin associated with terms such as desmosomopathy, cytoskeletalopathy, sarcomyopathy, or channelopathy. To incorporate different attributes into defining cardiomyopathies the MOGE(S) nosology system has been proposed, which describes morphofunctional phenotype, organ(s) involvement, genetic inheritance pattern, etiological annotation (including genetic defect or underlying disease/substrate), and the functional status [8].

Since the identification of the first disease-causing *MYH7* mutation in 1990 [9], there has been a huge progress in uncovering the genetic background of hereditary cardiomyopathies, proving them to be highly heterogenic diseases caused by a plethora of genes involved in generation, transmission, and regulation of the contractile force of heart muscle, its electric activity, and its energy supply. The introduction of NGS techniques provided the possibility of increasing the knowledge of the underlying genetic causes as well as easing their detection.

The application of a powerful tool, WES, allowed the discovery of several new genes associated with cardiomyopathy. For instance, WES led to the identification of mutations in *MRPL3*, *MRPL44*, and *AARS2* as the causes of mitochondrial cardiomyopathies [10–12] and to *GATAD1* being implicated in the autosomal recessive form of DCM [13].

The first attempt to employ a targeted NGS approach in screening patients with hereditary cardiomyopathy was described in 2011. The authors used a 47-gene panel to detect mutations in patients with DCM and HCM (*n* = 10). They found known pathogenic variants in six patients and identified 27 new possibly damaging mutations [14]. Since then NGS has entered the clinical diagnosis of cardiomyopathies, making identification of underlying genetic defects time- and cost-efficient [15].

In patients with cardiomyopathy, sequencing multiple genes in parallel is particularly attractive as many subjects have more than one causative variant, either in the same gene (compound heterozygotes) or in different genes (double heterozygotes) [16–18]. Such cases may account for 5% of HCM [19] and as much as 57% of ARVC [20].

Identification of a truncating mutation in the *TTN* gene as a leading cause of DCM is the hallmark example of the impact NGS had on uncovering the genetic background of cardiomyopathies [21]. Although mutations in *TTN*, which encodes the giant muscle filament titin, have been known to cause DCM in several families since the end of the twentieth century [22,23], the scale of this association remained underestimated for a long time because of the immense size of this gene (363 exons), making analysis by traditional Sanger sequencing difficult. In 2012 Herman and colleagues published the results of a multicenter study in which the full coding *TTN* sequence was assessed mostly by NGS among 312 individuals with DCM, 231 with HCM, and 249 controls. Mutations altering the full-length titin were discovered in 27% of subjects with dilated cardiomyopathy. The authors concluded that incorporation of NGS analyses of *TTN* into clinical genetic screens should increase the detection of DCM-causing mutations by ~50%, permitting earlier diagnosis and interventions to prevent disease progression [21]. This assumption was confirmed by an independent study in 2014 [24].

Another large NGS study including 223 unrelated patients with HCM analyzed coding, intronic, and regulatory sequences of 41 genes implied in the pathogenesis of HCM and DCM. The authors concentrated on the clinical implications of a targeted NGS strategy and quantitative analysis of the prevalence of variation in sarcomeric genes. To quantify the strength of evidence supporting the causality of detected variants sequencing, results were compared with published findings and data from a large-scale exome sequencing program. One hundred fifty-seven potentially causative variants within sarcomeric or associated genes (not including titin) were found in 64% of patients. Four sarcomeric genes (*MYH7, MYBPC3, TNNI3,* and *TNNT2*) showed an excess of rare nonsynonymous single-nucleotide polymorphisms in cases compared to controls, with 34% of patients carrying candidate variants in desmosomal and ion channel genes [25].

The largest study so far that tested the utility of broad gene panels in cardiomyopathy was conducted by Pugh et al. and included 766 DCM patients from the United States. As the testing was carried out over a period of 5.5 years and as the knowledge of the genetic background of DCM grew, gene panels of increasing size were used starting from a 5-gene Sanger panel through a microarray-based sequencing assay targeting 19 DCM genes to the NGS Pan Cardiomyopathy Panel covering 46 genes. As a result the clinical sensitivity for DCM more than tripled, from a range of 7.7–10% to a range of 27–37%. The major contribution to this increase was caused by the inclusion of the *TTN* gene. As a drawback, the percentage of patients receiving an inconclusive test result increased from 4.6–6.5% to 51–61%. Among the 20 genes accounting for 37% of cases carrying a clinically relevant variant, the largest contributor was *TTN* (up to 14%), followed by *LMNA* (4.1–4.5%), *MYH7* (3.4–4.9%), and *DSP* (2.4%). Loss-of-function variants in the *VCL* gene were for the first time reported as a cause of DCM but no pathogenic variants were found in 11 genes previously reported as likely or candidate DCM genes (*ANKRD1, CAV3, CRYAB, CTF1, DSC2, EMD, FHL2, LAMA4, LAMP2, MYH6,* and *PKP2*). Interestingly, having a family history of DCM did not substantially increase the yield of genetic testing in the examined cohort [26].

The most recent study (2014) enrolling smaller number of well-characterized DCM patients applied ultra-high-coverage NGS of 84 genes for 639 individuals with sporadic or familial DCM from eight European countries. Sixteen percent of patients were found to carry known mutations previously reported as DCM causing, while 46% had variants associated with broadly defined heart muscle diseases and channelopathies. Twenty-three percent carried 1 of 117 previously not annotated highly "likely" pathogenic variants in 26 genes and another 141 variants were classified as potentially disease causing. *SMYD1* has been identified as a new gene associated with DCM. The highest numbers of known cardiomyopathy mutations were found in *PKP2*, *MYBPC3*, and *DSP*. If yet unknown but predicted disease loci were considered, *TTN*, *PKP2*, *MYBPC3*, *DSP*, *RYR2*, *DSC2*, *DSG2*, and *SCN5A* were most commonly mutated. The percentage of variants found in the *TTN* gene (13%) was similar to that observed by Pugh et al. [26]; however, observations regarding *DSC2* and *PKP2* were diametrically different. After normalizing the number of variants to the size of each gene a rather even distribution was observed, disproving the existence of mutation hot spots in DCM genes. The overlap between DCM and specific mutations associated with other cardiomyopathies was considerable (ARVC—31%, HCM—16%, and channelopathies—6%). Remarkably, 38% of patients had compound or combined mutations and 12.8% had three or even more mutations. When comparing patients recruited in different countries no large differences in mutation frequencies or affected genes were found, indicating that genetic testing for DCM can be applied in a uniform setting across Europe. Although mutation rates in familial cases were significantly higher than in the sporadic ones, known and well-characterized disease mutations were also found in many cases of idiopathic DCM, indicating that in Western countries, the small size of contemporary families may obscure the genetic nature of the disease and it is important to consider that sporadic DCM cases can also be due to well-known mutations [27].

According to the Working Group on Myocardial and Pericardial Diseases of the European Society of Cardiology statement on genetic counseling and testing in cardiomyopathies, reasons to perform genetic testing include diagnosis, prognostic evaluation, and therapeutic decision-making, and it is considered reasonable in the following situations:

1. Genetic testing is appropriate for the diagnosis of a rare or particular cardiomyopathy, especially in the presence of atypical phenotypic features, in the setting of expert teams after detailed clinical and family assessment.

2. Genetic testing is not indicated for the diagnosis of a borderline or doubtful cardiomyopathy, except for selected cases in the setting of expert teams after detailed clinical and family assessment.

3. The main role of genetic testing is to provide predictive diagnosis in first-degree relatives.
   a. Genetic testing is appropriate for predictive diagnosis in asymptomatic relatives of a patient with a cardiomyopathy when the disease-causing mutation has been previously characterized in the family.
   b. Genetic testing is therefore indicated in the proband of a family (the first or most clearly affected patient with a cardiomyopathy) as a first condition for the proposal of predictive diagnosis within the family. NB1: Postmortem (necropsy) molecular analyses can be considered in the proband if he/she is the only patient with the cardiomyopathy within the family. NB2: This is appropriate, whatever may be the "familial" or "sporadic" context, in HCM and ARVC, but questionable in sporadic DCM and sporadic RCM (except in the presence of atypical associated phenotype or red flags).

c. Predictive diagnosis in children can be considered at the age at which cardiac examination is useful (10–12 years of age for most cardiomyopathies) [28].

Among the analyzed offers, we found six comprehensive cardiomyopathy panels consisting of 51–103 genes. Prices range widely between US$1500 and US$4580, and the average is US$2867. The price of the most expensive 76-gene panel includes deletion/duplication analysis.

HYPERTROPHIC CARDIOMYOPATHY

HCM is the most common inherited cardiac disease affecting 1 in 500 people, and it is characterized by the presence of increased ventricular wall thickness or mass in the absence of loading conditions (hypertension, valve disease) sufficient to cause the observed abnormality [7].

Isolated familial HCM is inherited as an autosomal dominant trait. Although the penetrance increases with age, it remains incomplete and the phenotypes may vary, ranging from asymptomatic LVH, through atrial and ventricular arrhythmias, to refractory heart failure. De novo mutations are observed sporadically. HCM is often called the "disease of the sarcomere" and, accordingly, over 1000 causative mutations identified so far almost exclusively reside within the genes involved in the structure and functioning of the sarcomere [29]. The disease-causing defect is found in approximately 50–60% of patients with a family history of HCM and 20–30% without. The leading causes of HCM are mutations in *MYH7* (β-myosin heavy chain) and *MYBPC3* (cardiac myosin-binding protein C), which are collectively responsible for around 80% of cases. *TNNT2* (troponin T) and *TNNI3* (troponin I) can be assigned to 5% cases each, while *TPM1* (tropomyosin α1 chain) causes around 2% and *MYL3* (myosin light chain 3) 1% of cases [30].

Although identifying the causative gene and mutation has limited prognostic implications, some *MYH7* variants are considered particularly malignant (e.g., R403Q, R453C, G716R, and R719W) [31] and several mutations in *TNNT2* associate with smaller thickness of the left-ventricular wall but high incidence of sudden death in the early age [32], while *MYBPC3* variants are linked with later onset of the disease [33]. In ~5% of patients with an identified genetic cause more than one mutation in sarcomeric genes is observed, which is associated with a more severe phenotype and poor prognosis [30,34]. Finally, patients whose genetic tests turned out to be negative are more likely to have a mild phenotype [35]. Furthermore, the HCM phenotype may be modified by common polymorphisms. For example, a deletion in *ACE* encoding angiotensin-converting enzyme has been associated with a faster progression of hypertrophy and higher incidence of sudden cardiac death [36,37].

Increased left-ventricular wall thickness can also be a secondary effect of more complex hereditary disease, such as Danon disease, caused by mutations in *LAMP2;* autosomal dominant conditions resulting from mutations in *PRKAG2* [38]; or Fabry disease—an X-linked trait caused by defects in the *GLA* gene [39]. The distinction between sarcomeric HCM and its phenocopies may be important when considering therapeutic implications.

The ESC Task Force for the Diagnosis and Management of Hypertrophic Cardiomyopathy has set guidelines regarding genetic testing. It is recommended in patients fulfilling diagnostic criteria for HCM when it enables cascade genetic screening of relatives (class of recommendation I/level of evidence B). It is also recommended that genetic testing be performed in certified diagnostic laboratories with expertise in the interpretation of cardiomyopathy-related mutations (I/C). In the presence of symptoms and signs of disease suggestive of specific causes of HCM, genetic testing is indicated to confirm the diagnosis (I/B). In patients with a borderline diagnosis of HCM, genetic testing should be performed only after detailed assessment by specialist teams (IIa/C). Postmortem genetic analysis of stored tissue or DNA should be considered in deceased patients with pathologically confirmed HCM, to enable cascade genetic screening of their relatives (IIa/C). First-degree relatives who do not have the same definite disease-causing mutation as the proband should be discharged from further follow-up but advised to seek reassessment if they develop symptoms or when new clinically relevant data emerge in the family (IIa/B) [40].

The number of genes in available NGS diagnostic panels ranges from 11 to 25. Depending on the laboratory, the list consists of only well-established sarcomeric genes or may include those with lesser evidence for pathogenicity. Most panels test also for the genes causing phenocopies (*LAMP2, PRKAG2,* and *GLA*). There are also broad panels including genes responsible for concomitant phenotypes, for example, a 35-gene panel for HCM with cardiac conduction defect.

DILATED CARDIOMYOPATHY

DCM is defined by the presence of left-ventricular dilatation and left-ventricular systolic dysfunction in the absence of abnormal loading conditions (hypertension, valve disease) or coronary artery disease sufficient to cause global systolic impairment. Right-ventricular dilation and dysfunction may be present but are not necessary for the diagnosis [7]. Disease may be initially asymptomatic followed by variable clinical presentations, including heart failure, arrhythmias with conduction system disease, and thromboembolic episodes. It is also a leading cause of heart transplantation [41].

Up to 30% of cases of DCM have a hereditary background, mostly with an autosomal dominant pattern of inheritance. Autosomal recessive, X-linked, and mitochondrial have also been observed but rather in syndromic variants such as Alström syndrome, Duchenne muscular dystrophy, or Kearns–Sayre syndrome [42]. The genetic basis of isolated DCM is highly heterogeneous, with over 30 implicated genes, which can be divided into three major groups: those affecting force generation and regulation (*MYH7, ACTC1, TNNT2, TNNI3, TNNC1, PLN*), force transmission and mechanosensing (*TTN, DMD, SGCD, DES*), or nuclear function including transcription (*LMNA, TMPO, ANKRD1*). The yield of genetic testing is estimated at 30–35% [41] and is highest among the patients with concomitant conduction disease and elevated serum creatine kinase [3]. The most commonly identified causative mutations reside within *TTN, LMNA, MYH7,* and *TNNT2* [41]. The unique role of *TTN* mutations in DCM as tested by NGS has been discussed earlier in this chapter.

The prognostic implications of genetic testing are limited to assessing higher risk of sudden cardiac death (SCD) among individuals with *LMNA* and *DES* mutations [43,44]. In the case of *LMNA* mutation carriers with DCM and cardiac conduction disease (CCD), there are also some therapeutic implications that need to be taken into account regarding early or preemptive use of an implantable cardiac defibrillator, rather than a pacemaker. For individuals with DCM and CCD comprehensive or targeted (*LMNA, SCN5A*) analysis is being recommended [3]. In the mentioned earlier multicenter study by Haas et al., ICD-carrier status in DCM was strongly associated with the presence of *RBM20* mutation, whereas *SMYD1* and *CRYAB* defects correlated with left-ventricular ejection fraction. Alterations in left ventricular end diastolic diameter was in association with *TBX20*, and heart transplantation was significantly associated with *MYPN* [27]. Another observation was that the age of onset of symptoms varies according to the defective gene. *LMNA* variants are more likely to present in adulthood, *DSP* variants were reported to be unique to the adult cohort, while *RBM20* variants were enriched in the pediatric group [26]. Mutations in *MYH6* and *ADRB3* have been associated with age at diagnosis in another study [27].

As a result of genetic heterogeneity, DCM diagnostic panels are broad, including 25 up to 48 genes, with prices starting at US$2500 and ranging to over US$7000. As often is the case with NGS, the size of the panel does not correlate with turnaround time. The most comprehensive tests also include a large number of mitochondrial genes.

ARRHYTHMOGENIC RIGHT VENTRICULAR CARDIOMYOPATHY

ARVC is defined by the presence of right-ventricular dysfunction (global or regional), with or without left-ventricular disease, in the presence of histological evidence for the disease and/or electrocardiographic abnormalities in accordance with published criteria [7,45].

In over 50% of ARVC cases the disease has a familial background with autosomal dominant inheritance, although rare syndromic, cardiocutaneous forms are inherited in autosomal recessive mode (Naxos disease, Caravajal syndrome). The penetrance is low and clinical features are highly variable even within families. As a substantial number of cases may result from double or compound heterozygosity (up to 57%), a Mendelian inheritance pattern is often obscured and large pedigrees with numerous affected individuals are rarely seen, which makes cosegregation analysis a challenge [20]. Despite these difficulties a number of ARVC-associated genes have been identified over the years and a causative genetic defect is currently identified in 50% of families [46]. There are eight established genes with mutations known to cause ARVC. Five of them are involved in function and structure of the desmosome, *DSP, PKP2, DSG2, DSC2,* and *JUP,* and the remaining ones are *TGFB3, RYR2,* and *TMEM43* [20]. Mutations in *PKP2* are responsible for 9–43% of cases according to various studies and are considered the most common genetic cause of ARVC, although one pathogenic *PKP2* variant may not be enough to cause full disease; another mutation in *PKP2* or in another desmosomal gene is required [3]. While new causative variants in desmosomal genes are being identified, others are put into doubt because of their prevalence in control populations, although this argument fails to take into account the frequent digenic character of the disease. A 2013 study identified two damaging mutations in *CTNNA3* among patients without defects in genes commonly mutated in ARVC [47]. The *PLN* R14del founder mutation is present in a substantial number of patients clinically diagnosed with ARVC or DCM in a population in the Netherlands [48]. The current state of knowledge on the genetics of ARVC is available in the database at www. arvcdatabase.info.

In ARVC, from the point of view of prognostic and therapeutic implications, genetic testing is of little use. The few genotype–phenotype relationships that have been observed include the suggestion that individuals with a *DSP* mutation and/or chain-termination variants in desmosomal genes are more likely to develop left-ventricle involvement [49] and that *TMEM43* mutations have a more lethal effect in males than in females [50]. Still, genetic testing may have an important confirmatory role in diagnosing borderline cases [20].

NGS-based panels are particularly attractive for ARVC diagnosis, as they allow simultaneous molecular analysis of all the disease-related genes, which is optimal to fully explain the phenotype [20].

The majority of offered ARVC panels consist of seven genes that include a combination of desmosomal genes plus *RYR2* and *TMEM43*. One 13-gene test that included *TGFB3, PKP4* (less established ARVC-causing gene), and a set of unconfirmed candidate genes (*PNN, RPSA,* and *TNN*) can also be found. The price range is between US$2500 and US$3800.

RESTRICTIVE CARDIOMYOPATHY

RCM is a rare disease defined as restrictive ventricular physiology in the presence of normal or reduced diastolic volumes (of one or both ventricles), normal or reduced systolic volumes, and normal ventricular-wall thickness [7].

Isolated RCM is rare, occurring usually as a secondary feature of other diseases, including hereditary ones such as the cardiomyopathies (especially HCM), Noonan syndrome, or amyloidosis. Individuals with RCM and HCM may coexist within one family. Another condition concurrent with the presence of RCM is skeletal myopathy [51].

Primary RCM generally has a genetic cause. So far RCM-causing mutations have been identified mostly in sarcomere genes: *MYH7, TNNT2, TNNI3*, and *ACTC1* [3]. In addition, *DES* mutations have been shown to cause RCM with conduction disease [52].

Diagnostic and prognostic implications of genetic testing apart from the usual confirmation of diagnosis and family evaluation are limited to potential identification of concomitant syndromic disorder. In the differential diagnosis of RCM, transthyretin amyloidosis should be taken into consideration in relation to mutations in the *TTR* gene. Establishing the diagnosis has therapeutic implications [53].

Among the analyzed offers only one RCM-dedicated panel was found. It consists of the following genes: *ACTC1, DES, MYH7, MYPN, TNNI3, TNNT2, TNNC1*, and *TTR*.

LEFT-VENTRICULAR NONCOMPACTION

LVNC is a form of inherited cardiomyopathy that is characterized by prominent left-ventricular trabeculae and deep intertrabecular recesses. The myocardial wall is often thickened with a thin, compacted epicardial layer and a thickened endocardial layer. In some patients, LVNC is associated with left-ventricular dilatation and systolic dysfunction, which can be transient in neonates [7].

There is substantial overlap between LVNC and other cardiac disorders, including congenital heart defects and other cardiomyopathies, HCM, DCM, and RCM, as a coexisting phenotype or occurring in the same families [54]. It may also be a feature of systemic disease such as mitochondrial myopathy, Noonan, or Barth syndrome—disorders that are likewise manifested by other cardiomyopathies. Unsurprisingly, LVNC has a genetic background similar to those of other cardiomyopathies, being caused by mutations in sarcomeric genes (*MYH7, ACTC1, TNNT2, MYBPC3*, and *TPM1*) [41]. Moreover, the same variants have been found to cause variable symptoms within one family. For example, the *MYH7* mutation R281T has been shown to cause LVNC in combination with other heart defects, such as Ebstein's anomaly and atrial septal defect (ASD) [55], and the *ACTC1* variant E101K has resulted in LVNC, HCM, or ASD [54]. Other genes reported to be associated with LVNC are *MIB1* [56], *LDB3* [57], *PRDM16* [58], *LMNA* [59], and *DTNA* [60]. Barth syndrome is an X-linked disease caused by *TAZ* gene defects [61].

The yield of genetic testing among LVNC patients is relatively small, estimated at approximately 15–20%, although it may depend on the number of tested genes. Using a panel of six sarcomeric genes, the authors identified causative defects in 17% of 63 cases [62], while in another study analyzing a 17-gene panel investigators managed to identify causative variants in 41% of 56 index cases. Apart from sarcomere defects, damaging mutations have been found in *CASQ2, CALR3, LMNA, LDB3*, and *TAZ* [63].

We have found three tests specifically designed to diagnose LVNC. They consist of 6–10 genes being combinations of *ACTC1, CASQ2, DTNA, LDB3, LMNA, MYBPC3, MYH7, TAZ, TNNT2, TPM1, LMNA*, and *VCL*. Prices range from US$2290 to US$3200.

ARRHYTHMIAS

Inherited arrhythmias are often referred to as channelopathies, since underlying genetic defects affect ion channel subunits or the proteins that regulate them. Although results of genotype–phenotype studies are slowly starting to influence patient management, the main goal of genetic testing remains the identification of causative mutations for early diagnosis of relatives at risk of frequently fatal conditions [64], so regardless of recommendations concerning the need for genetic testing of index patients, once the genetic cause of disease has been established, experts always recommend mutation-specific testing for family members [3].

Another application of comprehensive genetic testing is the determination of the cause of SCD. When an individual dies suddenly without an obvious cause, with inconclusive postmortem results and negative toxicology tests, the death is classified as SADS (sudden arrhythmic death syndrome). The term "genetic autopsy" is applied to the genetic testing in samples acquired after death to diagnose the cause of demise. Ion channelopathies account for up to 35% of sudden unexplained death cases when *KCNQ1, KCNH2, SCN5A,* and *RYR2* are tested. Following the results of positive genetic testing, the most frequent causes of SADS are catecholaminergic polymorphic ventricular tachycardia (20% of all cases), long QT syndrome (15%), and Brugada syndrome (<1%) [65].

Similar to cardiomyopathies, there is a substantial overlap in genes related to several arrhythmic phenotypes, as different mutations in one gene can cause distinct conditions. Suitably comprehensive arrhythmia panels are commercially available and are priced similar to the phenotype-specific panels mentioned in sections below. Six currently available comprehensive arrhythmia panels include 29–62 genes and cost from US$1450 to US$6218.

LONG QT SYNDROME

Hereditary long QT syndrome (LQTS) is a repolarization disorder characterized by prolongation of the QT interval that may result in syncopal episodes, potentially lethal *torsades de pointes* tachyarrhythmias, and SCD in young age [66].

There are currently 13 genes associated with LQTS: *KCNQ1, KCNH2, SCN5A, ANK2, KCNE1, KNCE2, KCNJ2, CACNA1C, CAV3, SCN4B, AKAP9, SNTA1,* and *KCNJ5*. Loss-of-function mutations in *KCNQ1* and *KCNH2* are together responsible for approximately 70% of LQTS and another 5–10% can be attributed to gain-of-function variants in *SCN5A*. Collectively these three genes comprise up to 80% of clinically definite LQTS cases, while the remaining 10 genes cover another 5% [3].

Romano–Ward syndrome is a term generally used to refer to LQTS caused by autosomal dominant mutations with the exception of *KCNJ2* and *CACNA1C* defects. Mutations in *KCNJ2* cause Andersen–Tawil syndrome associated with physical abnormalities in addition to arrhythmia [67] and defective variants in *CACNA1C* that cause Timothy syndrome characterized by physical malformations, neurological and developmental defects, and autism spectrum disorders [68]; however, a WES study identified a *CACNA1C* mutation as a cause of autosomal dominant LQTS in the absence of Timothy syndrome in a large family [69]. Autosomal recessive mutations in *KCNQ1* and *KCNE1* result in another rare disease called Jervell–Lange–Nielsen syndrome in which arrhythmia coincides with hearing loss [70].

Experts recommend genetic testing for any patient with strong suspicion of LQTS and asymptomatic patients with idiopathic QT prolongation (class I). It may also be considered for asymptomatic patients with idiopathic borderline QT values (IIb). Once the causative variant has been identified in the index case, mutation-specific testing is recommended for family members (class I). Although experts recommend sequencing of *KCNQ1, KCNH2,* and *SCN5A,* a more comprehensive analysis, including the remaining LQT genes, may increase testing yield as well as the risk of false positive results [3].

Results of genetic testing may have prognostic significance including course of disease, arrhythmogenic triggers, and response to pharmacotherapy [71–73]. Generally *SCN5A* mutation carriers are at higher risk of mortality per event compared to *KCNQ1* and *KCNH2* [72]. There are also risk estimations based on the mutation's location within the protein [74,75].

We found 11 diagnostic panels dedicated to LQTS. One is limited to three main LQT genes, but most consist of 12 or 13 genes. One panel includes a 14th gene—*NOS1AP,* whose polymorphisms have been reported to modify the LQTS phenotype [76]. Costs vary between US$1340 and US$4565. Broadening the sequencing-based tests by deletion/duplication analysis of *KCNQ1* and *KCNH2* may increase the yield by 5% [77,78].

BRUGADA SYNDROME

Brugada syndrome (BrS) is an autosomal dominant hereditary disease characterized by conduction delays and predisposition to ventricular tachycardia leading to ventricular fibrillation causing sudden (often nocturnal) cardiac death. The disease is recognized by precordial ST segment elevation in the electrocardiographic leads V1 and V2 [79]. BrS is the major cause of sudden unexplained death syndrome [80].

The yield of genetic testing is relatively low (approximately 25%). Among identified cases the leading causes of disease are mutations in *SCN5A* (75%) [81]. Remaining genes encode other ion channels (sodium, potassium, and calcium), each responsible for a small number of individual cases [82–87].

There are not many genotype–phenotype associations observed in BrS and the main value of genetic testing is the possibility of recommending appropriate precautions to asymptomatic mutation-positive relatives of the index case. HRS–EHRA experts consider that genetic testing "can be useful" (IIa) for patients with an established clinical index of suspicion for the disease [3]. According to a meta-analysis, the sole presence of an *SCN5A* gene mutation is of no use, considering the management of patients [88]; however, the comparison between missense and truncating mutations shows better prognosis for the first type [89], and phenotypes of individuals with BrS associated with *CACNA1C* and *CACNB2* mutation include a shortened QT interval [83].

Among the analyzed diagnostic offers there are seven BrS-dedicated panels consisting of 5–18 genes. Prices range from US$1474 to US$3375.

CATECHOLAMINERGIC POLYMORPHIC VENTRICULAR TACHYCARDIA

Catecholaminergic polymorphic ventricular tachycardia (CPVT) is a genetic disease characterized by adrenergically mediated ventricular arrhythmias causing syncope, cardiac arrest, and SCD in young individuals with structurally normal hearts. Symptoms are typically triggered by physical or emotional stress [3].

The causative genetic defect is identified in approximately 65% of cases. The main culprits in CPVT are mutations in the ryanodine receptor (*RYR2*) inherited in an autosomal dominant manner. The autosomal recessive mode of inheritance is associated with the *CASQ2* gene, which accounts for 3–5% of cases. Compound heterozygotes for mutation in this gene also have been reported [90]. Other genes related to the pathogenesis of CPVT are *CALM1* [91], *TRDN* [92], *KCNJ2* [93], and *ANK2* [94]. *KCNJ2* testing should be considered if abundant ectopy and a prominent U wave are found on 24-h ambulatory monitoring [3].

Comprehensive testing or *RYR2* and *CASQ2* testing is recommended for patients with a clinical index of suspicion of CPVT (class I). Considering the early age of manifestation (mean age of onset is 8 years), early CPVT genetic evaluation is important for all family members of the CPVT index case, even those with a negative clinical phenotype, as sudden death might be the first manifestation of the disease. Since CPVT is associated with sudden infant death syndrome, testing should also be performed at birth to enable early initiation of preventive therapy [3].

Taking into account the large size of the *RYR2* gene and the fact that time-efficient genetic testing is of significant importance in the case of CPVT, the application of NGS-based diagnostics seems particularly attractive. Six analyzed panels dedicated to CPVT consist of two to six genes and all include *RYR2* and *CASQ2*. Turnaround time ranges from 2 to 13 weeks. Prices vary between US$1608 and US$3375.

CARDIAC CONDUCTION DISEASE

CCD is characterized by a delay in electrical impulse conduction at the atrial, nodal, and ventricular levels. Inherited forms of CCD may be secondary to congenital heart defects and cardiomyopathies or may exist as an isolated condition with a structurally normal heart.

The major cause of isolated CCD is mutation in *SCN5A*; however, the 1795insD mutation in this gene has been reported to cause overlapping phenotypes of LQTS, BrS, and CCD in a single family [95]. Defects in *TRPM4* are responsible for approximately 25% of right bundle branch block and 10% of atrioventricular block [96,97]. A less frequently reported gene is *SCN1B* [84] and mutations in *HCN4* have been shown to cause idiopathic sinus node dysfunction [98].

In the case of the cardiomyopathy-associated form of disease, symptoms of CCD may precede development of DCM or LVNC, especially if caused by the *LMNA* or *DES* gene [44,99]. When CCD accompanies a congenital heart disease such as atrial septal defect, mutations in *NKX2.5* or *GATA4* are more likely [100,101].

According to an expert consensus statement genetic testing may be considered as a part of the diagnostic evaluation for patients with either isolated CCD or CCD with concomitant congenital heart disease, especially when there is documentation of a positive family history of CCD (class IIb) [3].

Two analyzed CCD panels consist of 2 and 17 genes. The first includes *SCN5A* and *LMNA*; the second one, *DES, EMD, GAA, GLA, HCN1, HCN4, KCNA5, KCNQ1, LAMP2, LMNA, MYH6, NKX2.5, PRKAG2, SCN5A, SCN1B, SCN4B*, and *TRPM4*.

SHORT QT SYNDROME

The short QT syndrome (SQTS) is a newly described cardiac channelopathy associated with a predisposition to atrial fibrillation and sudden cardiac death, characterized by the presence of a very short QT interval on electrocardiogram (ECG). The syndrome appears to be inherited in an autosomal dominant pattern. Age of onset can be very low [102].

Not surprisingly, SQTS is caused by defects in the same ion channels as LQTS. Mutations in *KCNH2* (SQT1), *KCNQ1* (SQT2), and *KCNJ2* (SQT3) account for 20% of index cases. Short QT intervals with concomitant BrS ECG pattern have been also reported in individuals with *CACNA1C* and

CACNB2B mutations [83]. As yet there are no available data regarding genotype–phenotype correlations. According to HRS–EHRA experts comprehensive or SQT1–3-targeted genetic testing may be considered for patients with a strong suspicion of SQTS (class IIb) [3].

Among the analyzed diagnostic offers we found five SQTS panels and one dedicated to both SQTS and LQTS. One panel includes only three established potassium channel genes. Others consist of six to eight genes and also include varying numbers of genes encoding calcium channel subunits (*CACNA1B, CACNA1C, CACNA1D, CACNA2D1,* and *CACNB2*). One panel includes *ABCC9,* recently associated with Brugada and early depolarization syndrome [103]. The LQTS/SQTS panel consists of 12 of 13 standard LQTS genes. The price range is US$1340–US$3375.

ATRIAL FIBRILLATION

Atrial fibrillation (AF) is the most common sustained arrhythmia, with a multifactorial background that includes genetics; however, families with a monogenic isolated form of AF inherited in an autosomal dominant pattern have been described.

The list of genes currently associated with familial AF includes *KCNQ1, KCNJ2, KCNE2, SCN5A, KCNA5, NPPA, GJA1, GJA5* [3], *ABCC9* [104], *SCN1B, SCN2B* [105], *NUP155* [106], *SCN3B* [107], and *SCN4B* [108] and is still growing as new unique mutations suspected of causing AF are being reported. It is hard to indicate frequent common causes of familial AF, and as few genotype–phenotype relationships have been observed, they are of no use in considering the prognostic or therapeutic impact of genetic testing.

Whereas at the time of this writing genetic testing is not indicated by experts, it is considered useful to refer individuals with an extensive history of familial AF to a research center [3].

Among the analyzed offers, three familial AF testing panels were found. One consists only of *KCNE2, KCNJ2,* and *KCNQ1*. The other two consist of 19 and 23 genes from the following list: *ABCC9, CACNA1D, GATA4, GATA6, GJA1, GJA5, HCN1, HCN4, KCNA5, KCND3, KCNE1, KCNE1L, KCNE2, KCNE4, KCNH2, KCNJ2, KCNJ5, KCNJ8, KCNQ1, LMNA, NPPA, NUP155, RANGRF, RYR2, SCN10A, SCN1B, SCN2B, SCN3B, SCN4B, SCN5A, SCNN1B,* and *SCNN1G*.

THORACIC AORTIC ANEURYSMS AND DISSECTIONS

Thoracic aortic aneurysm (TAA) is a permanent localized dilatation having at least a 50% increase in diameter localized at the ascending thoracic aorta at the level of the sinuses of Valsalva or the ascending aorta or both [109]. TAA is a dangerous but usually silent disease. In many cases, the first manifestation is acute aortic dissection (AAD) or rupture, resulting in high mortality despite progress in therapeutic surgical procedures. Importantly, the risk may be decreased by adequate management, such as a pharmacological approach aimed at reduction of aortic wall stress and prophylactic surgery, which at present is relatively safe. Thus, it is reasonable to carry out active screening to find people at risk at the early stages of disease [110].

Over 20% of cases of TAAD have a familial form (FTAAD) with a usually autosomal dominant pattern of inheritance and limited penetrance. An underlying genetic defect is found in ~20% of index cases. FTAAD may be part of a well-characterized genetic syndrome or nonsyndromic [111]. The correct interpretation of results obtained from molecular genetic testing requires basic knowledge of these different entities particularly since medical and surgical management may differ according to the underlying diagnosis, as is the case, for example, in Loyes–Dietz syndrome [112,113].

Marfan syndrome (MS) caused by *FBN1* mutations is a connective tissue disorder that manifests mainly in cardiovascular, ocular, and skeletal systems with a strong tendency for AAD [114].

The vascular type of Ehlers–Danlos syndrome (EDS) is caused by mutations in *COL3A1* and *COL1A2* [115,116] and associated with a high risk for aortic dissection and intestinal or uterine rupture. It is important to diagnose EDS. As this is often difficult on solely a clinical basis genetic testing is an attractive option [115].

Loeys–Dietz syndrome (LDS), caused by mutations in *TGFBR1* and *TGFBR2,* is associated with frequent coexisting aneurysms of other arteries and a more aggressive course of the aortic disease (acute aortic dissection has been described with aortic diameters of less than 4.5 cm) [117]. Aneurysm–osteoarthritis syndrome (AOS), caused by *SMAD3* mutations, is a form of TAAD characterized by the presence of arterial aneurysms and tortuosity; mild craniofacial, skeletal, and cutaneous anomalies; and early-onset osteoarthritis [118]. Together with some syndromic presentations related to *TGFB2* [119], LDS and AOS are generally referred to as TGF-β-related vasculopathies [112].

Other syndromic forms of TAAD are arterial tortuosity syndrome, caused by *SLC2A10* mutations [120], and cutis laxa syndromes, caused by autosomal recessive mutations in *FBLN4* [121] and autosomal dominant variants in *ELN* [122].

Most FTAAD patients are nonsyndromic, although they can have other coexisting cardiovascular conditions. The most frequent cause of FTAAD is mutations in the *ACTA2* gene, responsible for up to 14% of identified cases [123]. Apart from TAAD, *ACTA2* mutation carriers from the same family can have a diversity of cardiovascular diseases including premature coronary artery disease and ischemic stroke [124]. Defects in *TGFBR1* and *TGFBR2* can also be responsible for nonsyndromic forms of disease accounting for 3–5% of cases, similar to *SMAD3* (2%) and *TGFB2* [112]. Not all TAAD patients with an *FBN1* mutation fulfill the criteria for MS; those who do not often belong to a broader category termed type I fibrillinopathies associated with increased risk for AAD [125]. Less frequently found causative defects reside within *MYLK, MYH11*, and *PRKG1* [123]. Mutations in *NOTCH1* are identified in 10% patients with TAA coexisting with bicuspid aortic valve (BAV) [126].

AHA guidelines for genetic disorders associated with TAAD, including the principles of family screening, were published in 2010. Accordingly, sequencing of the *ACTA2* gene is reasonable in patients with a family history of TAAD to determine if mutations in this gene are responsible for the inherited predisposition (class IIa), and sequencing of other genes (*TGFBR1, TGFBR2, MYH11*) may be considered (IIb) [113]. The more recent ESC guidelines on the diagnosis and treatment of aortic diseases recommend referring patients with highly suspected familial form of TAAD to a geneticist for family investigation and molecular testing (Ic). Some authors advise adequate gene-targeted analysis in the case of syndromic disease and broader multiple NGS panel testing for individuals with nonsyndromic familial TAAD. The same approach may be considered in young subjects with sporadic disease but no additional risk factors [112].

There is evidence that rare CNVs significantly contribute to the pathogenesis of TAAD. Results of a case–control study suggest that rare CNVs are present in 13% of sporadic and 23% of familial TAAD cases without common causative genetic mutation. In both groups enriched CNVs reside within genes responsible for the regulation of cell adhesion or actin cytoskeleton. Although genes already associated with FTAAD belong to this group, none of the discovered particular loci have been previously reported to play a role in TAAD apart from *MYH11* [127]. These findings need further investigation in order to implement them into diagnostic practice, and currently CNV analysis is offered only for the standard set of FTAAD genes.

Analysis of commercially available diagnostic offers revealed a variety of aortic aneurysm-oriented NGS testing panels dedicated to particular syndromes or aortopathy in general. The overall 23 panels consist of 2–18 genes in the price range US$1450–US$5090.

CONGENITAL HEART DISEASE

CHD is defined as structural or functional abnormalities that occur in utero and are present at birth, even if discovered later in life [128]. Until recently the heredity of CHD has been significantly underestimated because most patients have a negative family history and as such are considered sporadic cases. Big multigeneration families with multiple affected members are scarcely observed. One of the reasons for this has been the high mortality of affected infants, but surgical interventions have improved enough to enable survival into adulthood and procreation. Based on estimations from the year 2000, there were 787,000 adults living with CHD in the United States alone and this number increases by 5% per year [129]. The 2010 estimation for Europe is 2.3 million and the number of adults with CHD surpasses the number of affected children [130]. The first parent-to-child recurrence of total anomalous pulmonary venous connection was described in 1991 with more cases reported subsequently [131]. Over the years more lesions proved to be highly hereditary. Taking into account the growing number of CHD families it becomes crucial to improve our understanding of the genetics of CHD to conduct more effective genetic counseling.

Examples of CHD that may be heritable include septal defects such as ASD, ventricular septal defect (VSD), and atrioventricular septal defect (AVSD); valvular and vascular defects of the cardiac outflow tract such as BAV, aortic valve stenosis (AS), coarctation of the aorta (CoA), and patent ductus arteriosus; and more complex defects such as tetralogy of Fallot (TOF), hypoplastic left heart syndrome, double-outlet right ventricle, or transposition of great arteries.

As of this writing the genetic cause of the disease has been found in only 10% of cases. Around 10% of those recognized cases are part of Down or diGeorge syndrome and other chromosomal abnormalities [132]. Although most CHDs have complex inheritance owing to multifactorial mechanisms, since 1995 positional cloning and family-based studies have revealed single-gene mutations causing both isolated and syndromic forms of CHD [100,101,133,134]. At present the burden of unraveling genetic "dark matter" rests on NGS techniques, which have already made some contribution to identifying causative variants among CHDs. The process is particularly challenging because of the genetic heterogeneity, lack of large pedigrees combined with variable expressivity, and reduced penetrance. The first use of NGS to identify a genetic cause of CHD revealed a novel *MYH6* mutation in a large family with pleiotropic CHD by performing exome sequencing on DNA samples from two cousins [135]. The results of another WES study analyzing de novo variants in 13 parent–offspring trios as well as 112 unrelated individuals with nonsyndromic AVSD identified the new CHD-associated gene *NR2F2* with several disease-causing variants, which turned out to be responsible for other forms of CHD, including TOF, AS, VSD, and CoA [136]. In 2013 a notable study was published in which the authors combined a case–control study with WES of trios. Comparison of the number of *de novo* damaging variants in genes with the highest heart expression between 362 trios of healthy parents and affected offspring and 264 control trios without heart defect revealed a novel set of potentially causal genes involved in histone methylation. Furthermore, it was estimated that *de novo* point and small insertion/deletion mutations may contribute to approximately 10% of severe CHD cases [137].

The 2007 AHA Scientific Statement emphasizes the importance of genetic testing in CHD patients to assess risk for reproduction and other family members [129] as well as predicting involvement of another organ and/or future complications. For example, patients with the *NKX2.5* mutation who had successful surgery for cardiac defect may still be at risk of cardiac block [138]. The authors presented a detailed algorithm for diagnostic procedures based on the initial presentation. For patients with multiple congenital anomalies, especially with dysmorphic facial features and mental retardation, they recommend karyotype analysis and other cytogenetic techniques (FISH and subtelomere FISH); however other authors also recommend array CGH as a useful technique in CNV detection [1,139,140].

The AHA statement concludes that most of the genes for isolated CHD were unavailable except on the research level at that time, although some of these genes were already transitioning to clinical availability. They advised clinicians to check the GeneTests Web site, or other publicly funded medical genetics information resource, for updates on current tests.

The majority (nine) of NGS-based panels dedicated to congenital defects listed on GeneTests were related to neurocardiofacial–cutaneous syndromes in general or chosen diseases (Noonan, Costello, and Leopard syndromes, RASopathies) and, according to a number of covered phenotypes, consisted of 3–12 genes from the following list: *BRAF, CBL, HRAS, KRAS, M2K1, MAP2K1, MAP2K2, NRAS, PTPN11, RAF1, SHOC2, SOS1,* and *SPRED1.* Two heterotaxy panels consisted of 6 and 10 genes: *ACVR2B, CCDC39, CCDC40, CFC1, CRELD1, FOXH1, GDF1, GJA1, LEFTY2, NKX2-5, NODAL,* and *ZIC3 (FOXH1, GDF, NODAL,* and *ZIC3* the only common ones). Two tests described as congenital heart disease/defect panels consisted of 6 and 19 genes: *ACTC1, BMPR2, CFC1, CHD7, ELN, G6PC3, GATA4, GATA6, GDF1, GJA1, HCN4, JAG1, MYH6, NKX23, NKX25, NKX26, RBM10, TBX1, TBX20,* and *TLL1 (ELN, JAG1, GATA4, NKX2-5,* and *TBX5* appeared in both panels). Finally, one panel dedicated to bicuspid aortic valve and arterial tortuosity consisted of *EFEMP2, FBLN5, NOTCH1,* and *SLC2A10.* Prices oscillate in the range of US$950 to US$3375.

FAMILIAL HYPERCHOLESTEROLEMIA

Familial hypercholesterolemia (FH) is a common hereditary condition characterized by severely elevated low-density lipoprotein cholesterol (LDL-C) levels from birth on, enhanced atherosclerosis progression, and premature cardiovascular events. Early diagnosis and treatment of FH are pivotal because therapy with lipid-lowering agents strongly decreases the risk for such events [141]. The prevalence of FH is thought to be between 1/500 and 1/200 [142].

Autosomal dominant FH is caused by mutations in the *LDLR, APOB,* and *PCSK9* genes. The yield of genetic testing including these three genes varies strongly according to different reports from 28 to 88% [143]. The *APOE* gene, which has been formerly associated with LDL-C levels [144], has been shown to be implicated in FH by two independent studies using WES [145,146].

The autosomal recessive form of the disease is caused by mutations in *LDLRAP1* [147]. Sitosterolemia is an autosomal recessive disorder characterized by increased intestinal absorption and decreased biliary excretion of dietary sterols, hypercholesterolemia, and premature coronary atherosclerosis caused by mutations in the *ABCG5* and *ABCG8* genes [148]. Although sitosterolemia and FH are etiologically different entities, they overlap phenotypically. An interesting case illustrating the value of broad NGS analysis in diagnosing sitosterolemia has been published [149]. WGS was performed to diagnose an 11-month-old breast-fed girl with xanthomas and very high plasma cholesterol levels.

Her parents were unaffected and no family history of hypercholesterolemia was reported. Known genetic causes of severe hypercholesterolemia were previously ruled out by sequencing the responsible genes (*LDLRAP, LDLR, PCSK9, APOE,* and *APOB*). Although sitosterolemia was ruled out by a normal plasma sitosterol:cholesterol ratio, WGS showed two nonsense mutations in *ABCG5* indicating the presence of this disease, albeit with an atypical presentation. Diagnosis was finally confirmed by the finding of severe sitosterolemia in a blood sample obtained after the infant had been weaned [149].

In cases of probable or definite FH the Consensus Statement of the European Atherosclerosis Society recommends referring subjects for genetic testing if available, with subsequent cascade testing in the family if a causative mutation is found [150].

Among the analyzed commercial offers one NGS panel test dedicated to FH was found, which included only *LDLR, APOB,* and *PCSK9,* as well as another broader hyperlipidemia panel including *ABCG5, ABCG8, APOB, APOE, LDLR, LDLRAP1, LPL*, and *PCSK9*.

CONCLUSIONS

NGS has entered into clinical molecular diagnostics of genetic cardiac diseases. Considering the large amount of data from a single experiment and the difficulties in interpretation, the results often produce more questions than answers. Segregation analysis in families with multiple members and long follow-up of carriers is necessary to assess the pathogenicity of possibly causative and often unique variants. Nevertheless, NGS is starting to have an impact on cardiogenetics and it is important for cardiologists and other physicians involved in the diagnosis and treatment of inherited cardiac diseases to be aware of the possibilities as well as the limitations of this new technology.

LIST OF ACRONYMS AND ABBREVIATIONS

AAD Acute aortic dissection
AF Atrial fibrillation
AHA American Heart Association
AOS Aneurysm–osteoarthritis syndrome
ARVC Arrhythmogenic right-ventricular cardiomyopathy
AS Aortic valve stenosis
ASD Atrial septal defect
AVSD Atrioventricular septal defect
BAV Bicuspid aortic valve
BrS Brugada syndrome
CCD Cardiac conduction disease
CGH Comparative genomic hybridization
CHD Congenital heart disease/defect
CNV Copy number variation
CoA Coarctation of the aorta
CPVT Catecholaminergic polymorphic ventricular tachycardia
DCM Dilated cardiomyopathy
EDS Ehlers–Danlos syndrome

EHRA European Heart Rhythm Association
ESC European Society of Cardiology
FH Familial hypercholesterolemia
FISH Fluorescence in situ hybridization
FTAAD Familial thoracic aortic aneurysms and dissections
HCM Hypertrophic cardiomyopathy
HRS Heart Rhythm Society
ICD Implantable cardioverter defibrillator
LDS Loyes–Dietz syndrome
LQTS Long QT syndrome
LVH Left-ventricular hypertrophy
LVNC Left-ventricular noncompaction
MLPA Multiplex ligation-dependent probe amplification
MS Marfan syndrome
NGS Next generation sequencing
Q-PCR Quantitative real-time polymerase chain reaction
RCM Restrictive cardiomyopathy
SADS Sudden arrhythmic death syndrome
SCD Sudden cardiac death
SQTS Short QT syndrome
TAA Thoracic aortic aneurysm
TAAD Thoracic aortic aneurysms and dissections
TOF Tetralogy of Fallot
VSD Ventricular septal defect
WES Whole exome sequencing
WGS Whole-genome sequencing

ACKNOWLEDGMENTS

Supported by Polish National Science Center (NCN) Grant 2011/01/B/NZ4/03455.

REFERENCES

[1] Jongbloed JD, Posafalvi A, Kerstjens-Frederikse WS, Sinke RJ, van Tintelen JP. New clinical molecular diagnostic methods for congenital and inherited heart disease. Expert Opin Med Diagn 2011;5(1):9–24.

[2] Ploski R, Pollak A, Muller S, Franaszczyk M, Michalak E, Kosinska J, et al. Does p.Q247X in TRIM63 cause human hypertrophic cardiomyopathy? Circ Res 2014;114(2):e2–5.

[3] Ackerman MJ, Priori SG, Willems S, Berul C, Brugada R, Calkins H, et al. HRS/EHRA expert consensus statement on the state of genetic testing for the channelopathies and cardiomyopathies this document was developed as a partnership between the Heart Rhythm Society (HRS) and the European Heart Rhythm Association (EHRA). Heart Rhythm 2011;8(8):1308–39.

[4] Green RC, Berg JS, Grody WW, Kalia SS, Korf BR, Martin CL, et al. ACMG recommendations for reporting of incidental findings in clinical exome and genome sequencing. Genet Med 2013;15(7):565–74.

[5] Platt J, Cox R, Enns GM. Points to consider in the clinical use of NGS panels for mitochondrial disease: an analysis of gene inclusion and consent forms. J Genet Couns 2014;23(4):594–603.

[6] Greenway SC, Pereira AC, Lin JC, DePalma SR, Israel SJ, Mesquita SM, et al. De novo copy number variants identify new genes and loci in isolated sporadic tetralogy of Fallot. Nat Genet 2009;41(8):931–5.

[7] Elliott P, Andersson B, Arbustini E, Bilinska Z, Cecchi F, Charron P, et al. Classification of the cardiomyopathies: a position statement from the European Society of Cardiology Working Group on Myocardial and Pericardial Diseases. Eur Heart J 2008;29(2):270–6.

[8] Arbustini E, Narula N, Tavazzi L, Serio A, Grasso M, Favalli V, et al. The MOGE(S) classification of cardiomyopathy for clinicians. J Am Coll Cardiol 2014;64(3):304–18.

[9] Geisterfer-Lowrance AA, Kass S, Tanigawa G, Vosberg HP, McKenna W, Seidman CE, et al. A molecular basis for familial hypertrophic cardiomyopathy: a beta cardiac myosin heavy chain gene missense mutation. Cell 1990;62(5):999–1006.

[10] Galmiche L, Serre V, Beinat M, Assouline Z, Lebre AS, Chretien D, et al. Exome sequencing identifies MRPL3 mutation in mitochondrial cardiomyopathy. Hum Mutat 2011;32(11):1225–31.

[11] Carroll CJ, Isohanni P, Poyhonen R, Euro L, Richter U, Brilhante V, et al. Whole-exome sequencing identifies a mutation in the mitochondrial ribosome protein MRPL44 to underlie mitochondrial infantile cardiomyopathy. J Med Genet 2013;50(3):151–9.

[12] Gotz A, Tyynismaa H, Euro L, Ellonen P, Hyotylainen T, Ojala T, et al. Exome sequencing identifies mitochondrial alanyl-tRNA synthetase mutations in infantile mitochondrial cardiomyopathy. Am J Hum Genet 2011;88(5):635–42.

[13] Theis JL, Sharpe KM, Matsumoto ME, Chai HS, Nair AA, Theis JD, et al. Homozygosity mapping and exome sequencing reveal GATAD1 mutation in autosomal recessive dilated cardiomyopathy. Circ Cardiovasc Genet 2011;4(6):585–94.

[14] Meder B, Haas J, Keller A, Heid C, Just S, Borries A, et al. Targeted next-generation sequencing for the molecular genetic diagnostics of cardiomyopathies. Circ Cardiovasc Genet 2011;4(2):110–22.

[15] Punetha J, Hoffman EP. Short read (next-generation) sequencing: a tutorial with cardiomyopathy diagnostics as an exemplar. Circ Cardiovasc Genet 2013;6(4):427–34.

[16] Lekanne Deprez RH, Muurling-Vlietman JJ, Hruda J, Baars MJ, Wijnaendts LC, Stolte-Dijkstra I, et al. Two cases of severe neonatal hypertrophic cardiomyopathy caused by compound heterozygous mutations in the MYBPC3 gene. J Med Genet 2006;43(10):829–32.

[17] Xu T, Yang Z, Vatta M, Rampazzo A, Beffagna G, Pilichou K, et al. Compound and digenic heterozygosity contributes to arrhythmogenic right ventricular cardiomyopathy. J Am Coll Cardiol 2010;55(6):587–97.

[18] Girolami F, Ho CY, Semsarian C, Baldi M, Will ML, Baldini K, et al. Clinical features and outcome of hypertrophic cardiomyopathy associated with triple sarcomere protein gene mutations. J Am Coll Cardiol 2010;55(14):1444–53.

[19] Tsoutsman T, Kelly M, Ng DC, Tan JE, Tu E, Lam L, et al. Severe heart failure and early mortality in a double-mutation mouse model of familial hypertrophic cardiomyopathy. Circulation 2008;117(14):1820–31.

[20] McNally E, MacLeod H, Dellefave-Castillo L, editors. Arrhythmogenic right ventricular dysplasia/cardiomyopathy. Seattle: University of Washington; 2014.

[21] Herman DS, Lam L, Taylor MR, Wang L, Teekakirikul P, Christodoulou D, et al. Truncations of titin causing dilated cardiomyopathy. N Engl J Med 2012;366(7):619–28.

[22] Siu BL, Niimura H, Osborne JA, Fatkin D, MacRae C, Solomon S, et al. Familial dilated cardiomyopathy locus maps to chromosome 2q31. Circulation 1999;99(8):1022–6.

[23] Gerull B, Gramlich M, Atherton J, McNabb M, Trombitas K, Sasse-Klaassen S, et al. Mutations of TTN, encoding the giant muscle filament titin, cause familial dilated cardiomyopathy. Nat Genet 2002;30(2):201–4.

[24] Johnson R, Stockhammer K, Soka M, Ohanian M, Lam L, Fatkin D. Truncating Titin mutations: are they a primary cause of dilated cardiomyopathy or a susceptibility factor? Heart Lung Circ 2014;23(Suppl. 2):e11–2.

[25] Lopes LR, Zekavati A, Syrris P, Hubank M, Giambartolomei C, Dalageorgou C, et al. Genetic complexity in hypertrophic cardiomyopathy revealed by high-throughput sequencing. J Med Genet 2013;50(4):228–39.

[26] Pugh TJ, Kelly MA, Gowrisankar S, Hynes E, Seidman MA, Baxter SM, et al. The landscape of genetic variation in dilated cardiomyopathy as surveyed by clinical DNA sequencing. Genet Med 2014;16(8):601–8.

[27] Haas J, Frese KS, Peil B, Kloos W, Keller A, Nietsch R, et al. Atlas of the clinical genetics of human dilated cardiomyopathy. Eur Heart J 2014.

[28] Charron P, Arad M, Arbustini E, Basso C, Bilinska Z, Elliott P, et al. Genetic counselling and testing in cardiomyopathies: a position statement of the European Society of Cardiology Working Group on Myocardial and Pericardial Diseases. Eur Heart J 2010;31(22):2715–26.

[29] Cirino AL, Ho C. Hypertrophic cardiomyopathy overview. 1993.

[30] Richard P, Charron P, Carrier L, Ledeuil C, Cheav T, Pichereau C, et al. Hypertrophic cardiomyopathy: distribution of disease genes, spectrum of mutations, and implications for a molecular diagnosis strategy. Circulation 2003;107(17):2227–32.

[31] Ackerman MJ, VanDriest SL, Ommen SR, Will ML, Nishimura RA, Tajik AJ, et al. Prevalence and age-dependence of malignant mutations in the beta-myosin heavy chain and troponin T genes in hypertrophic cardiomyopathy: a comprehensive outpatient perspective. J Am Coll Cardiol 2002;39(12):2042–8.

[32] Watkins H, McKenna WJ, Thierfelder L, Suk HJ, Anan R, O'Donoghue A, et al. Mutations in the genes for cardiac troponin T and alpha-tropomyosin in hypertrophic cardiomyopathy. N Engl J Med 1995;332(16):1058–64.

[33] Niimura H, Bachinski LL, Sangwatanaroj S, Watkins H, Chudley AE, McKenna W, et al. Mutations in the gene for cardiac myosin-binding protein C and late-onset familial hypertrophic cardiomyopathy. N Engl J Med 1998;338(18):1248–57.

[34] Ingles J, Sarina T, Yeates L, Hunt L, Macciocca I, McCormack L, et al. Clinical predictors of genetic testing outcomes in hypertrophic cardiomyopathy. Genet Med 2013;15(12):972–7.

[35] Olivotto I, Girolami F, Ackerman MJ, Nistri S, Bos JM, Zachara E, et al. Myofilament protein gene mutation screening and outcome of patients with hypertrophic cardiomyopathy. Mayo Clin Proc 2008;83(6): 630–8.

[36] Marian AJ, Yu QT, Workman R, Greve G, Roberts R. Angiotensin-converting enzyme polymorphism in hypertrophic cardiomyopathy and sudden cardiac death. Lancet 1993;342(8879):1085–6.

[37] Doolan G, Nguyen L, Chung J, Ingles J, Semsarian C. Progression of left ventricular hypertrophy and the angiotensin-converting enzyme gene polymorphism in hypertrophic cardiomyopathy. Int J Cardiol 2004;96(2):157–63.

[38] Arad M, Maron BJ, Gorham JM, Johnson Jr WH, Saul JP, Perez-Atayde AR, et al. Glycogen storage diseases presenting as hypertrophic cardiomyopathy. N Engl J Med 2005;352(4):362–72.

[39] Sachdev B, Takenaka T, Teraguchi H, Tei C, Lee P, McKenna WJ, et al. Prevalence of Anderson-Fabry disease in male patients with late onset hypertrophic cardiomyopathy. Circulation 2002;105(12):1407–11.

[40] Elliott PM, Anastasakis A, Borger MA, Borggrefe M, Cecchi F, Charron P, et al. 2014 ESC Guidelines on diagnosis and management of hypertrophic cardiomyopathy: the Task Force for the Diagnosis and Management of Hypertrophic Cardiomyopathy of the European Society of Cardiology (ESC). Eur Heart J 2014.

[41] Cahill TJ, Ashrafian H, Watkins H. Genetic cardiomyopathies causing heart failure. Circ Res 2013;113(6): 660–75.

[42] Hershberger REM A, editor. Dilated cardiomyopathy overview. Seattle: University of Washington; 2007. [Updated 2013].

[43] Hershberger RE, Morales A. LMNA-related dilated cardiomyopathy. In: Pagon RA, Adam MP, Ardinger HH, et al., editors. GeneReviews®. 2010/03/20 ed. Seattle: University of Washington; 2013.

[44] van Spaendonck-Zwarts KY, van Hessem L, Jongbloed JD, de Walle HE, Capetanaki Y, van der Kooi AJ, et al. Desmin-related myopathy. Clin Genet 2011;80(4):354–66.

[45] Marcus FI, McKenna WJ, Sherrill D, Basso C, Bauce B, Bluemke DA, et al. Diagnosis of arrhythmogenic right ventricular cardiomyopathy/dysplasia: proposed modification of the task force criteria. Circulation 2010;121(13):1533–41.

[46] Quarta G, Muir A, Pantazis A, Syrris P, Gehmlich K, Garcia-Pavia P, et al. Familial evaluation in arrhythmogenic right ventricular cardiomyopathy: impact of genetics and revised task force criteria. Circulation 2011;123(23):2701–9.

[47] van Hengel J, Calore M, Bauce B, Dazzo E, Mazzotti E, De Bortoli M, et al. Mutations in the area composita protein alphaT-catenin are associated with arrhythmogenic right ventricular cardiomyopathy. Eur Heart J 2013;34(3):201–10.

[48] van der Zwaag PA, van Rijsingen IA, Asimaki A, Jongbloed JD, van Veldhuisen DJ, Wiesfeld AC, et al. Phospholamban R14del mutation in patients diagnosed with dilated cardiomyopathy or arrhythmogenic right ventricular cardiomyopathy: evidence supporting the concept of arrhythmogenic cardiomyopathy. Eur J Heart Fail 2012;14(11):1199–207.

[49] Sen-Chowdhry S, Syrris P, Ward D, Asimaki A, Sevdalis E, McKenna WJ. Clinical and genetic characterization of families with arrhythmogenic right ventricular dysplasia/cardiomyopathy provides novel insights into patterns of disease expression. Circulation 2007;115(13):1710–20.

[50] Merner ND, Hodgkinson KA, Haywood AF, Connors S, French VM, Drenckhahn JD, et al. Arrhythmogenic right ventricular cardiomyopathy type 5 is a fully penetrant, lethal arrhythmic disorder caused by a missense mutation in the TMEM43 gene. Am J Hum Genet 2008;82(4):809–21.

[51] Stollberger C, Finsterer J. Extracardiac medical and neuromuscular implications in restrictive cardiomyopathy. Clin Cardiol 2007;30(8):375–80.

[52] Arbustini E, Pasotti M, Pilotto A, Pellegrini C, Grasso M, Previtali S, et al. Desmin accumulation restrictive cardiomyopathy and atrioventricular block associated with desmin gene defects. Eur J Heart Fail 2006;8(5):477–83.

[53] Hund E, Linke RP, Willig F, Grau A. Transthyretin-associated neuropathic amyloidosis. Pathogenesis and treatment. Neurology 2001;56(4):431–5.

[54] Monserrat L, Hermida-Prieto M, Fernandez X, Rodriguez I, Dumont C, Cazon L, et al. Mutation in the alpha-cardiac actin gene associated with apical hypertrophic cardiomyopathy, left ventricular noncompaction, and septal defects. Eur Heart J 2007;28(16):1953–61.

[55] Budde BS, Binner P, Waldmuller S, Hohne W, Blankenfeldt W, Hassfeld S, et al. Noncompaction of the ventricular myocardium is associated with a de novo mutation in the beta-myosin heavy chain gene. PLoS One 2007;2(12):e1362.

[56] Luxan G, Casanova JC, Martinez-Poveda B, Prados B, D'Amato G, MacGrogan D, et al. Mutations in the NOTCH pathway regulator MIB1 cause left ventricular noncompaction cardiomyopathy. Nat Med 2013;19(2): 193–201.

[57] Vatta M, Mohapatra B, Jimenez S, Sanchez X, Faulkner G, Perles Z, et al. Mutations in Cypher/ZASP in patients with dilated cardiomyopathy and left ventricular non-compaction. J Am Coll Cardiol 2003;42(11): 2014–27.

[58] Arndt AK, Schafer S, Drenckhahn JD, Sabeh MK, Plovie ER, Caliebe A, et al. Fine mapping of the 1p36 deletion syndrome identifies mutation of PRDM16 as a cause of cardiomyopathy. Am J Hum Genet 2013;93(1): 67–77.

[59] Hermida-Prieto M, Monserrat L, Castro-Beiras A, Laredo R, Soler R, Peteiro J, et al. Familial dilated cardiomyopathy and isolated left ventricular noncompaction associated with lamin A/C gene mutations. Am J Cardiol 2004;94(1):50–4.

[60] Ichida F, Tsubata S, Bowles KR, Haneda N, Uese K, Miyawaki T, et al. Novel gene mutations in patients with left ventricular noncompaction or Barth syndrome. Circulation 2001;103(9):1256–63.

[61] Bione S, D'Adamo P, Maestrini E, Gedeon AK, Bolhuis PA, Toniolo D. A novel X-linked gene, G4.5. is responsible for Barth syndrome. Nat Genet 1996;12(4):385–9.

[62] Klaassen S, Probst S, Oechslin E, Gerull B, Krings G, Schuler P, et al. Mutations in sarcomere protein genes in left ventricular noncompaction. Circulation 2008;117(22):2893–901.

[63] Hoedemaekers YM, Caliskan K, Michels M, Frohn-Mulder I, van der Smagt JJ, Phefferkorn JE, et al. The importance of genetic counseling, DNA diagnostics, and cardiologic family screening in left ventricular noncompaction cardiomyopathy. Circ Cardiovasc Genet 2010;3(3):232–9.

[64] Bai R, Napolitano C, Bloise R, Monteforte N, Priori SG. Yield of genetic screening in inherited cardiac channelopathies: how to prioritize access to genetic testing. Circ Arrhythm Electrophysiol 2009;2(1):6–15.

[65] Semsarian C, Hamilton RM. Key role of the molecular autopsy in sudden unexpected death. Heart Rhythm 2012;9(1):145–50.

[66] Ackerman MJ, Khositseth A, Tester DJ, Schwartz PJ. Congenital long QT syndrome. Electrical diseases of the heart. New York: Springer Publishing; 2008. p. 462–482.

[67] Tristani-Firouzi M, Jensen JL, Donaldson MR, Sansone V, Meola G, Hahn A, et al. Functional and clinical characterization of KCNJ2 mutations associated with LQT7 (Andersen syndrome). J Clin Invest 2002;110(3): 381–8.

[68] Splawski I, Timothy KW, Sharpe LM, Decher N, Kumar P, Bloise R, et al. Ca(V)1.2 calcium channel dysfunction causes a multisystem disorder including arrhythmia and autism. Cell 2004;119(1):19–31.

[69] Boczek NJ, Best JM, Tester DJ, Giudicessi JR, Middha S, Evans JM, et al. Exome sequencing and systems biology converge to identify novel mutations in the L-type calcium channel, CACNA1C, linked to autosomal dominant long QT syndrome. Circ Cardiovasc Genet 2013;6(3):279–89.

[70] Tranebjaerg L, Samson RA, Green GE. Jervell and Lange-Nielsen syndrome. In: Pagon RA, Adam MP, Ardinger HH, editors. GeneReviews® Seattle. University of Washington; 2012.

[71] Zhang L, Timothy KW, Vincent GM, Lehmann MH, Fox J, Giuli LC, et al. Spectrum of ST-T-wave patterns and repolarization parameters in congenital long-QT syndrome: ECG findings identify genotypes. Circulation 2000;102(23):2849–55.

[72] Schwartz PJ, Priori SG, Spazzolini C, Moss AJ, Vincent GM, Napolitano C, et al. Genotype-phenotype correlation in the long-QT syndrome: gene-specific triggers for life-threatening arrhythmias. Circulation 2001;103(1):89–95.

[73] Choi G, Kopplin LJ, Tester DJ, Will ML, Haglund CM, Ackerman MJ. Spectrum and frequency of cardiac channel defects in swimming-triggered arrhythmia syndromes. Circulation 2004;110(15):2119–24.

[74] Moss AJ, Shimizu W, Wilde AA, Towbin JA, Zareba W, Robinson JL, et al. Clinical aspects of type-1 long-QT syndrome by location, coding type, and biophysical function of mutations involving the KCNQ1 gene. Circulation 2007;115(19):2481–9.

[75] Shimizu W, Moss AJ, Wilde AA, Towbin JA, Ackerman MJ, January CT, et al. Genotype-phenotype aspects of type 2 long QT syndrome. J Am Coll Cardiol 2009;54(22):2052–62.

[76] Crotti L, Monti MC, Insolia R, Peljto A, Goosen A, Brink PA, et al. NOS1AP is a genetic modifier of the long-QT syndrome. Circulation 2009;120(17):1657–63.

[77] Eddy CA, MacCormick JM, Chung SK, Crawford JR, Love DR, Rees MI, et al. Identification of large gene deletions and duplications in KCNQ1 and KCNH2 in patients with long QT syndrome. Heart Rhythm 2008;5(9):1275–81.

[78] Tester DJ, Benton AJ, Train L, Deal B, Baudhuin LM, Ackerman MJ. Prevalence and spectrum of large deletions or duplications in the major long QT syndrome-susceptibility genes and implications for long QT syndrome genetic testing. Am J Cardiol 2010;106(8):1124–8.

[79] Priori SG, Wilde AA, Horie M, Cho Y, Behr ER, Berul C, et al. Executive summary: HRS/EHRA/APHRS expert consensus statement on the diagnosis and management of patients with inherited primary arrhythmia syndromes. Heart Rhythm 2013;10(12):e85–108.

[80] Vatta M, Dumaine R, Varghese G, Richard TA, Shimizu W, Aihara N, et al. Genetic and biophysical basis of sudden unexplained nocturnal death syndrome (SUNDS), a disease allelic to Brugada syndrome. Hum Mol Genet 2002;11(3):337–45.

[81] Kapplinger JD, Tester DJ, Alders M, Benito B, Berthet M, Brugada J, et al. An international compendium of mutations in the SCN5A-encoded cardiac sodium channel in patients referred for Brugada syndrome genetic testing. Heart Rhythm 2010;7(1):33–46.

[82] Weiss R, Barmada MM, Nguyen T, Seibel JS, Cavlovich D, Kornblit CA, et al. Clinical and molecular heterogeneity in the Brugada syndrome: a novel gene locus on chromosome 3. Circulation 2002;105(6):707–13.

[83] Antzelevitch C, Pollevick GD, Cordeiro JM, Casis O, Sanguinetti MC, Aizawa Y, et al. Loss-of-function mutations in the cardiac calcium channel underlie a new clinical entity characterized by ST-segment elevation, short QT intervals, and sudden cardiac death. Circulation 2007;115(4):442–9.

[84] Watanabe H, Koopmann TT, Le Scouarnec S, Yang T, Ingram CR, Schott JJ, et al. Sodium channel beta1 subunit mutations associated with Brugada syndrome and cardiac conduction disease in humans. J Clin Invest 2008;118(6):2260–8.

[85] Delpon E, Cordeiro JM, Nunez L, Thomsen PE, Guerchicoff A, Pollevick GD, et al. Functional effects of KCNE3 mutation and its role in the development of Brugada syndrome. Circ Arrhythm Electrophysiol 2008;1(3):209–18.

[86] Hu D, Barajas-Martinez H, Burashnikov E, Springer M, Wu Y, Varro A, et al. A mutation in the beta 3 subunit of the cardiac sodium channel associated with Brugada ECG phenotype. Circ Cardiovasc Genet 2009;2(3):270–8.

[87] Ueda K, Hirano Y, Higashiuesato Y, Aizawa Y, Hayashi T, Inagaki N, et al. Role of HCN4 channel in preventing ventricular arrhythmia. J Hum Genet 2009;54(2):115–21.

[88] Gehi AK, Duong TD, Metz LD, Gomes JA, Mehta D. Risk stratification of individuals with the Brugada electrocardiogram: a meta-analysis. J Cardiovasc Electrophysiol 2006;17(6):577–83.

[89] Meregalli PG, Tan HL, Probst V, Koopmann TT, Tanck MW, Bhuiyan ZA, et al. Type of SCN5A mutation determines clinical severity and degree of conduction slowing in loss-of-function sodium channelopathies. Heart Rhythm 2009;6(3):341–8.

[90] di Barletta MR, Viatchenko-Karpinski S, Nori A, Memmi M, Terentyev D, Turcato F, et al. Clinical phenotype and functional characterization of CASQ2 mutations associated with catecholaminergic polymorphic ventricular tachycardia. Circulation 2006;114(10):1012–9.

[91] Toutenhoofd SL, Foletti D, Wicki R, Rhyner JA, Garcia F, Tolon R, et al. Characterization of the human CALM2 calmodulin gene and comparison of the transcriptional activity of CALM1, CALM2 and CALM3. Cell Calcium 1998;23(5):323–38.

[92] Roux-Buisson N, Cacheux M, Fourest-Lieuvin A, Fauconnier J, Brocard J, Denjoy I, et al. Absence of triadin, a protein of the calcium release complex, is responsible for cardiac arrhythmia with sudden death in human. Hum Mol Genet 2012;21(12):2759–67.

[93] Vega AL, Tester DJ, Ackerman MJ, Makielski JC. Protein kinase A-dependent biophysical phenotype for V227F-KCNJ2 mutation in catecholaminergic polymorphic ventricular tachycardia. Circ Arrhythm Electrophysiol 2009;2(5):540–7.

[94] Mohler PJ, Splawski I, Napolitano C, Bottelli G, Sharpe L, Timothy K, et al. A cardiac arrhythmia syndrome caused by loss of ankyrin-B function. Proc Natl Acad Sci USA 2004;101(24):9137–42.

[95] Postema PG, Van den Berg M, Van Tintelen JP, Van den Heuvel F, Grundeken M, Hofman N, et al. Founder mutations in the Netherlands: SCN5a 1795insD, the first described arrhythmia overlap syndrome and one of the largest and best characterised families worldwide. Neth Heart J 2009;17(11):422–8.

[96] Kruse M, Schulze-Bahr E, Corfield V, Beckmann A, Stallmeyer B, Kurtbay G, et al. Impaired endocytosis of the ion channel TRPM4 is associated with human progressive familial heart block type I. J Clin Invest 2009;119(9):2737–44.

[97] Liu H, El Zein L, Kruse M, Guinamard R, Beckmann A, Bozio A, et al. Gain-of-function mutations in TRPM4 cause autosomal dominant isolated cardiac conduction disease. Circ Cardiovasc Genet 2010;3(4):374–85.

[98] Stieber J, Hofmann F, Ludwig A. Pacemaker channels and sinus node arrhythmia. Trends Cardiovasc Med 2004;14(1):23–8.

[99] van Tintelen JP, Hofstra RM, Katerberg H, Rossenbacker T, Wiesfeld AC, du Marchie Sarvaas GJ, et al. High yield of LMNA mutations in patients with dilated cardiomyopathy and/or conduction disease referred to cardiogenetics outpatient clinics. Am Heart J 2007;154(6):1130–9.

[100] Schott JJ, Benson DW, Basson CT, Pease W, Silberbach GM, Moak JP, et al. Congenital heart disease caused by mutations in the transcription factor NKX2-5. Science 1998;281(5373):108–11.

[101] Garg V, Kathiriya IS, Barnes R, Schluterman MK, King IN, Butler CA, et al. GATA4 mutations cause human congenital heart defects and reveal an interaction with TBX5. Nature 2003;424(6947):443–7.

[102] Gollob MH, Redpath CJ, Roberts JD. The short QT syndrome: proposed diagnostic criteria. J Am Coll Cardiol 2011;57(7):802–12.

[103] Hu D, Barajas-Martinez H, Terzic A, Park S, Pfeiffer R, Burashnikov E, et al. ABCC9 is a novel Brugada and early repolarization syndrome susceptibility gene. Int J Cardiol 2014;171(3):431–42.

[104] Olson TM, Alekseev AE, Moreau C, Liu XK, Zingman LV, Miki T, et al. KATP channel mutation confers risk for vein of Marshall adrenergic atrial fibrillation. Nat Clin Pract Cardiovasc Med 2007;4(2):110–6.

[105] Watanabe H, Darbar D, Kaiser DW, Jiramongkolchai K, Chopra S, Donahue BS, et al. Mutations in sodium channel beta1- and beta2-subunits associated with atrial fibrillation. Circ Arrhythm Electrophysiol 2009;2(3):268–75.

[106] Nagase T, Ishikawa K, Suyama M, Kikuno R, Miyajima N, Tanaka A, et al. Prediction of the coding sequences of unidentified human genes. XI. The complete sequences of 100 new cDNA clones from brain which code for large proteins in vitro. DNA Res 1998;5(5):277–86.

[107] Olesen MS, Jespersen T, Nielsen JB, Liang B, Moller DV, Hedley P, et al. Mutations in sodium channel beta-subunit SCN3B are associated with early-onset lone atrial fibrillation. Cardiovasc Res 2011;89(4):786–93.

[108] Li RG, Wang Q, Xu YJ, Zhang M, Qu XK, Liu X, et al. Mutations of the SCN4B-encoded sodium channel beta4 subunit in familial atrial fibrillation. Int J Mol Med 2013;32(1):144–50.

[109] Hiratzka LF, Bakris GL, Beckman JA, Bersin RM, Carr VF, Casey Jr DE, et al. 2010 ACCF/AHA/AATS/ACR/ASA/SCA/SCAI/SIR/STS/SVM guidelines for the diagnosis and management of patients with thoracic aortic disease. A report of the American College of Cardiology Foundation/American Heart Association Task Force on Practice Guidelines, American Association for Thoracic Surgery, American College of Radiology, American Stroke Association, Society of Cardiovascular Anesthesiologists, Society for Cardiovascular Angiography and Interventions, Society of Interventional Radiology, Society of Thoracic Surgeons, and Society for Vascular Medicine Circulation 2010;121(13):e266–369.

[110] Elefteriades JA. Thoracic aortic aneurysm: reading the enemy's playbook. Curr Probl Cardiol 2008;33(5):203–77.

[111] Milewicz DM, Regalado E. Thoracic aortic aneurysms and aortic dissections. In: Pagon RA, Adam MP, Ardinger HH, Bird TD, Dolan CR, Fong CT, et al., editors. GeneReviews. 2010/03/20 ed. Seattle: University of Washington; 2012.

[112] De Backer J, Campens L, De Paepe A. Genes in thoracic aortic aneurysms/dissections – do they matter? Ann Cardiothorac Surg 2013;2(1):73–82.

[113] Hiratzka LF, Bakris GL, Beckman JA, Bersin RM, Carr VF, Casey Jr DE, et al. 2010 ACCF/AHA/AATS/ACR/ASA/SCA/SCAI/SIR/STS/SVM guidelines for the diagnosis and management of patients with thoracic aortic disease. A Report of the American College of Cardiology Foundation/American Heart Association Task Force on Practice Guidelines, American Association for Thoracic Surgery, American College of Radiology, American Stroke Association, Society of Cardiovascular Anesthesiologists, Society for Cardiovascular Angiography and Interventions, Society of Interventional Radiology, Society of Thoracic Surgeons, and Society for Vascular Medicine J Am Coll Cardiol 2010;55(14):e27–129.

[114] Judge DP, Dietz HC. Marfan's syndrome. Lancet 2005;366(9501):1965–76.

[115] Pepin M, Schwarze U, Superti-Furga A, Byers PH. Clinical and genetic features of Ehlers-Danlos syndrome type IV, the vascular type. N Engl J Med 2000;342(10):673–80.

[116] Malfait F, Symoens S, De Backer J, Hermanns-Le T, Sakalihasan N, Lapiere CM, et al. Three arginine to cysteine substitutions in the pro-alpha (I)-collagen chain cause Ehlers-Danlos syndrome with a propensity to arterial rupture in early adulthood. Hum Mutat 2007;28(4):387–95.

[117] Loeys BL, Chen J, Neptune ER, Judge DP, Podowski M, Holm T, et al. A syndrome of altered cardiovascular, craniofacial, neurocognitive and skeletal development caused by mutations in TGFBR1 or TGFBR2. Nat Genet 2005;37(3):275–81.

[118] van de Laar IM, van der Linde D, Oei EH, Bos PK, Bessems JH, Bierma-Zeinstra SM, et al. Phenotypic spectrum of the SMAD3-related aneurysms-osteoarthritis syndrome. J Med Genet 2012;49(1):47–57.

[119] Lindsay ME, Schepers D, Bolar NA, Doyle JJ, Gallo E, Fert-Bober J, et al. Loss-of-function mutations in TGFB2 cause a syndromic presentation of thoracic aortic aneurysm. Nat Genet 2012;44(8):922–7.

[120] Coucke PJ, Willaert A, Wessels MW, Callewaert B, Zoppi N, De Backer J, et al. Mutations in the facilitative glucose transporter GLUT10 alter angiogenesis and cause arterial tortuosity syndrome. Nat Genet 2006;38(4):452–7.

[121] Renard M, Holm T, Veith R, Callewaert BL, Ades LC, Baspinar O, et al. Altered TGFbeta signaling and cardiovascular manifestations in patients with autosomal recessive cutis laxa type I caused by fibulin-4 deficiency. Eur J Hum Genet 2010;18(8):895–901.

[122] Szabo Z, Crepeau MW, Mitchell AL, Stephan MJ, Puntel RA, Yin Loke K, et al. Aortic aneurysmal disease and cutis laxa caused by defects in the elastin gene. J Med Genet 2006;43(3):255–8.

[123] Campens L, Renard M, Callewaert B, Coucke P, De Backer J, De Paepe A. New insights into the molecular diagnosis and management of heritable thoracic aortic aneurysms and dissections. Pol Arch Med Wewn 2013;123(12):693–700.

[124] Guo DC, Papke CL, Tran-Fadulu V, Regalado ES, Avidan N, Johnson RJ, et al. Mutations in smooth muscle alpha-actin (ACTA2) cause coronary artery disease, stroke, and Moyamoya disease, along with thoracic aortic disease. Am J Hum Genet 2009;84(5):617–27.

[125] Stheneur C, Collod-Beroud G, Faivre L, Buyck JF, Gouya L, Le Parc JM, et al. Identification of the minimal combination of clinical features in probands for efficient mutation detection in the FBN1 gene. Eur J Hum Genet 2009;17(9):1121–8.

[126] McKellar SH, Tester DJ, Yagubyan M, Majumdar R, Ackerman MJ, Sundt 3rd TM. Novel NOTCH1 mutations in patients with bicuspid aortic valve disease and thoracic aortic aneurysms. J Thorac Cardiovasc Surg 2007;134(2):290–6.

[127] Prakash SK, LeMaire SA, Guo DC, Russell L, Regalado ES, Golabbakhsh H, et al. Rare copy number variants disrupt genes regulating vascular smooth muscle cell adhesion and contractility in sporadic thoracic aortic aneurysms and dissections. Am J Hum Genet 2010;87(6):743–56.

[128] Martin LAB. D.W. congenital heart disease. In: Ginsburg Gsw HF, editor. Genomic and personalized medicine. Elsevier Inc.; 2013. p. 624–34.

[129] Pierpont ME, Basson CT, Benson Jr DW, Gelb BD, Giglia TM, Goldmuntz E, et al. Genetic basis for congenital heart defects: current knowledge: a scientific statement from the American Heart Association Congenital Cardiac Defects Committee, Council on Cardiovascular Disease in the Young: endorsed by the American Academy of Pediatrics. Circulation 2007;115(23):3015–38.

[130] Moons P, Meijboom FJ, Baumgartner H, Trindade PT, Huyghe E, Kaemmerer H. Structure and activities of adult congenital heart disease programmes in Europe. Eur Heart J 2010;31(11):1305–10.

[131] Raisher BD, Dowton SB, Grant JW. Father and two children with total anomalous pulmonary venous connection. Am J Med Genet 1991;40(1):105–6.

[132] Ferencz C, Neill CA, Boughman JA, Rubin JD, Brenner JI, Perry LW. Congenital cardiovascular malformations associated with chromosome abnormalities: an epidemiologic study. J Pediatr 1989;114(1):79–86.

[133] Kodo K, Nishizawa T, Furutani M, Arai S, Yamamura E, Joo K, et al. GATA6 mutations cause human cardiac outflow tract defects by disrupting semaphorin-plexin signaling. Proc Natl Acad Sci USA 2009;106(33):13933–8.

[134] Garg V, Muth AN, Ransom JF, Schluterman MK, Barnes R, King IN, et al. Mutations in NOTCH1 cause aortic valve disease. Nature 2005;437(7056):270–4.

[135] Arrington CB, Bleyl SB, Matsunami N, Bonnell GD, Otterud BE, Nielsen DC, et al. Exome analysis of a family with pleiotropic congenital heart disease. Circ Cardiovasc Genet 2012;5(2):175–82.

[136] Al Turki S, Manickaraj AK, Mercer CL, Gerety SS, Hitz MP, Lindsay S, et al. Rare variants in NR2F2 cause congenital heart defects in humans. Am J Hum Genet 2014;94(4):574–85.

[137] Zaidi S, Choi M, Wakimoto H, Ma L, Jiang J, Overton JD, et al. De novo mutations in histone-modifying genes in congenital heart disease. Nature 2013;498(7453):220–3.

[138] Benson DW. Genetic origins of pediatric heart disease. Pediatr Cardiol 2010;31(3):422–9.

[139] Thienpont B, Mertens L, de Ravel T, Eyskens B, Boshoff D, Maas N, et al. Submicroscopic chromosomal imbalances detected by array-CGH are a frequent cause of congenital heart defects in selected patients. Eur Heart J 2007;28(22):2778–84.

[140] Li F, Lisi EC, Wohler ES, Hamosh A, Batista DA. 3q29 interstitial microdeletion syndrome: an inherited case associated with cardiac defect and normal cognition. Eur J Med Genet 2009;52(5):349–52.

[141] van der Graaf A, Avis HJ, Kusters DM, Vissers MN, Hutten BA, Defesche JC, et al. Molecular basis of autosomal dominant hypercholesterolemia: assessment in a large cohort of hypercholesterolemic children. Circulation 2011;123(11):1167–73.

[142] Brice P, Burton H, Edwards CW, Humphries SE, Aitman TJ. Familial hypercholesterolaemia: a pressing issue for European health care. Atherosclerosis 2013;231(2):223–6.

[143] Varret M, Abifadel M, Rabes JP, Boileau C. Genetic heterogeneity of autosomal dominant hypercholester-olemia. Clin Genet 2008;73(1):1–13.

[144] Asselbergs FW, Guo Y, van Iperen EP, Sivapalaratnam S, Tragante V, Lanktree MB, et al. Large-scale gene-centric meta-analysis across 32 studies identifies multiple lipid loci. Am J Hum Genet 2012;91(5):823–38.

[145] Marduel M, Ouguerram K, Serre V, Bonnefont-Rousselot D, Marques-Pinheiro A, Erik Berge K, et al. Description of a large family with autosomal dominant hypercholesterolemia associated with the APOE p.Leu167del mutation. Hum Mutat 2013;34(1):83–7.

[146] Awan Z, Choi HY, Stitziel N, Ruel I, Bamimore MA, Husa R, et al. APOE p.Leu167del mutation in familial hypercholesterolemia. Atherosclerosis 2013;231(2):218–22.

[147] Soutar AK, Naoumova RP. Autosomal recessive hypercholesterolemia. Semin Vasc Med 2004;4(3):241–8.

[148] Berge KE, Tian H, Graf GA, Yu L, Grishin NV, Schultz J, et al. Accumulation of dietary cholesterol in sitosterolemia caused by mutations in adjacent ABC transporters. Science 2000;290(5497):1771–5.

[149] Rios J, Stein E, Shendure J, Hobbs HH, Cohen JC. Identification by whole-genome resequencing of gene defect responsible for severe hypercholesterolemia. Hum Mol Genet 2010;19(22):4313–8.

[150] Nordestgaard BG, Chapman MJ, Humphries SE, Ginsberg HN, Masana L, Descamps OS, et al. Familial hypercholesterolaemia is underdiagnosed and undertreated in the general population: guidance for clinicians to prevent coronary heart disease: consensus statement of the European Atherosclerosis Society. Eur Heart J 2013;34(45):3478-90a.

NEXT GENERATION SEQUENCING IN PHARMACOGENOMICS

11

Urszula Demkow

Department of Laboratory Diagnostics and Clinical Immunology of Developmental Age, Medical University of Warsaw, Warsaw, Poland

Right drug, for the right patient, in the right dose, at the right time.

CHAPTER OUTLINE

INTRODUCTION

Pharmacogenomics refers to the relation between the human genetic profile and individual differences in clinical drug response. The aim of pharmacogenomics is to optimize treatment outcome and individualize therapy (maximal efficacy, increased safety of the medication, and decreased risk of drug interactions) [1]. Both single-nucleotide variants and structural alterations such as inversions, insertions, deletions, or copy number variations (CNVs) contribute to the drug response. Two approaches are essential in pharmacogenomics. The first is *pharmacokinetics,* exploring the metabolic profile of the drug, that is, the specific panel of enzymes catalyzing absorption, distribution, metabolism, and excretion (ADME). The second is *pharmacodynamics,* referring to the drug concentration at the site of action. Pharmacodynamics determines how the drug interacts with its transporters, receptors, ion channels, enzymes, and signal transduction pathways in the target organ [2]. Pharmacogenomics significantly increased our understanding of variability in treatment response. The interindividual variation in ADME enzyme activity and receptor/transporter binding can be predicted by patient genotype. Personalized pharmacotherapy is especially advantageous where the therapeutic index of the drug is narrow and medications are administered at maximally tolerated doses, balancing between low efficacy and toxic effects of the drug.

Rapid advances in sequencing technologies, especially next generation sequencing (NGS), have facilitated the clinical implementation of multigene panels and subsequently complex genome-wide approaches largely replacing single-pharmacogene studies [3]. NGS as a diagnostic tool in pharmacogenomics can be used in a targeted manner or can be applied to study whole exomes or even whole genomes [4]. Hundreds of pharmacogenes have been identified and the list is still growing. Registers of some of these variants are now available online (i.e., for cytochrome P450) [5]. Numerous pharmacogenomic panels are available to personalize therapy of various diseases. The main limitation of targeted analysis is the fact that the metabolism of only a small fraction of all drugs is controlled by a few genomic variants, whereas for the majority of the drugs, smaller effects of multiple variants (most unknown to date) can be expected.

CANCER THERAPY

An increasing emphasis is being placed on the use of pharmacogenomics in cancer therapy. Unlike in other diseases, the pharmacogenomic approach in oncology includes both inherited (germ-line) and acquired (somatic) variants, affecting the outcome of the therapy in parallel. NGS contributes to the identification of key genetic variants in oncology (both germ-line and somatic) with a single-base approach and to the development of an integrated molecular blueprint of the tumor [6]. Generally somatic variants relate to the treatment efficacy of both conventional chemotherapy and targeted molecular drugs, whereas germ-line variants, by their association with ADME enzymes, point to the risk of developing severe adverse effects [5,7,8]. Targeted therapies based on a single somatic molecular marker are already applied as standards of care in several cancer types, including solid tumors such as melanoma, breast, colon, lung, gastrointestinal, and ovarian cancer as well as lymphoma and leukemia.

Well-established applications of targeted drugs based on pharmacogenomic markers (somatic) include several marker–drug associations [5,7,9,10]:

- crizotinib for breast cancer, lung cancer, and melanoma in *ALK*-positive tumors,
- nilotinib, bosutinib, dasatinib, imatinib, and ponatinib (*ABL*-positive neoplasms),
- vemurafenib (*BRAF*-positive tumors),
- cetuximab, erlotinib, afatinib, panitumumab, and vandetinib (*EGFR* mutations) in gastrointestinal tumors,

- trastuzumab, lapatinib, and pertuzumab (*HER2* overexpression),
- cetuximab and panitumumab (*KRAS*-positive cancers), and
- imatinib (*KIT*-positive tumors).

Additionally, imatinib was the first tyrosine kinase inhibitor applied in chronic myeloid leukemia (CML) to inhibit the fusion protein BCR–ABL, now being replaced by new tyrosine kinase inhibitors such as nilotinib, dasatinib, or ponatinib in new cases of CML or as an alternate treatment for patients with imatinib-resistant or -intolerant disease [11].

Common germ-line pharmacogenetic markers in oncology identifying patients at high risk of toxicity include [12]:

- *CYP2B6* (cyclophosphamide) and
- *TPMT* (mercaptopurine, thioguanine, cisplatin).

Some germ-line mutations may also influence the drug efficacy [13,14]:

- *CYP2B6*—tamoxifen and
- *BIM*—imatinib.

A large panel of drugs approved by the Food and Drug Administration (FDA) includes pharmacogenetic information in the drug label, many with cancer indications referencing germ-line variations [7,15].

As of this writing, large observational and clinical studies in oncology combining testing for both somatic and germ-line variations are limited. However, there is ample evidence that by focusing only on somatic or germ-line variations, key actionable mutations can be missed [16]. The advent of NGS facilitated sequencing of vast panels of genes or even entire cancer genomes to develop targeted molecular drugs replacing or complementing conventional chemotherapy [10,17].

Several multimarker tumor panels based on NGS allowing parallel evaluation of multiple genomic alterations driving tumorigenesis are commercially available [16]. Those NGS tests are dedicated to detecting a broad spectrum of genomic variants through targeted sequencing of selected genes with the intention of providing clinically actionable therapeutic targets. Multigene tests are believed to guide the treatment decision process more efficiently than a single-marker–single-drug approach, especially for patients with metastatic disease who have exhausted available classical therapies [16,18].

Another key issue underlying this approach is that similar molecular profiles in entirely different cancer types may require similar specific therapies. It can be true or not depending on the presence of additional mutations downstream from the primary molecular target that may modify the response to targeted therapy differentially across cancer types [16,19]. Additionally, molecular heterogeneity within both primary and metastatic tumors may further limit effectiveness of specific therapies selected on the basis of a single tumor biopsy [16,19].

Examples of commercially available NGS panels for personalized therapy in oncology are presented in Table 1. All panels include a selection of relevant variants known to be somatically altered in solid cancers and/or in common nonsolid tumors [16,20–22].

Furthermore, numerous academic and local cancer treatment centers offer their own in-house NGS-based oncogene panels. These tests could be used to discover variants matching with molecular drugs. All available NGS arrays are designed to provide physicians with clinically valuable information to guide the therapy of tumors, theoretically independent of the affected organ or tumor histology. The application of multigene NGS panels has so far primarily focused on patients with advanced metastatic

Table 1 Commercially Available Pharmacogenomic NGS Tests in Oncology [20–22]

Company	Panel	Target Genes
Illumina	TruSight™ tumor sequencing panel	*AKT1, ALK, APC, BRAF, CDH1, CTNNB1, EGFR, ERBB2, FBXW7, FGFR2, FOXL2, GNAQ, GNAS, KIT, KRAS, MAP2K1, MET, MSH6, NRAS, PDGFRA, PIK3CA, PTEN, SMAD4, SRC, STK11, TP53*
Arub Laboratories	Solid tumor mutation panel	*ABL1, AKT1, ALK, APC, ATM, BRAF, CDH1, CDKN2A, CSF1R, CTNNB1, EGFR, ERBB2, ERBB4, EZH2, FBXW7, FGFR1, FGFR2, FGFR3, GNA11, GNAQ, GNAS, HNF1A, HRAS, IDH1, IDH2, JAK3, KDR, KIT, KRAS, MET, MLH1, MPL, NOTCH1, NPM1, NRAS, PDGFRA, PIK3CA, PTEN, PTPN11, RB1, RET, SMAD4, SMARCB1, SMO, SRC, STK11, TP53, VHL*
LabCorb Specialty Testing Group	IntelliGEN oncology therapeutic panel	*ABL1, AKT1, ALK, APC, ATM, BRAF, CDH1, CDKN2A, CSF1R, CTNNB1, EGFR, ERBB2, ERBB4, EZH2, FBXW7, FGFR1, FGFR2, FGFR3, FLT3, GNA11, GNAQ, GNAS, HNF1A, HRAS IDH1, IDH2, JAK2, JAK3, KDR, KIT, KRAS, MET, MLH1 MPL, NOTCH1, NPM1, NRAS, PDGFRA, PIK3CA, PTEN, PTPN11, RB1, RET, SMAD4, SMARCB1, SMO, SRC, STK11, TP53, VHL*
AsuraGen	SuraSeq 7500	*ABL1, DNMT3A, GNAQ, MET, PTCH1, TP53, AKT1, EGFR, HIF1A, MPL, PTEN, VHL, AKT2, ERBB2, HRAS, NF2, PTPN11, BRAF, FES, IDH1, NOTCH1, RB1, CDH1, FGFR1, IDH2, NPM1, RET, CDK4, FGFR3, IKBKB, NRAS, SMAD4, CDKN2A, FLT3, JAK2, PAX5, SMARCB1, CEBPA, FOXL2, KIT, PDGFRA, SMO, CREBBP, GATA1, KRAS, PIK3CA, SRC, CTNNB1, GNA11, MEN1, PIK3R1, STK11*
Illumina	TruSight™ myeloid sequencing panel	*ABL1, ASXL1, ATRX, BCOR, BCORL1, BRAF, CARL, CBL, CBCL, CBLC, CDKN2A, CEBPA, CSF3R, CUX1, DNMT3A, ETV6/TEL, EZH2, FBXW7, FLT3, GATA1, GATA2, GNAS, HRAS IDH1, IDH2, IKZF1, JAK2, JAK3, KDM6A, KIT, KRAS, MLL, MPL, MYD88, NOTCH1, NPM1, NRAS, PDGFRA, PHF6, PTEN, PTPN11, RAD21, RUNX1, SETBP1, SF3B1, SMC1A, SMC3, SRSF2, STAG2, TET2, TP53, U2AF1, WT1, ZRSR2*
Foundation Medicine	FoundationOne	Large panel of 236 cancer-related genes and 47 introns from 19 genes often rearranged or altered in solid tumor cancers. Somatic alterations, including single base pair changes, insertions, deletions, copy number alterations, and selected fusions.
Foundation Medicine	FoundationOne Heme	Panel of 405 cancer-related genes (all kinds of leukemias, lymphoma, myeloma, myelodysplastic syndrome, sarcomas, and pediatric cancers). All classes of genomic alterations, including base pair substitutions, insertions and deletions, copy number variations, and select gene rearrangements. Additionally RNA sequencing across 265 genes to capture a broad range of gene fusions.
Ambry Genetics	CancerNext™	*APC, ATM, BARD1, BRCA1, BRCA2, BRIP1, BMPR1A, CDH1, CDK4, CDKN2A, CHEK2, EPCAM, MLH1, MRE11A, MSH2, MSH6, MUTYH, NBN, NF1, PALB2, PMS2, PTEN, RAD50, RAD51C, RAD51D, SMAD4, STK11, TP53*

disease. In such cases, these tests may provide new therapies, including off-label options, against a disease that does not respond to attempted forms of treatment.

The largest commercially available NGS pharmacogenomic cancer panel is FoundationOne (see Table 1). This assay was reported to detect at least one clinically actionable variant in 76% of samples ($n = 2200$), with an average of 1.57 clinically actionable variants detected per sample (range: 0–16) [16,23].

CLINICAL TRIALS IN ONCOLOGY

Large-scale clinical trials are needed for evaluating the clinical relevance and cost-effectiveness of genetic testing and personalized medicine. It is very challenging and costly to conduct big, prospective studies to establish causal associations between genetic variations and drug response. Even though NGS panels of oncopharmacogenes are widely available and affordable, the evidence that individual genetic profiling can help to guide personalized therapy and enhance cancer treatment outcome is still lacking across all solid tumors. Therefore implementation of such testing panels into oncology practice is delayed, in part owing to the paucity of evidence-based guidelines and contradictory recommendations [7]. Randomized clinical trials are considered to be the gold standard for demonstrating the effectiveness of therapeutic interventions. However, randomized controlled trials are expensive and time-consuming and need to be based on a large sample size, and therefore their application for the validation of many biomarkers is limited [7]. Alternatively, observational studies have some advantages over clinical trials (larger sample size for lower cost, selection of well-defined genomic subgroups, off-label usage of drugs) but are more prone to a number of biases [24]. Both carefully designed, prospective trials with random allocation of patients and well-designed retrospective and prospective case–control and cohort studies based on big, robust databases are crucial to confirm genotype–phenotype associations and to fully implement pharmacogenomics into clinical care. The management of the large amount of data per patient and handling large data at a patient-population level is an emerging issue [6].

A further problem in oncology is the variability of treatment response observed in early-stage clinical trials. Drugs can fail to induce disease regression or to prolong median progression-free survival in most patients but may exhibit profound activity in a small number of patients [24]. This effect can be due to rare variants. In contemporary oncology there is an increasing emphasis on simultaneous evaluation of multiple genomic variants, taking into consideration the molecular heterogeneity of examined samples and including less common variants within the biological pathways driving development of the tumor.

The 1000 Genomes Project can be a source of information about rare genetic variants [25,26]. Nevertheless, studying infrequent variants using conventional approaches requires an enormous sample size and may not be feasible in conventional clinical trials. Such variants are also not included in widely used NGS-based tests. The Blue Cross and Blue Shield (BCBS) Technology Evaluation Center provides some guidelines derived from three observational studies for the implementation of multigene panels in the decision-making process in oncology [16,27–30]. According to BCBS there is a need to develop consensus standards for the content of such panels, for classification of variants according to their clinical utility and validity, and to optimize genomic testing for clinical implementation into the process of clinical decision-making in oncology practice. A further important point in question is to prioritize specific therapies when multiple variants with associated targeted therapies are revealed, as there may be overlapping toxicities and drug interactions [18,27]. Largely available databases such as CancerLinQ™ that include both genetic and clinical data can help to support clinical decisions [31]. Such projects are likely to result in the discovery of unexpected patterns of disease or unusual responses to therapy that would otherwise not be readily apparent [31]. The access to such information and databases must be greatly facilitated for clinicians. Mechanisms to disseminate the findings and associated knowledge are also important. Finally, a wider educational effort is required to better prepare clinicians to be able to appropriately interpret and use data resulting from multigene NGS cancer panel testing.

Several ongoing clinical trials are investigating both the feasibility and the utility of NGS technology, including commercially available tests (i.e., IMAGE and WINTHER), in oncology clinical practice (Table 2). The aim of these studies is to define participants with different solid tumors that could benefit from molecularly targeted therapy [16,32–34].

The results of several completed clinical studies have already been published. The CAPRI-GOIM trial demonstrated that NGS analysis in metastatic colorectal cancer is feasible, reveals significant intra- and intertumor heterogeneity, and identifies patients that might benefit from FOLFIRI plus cetuximab treatment [36]. Another study was aimed at testing the feasibility of incorporating the analysis of somatic mutations within exons of 19 genes into patient management. It was demonstrated that the use of NGS for genomic profiling in advanced cancer patients is feasible. Additionally, actionable mutations identified in this study were relatively stable between archival and biopsy samples, implying that cancer mutations that are good predictors of drug response may remain constant across clinical stages [37,38].

ETHICAL ISSUES IN ONCOPHARMACOGENOMICS

Pharmacogenomic tumor NGS testing may also cause ethical dilemmas [39–41]. All tests need patient consent and indications of how incidental findings should be reported. As these tests are intended to be implemented at the point of care, physicians have to face ethical difficulties at the bedside. There are no recommendations on how to provide information about incidental findings from somatic sequencing and how they apply to germ-line variations. An emerging issue is the off-label use of expensive drugs with well-known serious adverse effects for newly identified potential molecular targets, as such therapy can be ineffective, costly, and/or harmful [16].

MULTICENTER COLLABORATIONS IN ONCOPHARMACOGENOMICS

Join efforts across large pharmacogenomic networks in oncology may overcome the current challenges on the way to personalized medicine. The American Society of Clinical Oncology, the Association for Molecular Pathology, and the College of American Pathologists collectively increased efforts to develop consensus standards for multiplex cancer genomic testing for optimal integration of genetics data into clinical practice [42]. A major issue is to include into testing panels only genes or their regions that are frequently mutated in human cancers to maximize the coverage of actionable genetic variants while minimizing the chances of discovering variants with little or unknown clinical utility. There is a benefit from the laboratory and clinical perspective to developing a single, multiuse testing platform. Although

Table 2 Ongoing Clinical Studies Based on NGS Found in the Database "ClinicalTrials.gov"—A National Institutes of Health Registry of Publicly and Privately Supported Clinical Studies Conducted Around the World [35]

Clinical Trial	Purpose of the Trial
A prospective randomized trial comparing the effectiveness of physician discretion-guided therapy versus physician discretion-guided plus NGS-directed therapy	Randomized clinical trial study relating genomic variations detected by NGS to clinical phenotypes and to the effects of targeted therapy.
Decision impact analysis of foundation medicine's NGS test in advanced solid tumor malignancies	Pilot clinical trial study of comprehensive gene sequencing in guiding treatment recommendations in patients with metastatic or recurrent solid tumors.
NGS to evaluate breast cancer subtypes and genomic predictors of response to therapy in the preoperative setting for stage II–III breast cancer	Phase II trial study evaluating the effects of trastuzumab or bevacizumab with combination chemotherapy in patients with stage II–III breast cancer.
A prospective observational study to examine, in routine clinical practice in the USA, practice patterns and impact on clinical decision-making associated with the Foundation-One™ NGS test	A prospective, multicenter, observational study to characterize utilization patterns of the FoundationOne™ test by oncologists under conditions of routine clinical practice in the USA. The study will also examine the impact of test results on subsequent clinical decisions regarding choice of therapy. The planned duration of the study is at least 2 years with 1 year for patient recruitment and a minimum 1-year follow-up period for each patient.
FOrMAT—feasibility of a molecular characterization approach to treatment	This study will assess the feasibility of sequencing locally advanced/metastatic gastrointestinal cancers in real time to enable future treatment stratification by molecular characteristics. Targeted NGS of a panel of genes will be performed on tumor specimens and results will be discussed by a sequencing tumor board to establish if a patient is potentially suitable for a targeted therapy.
Identification of markers of primary or acquired resistance to anti-tumor treatment in patients with lung cancer or melanoma	The purpose of this study is to have a better understanding of why patients with lung cancer and melanoma are relapsing, using NGS to identify rare mutations and assess their predictive value.
WINTHER: a study to select rational therapeutics based on the analysis of matched tumor and normal biopsies in subjects with advanced malignancies	An open nonrandomized study using biology-driven selection of therapies. WINTHER will explore matched tumoral and normal tissue biopsies and use a novel method for predicting efficacy of drugs. The aim is to provide a rational personalized therapeutic choice to all (100%) patients enrolled in the study harboring oncogenic events (mutations/translocations/amplifications, etc.) or not. The total number of patients enrolled in the study will be 200 across all participating cancer centers (European countries France and Spain, Israel, USA, and Canada). All centers will realize the same study independently.
IMAGE study: individualized molecular analyses guide efforts in breast cancer—personalized molecular profiling in cancer treatment at Johns Hopkins	This study will test the feasibility of identifying patients who could benefit from tumor molecular profiling, of analyzing the patient's tumors in a timely (28 day) fashion, and of the identification of possible actionable mutations that are not just biologically interesting but also clinically relevant. The investigators will also examine the outcome data from patients who followed the molecular profiling tumor board suggestion compared with those who did not.
Tumor genomic profiling in patients evaluated for targeted cancer therapy	The goal of this study is to test cancer for certain mutations using leftover tumor tissue from a previous surgery or biopsy. The purpose of Part B of this study is to understand how genetic changes in tumor affect the chance of responding to experimental cancer treatment and how the genes in the tumor change over time in response to targeted cancer treatment.

numerous available databases of such recurrently altered genes exist, there is no cohesive initiative to standardize the information, and as of this writing this process is left to individual laboratories. Because of the ability of certain genes to be mutated differentially in different cancer types, separate gene panels for solid and nonsolid tumors seem appropriate. Moreover, a mechanism must exist by which the composition of each panel can be updated to implement new knowledge into clinical practice [42].

NGS IN NONCANCER PHARMACOGENOMICS

NGS can be used as cutting edge technology to analyze clinically important genomic variants to adjust the drug dose and balance the risk of adverse effects and drug effectiveness. As up to 80% of drugs are metabolized by polymorphic enzymes, adverse drug reactions as well as ultrafast drug metabolism and lowered drug efficacy may be due to genetic polymorphisms in drug-metabolizing enzymes (Table 3) [43]. While multiple genes are involved in drug metabolism, there are numerous points at which the system can be impaired.

CYTOCHROME P450

A major group of drug-metabolizing enzymes is the microsomal cytochrome P450 (CYP450) family [44]. CYP450s comprise an archaic superfamily of hemoproteins originating from an old gene that existed in very primitive organisms [43,44]. Repeated gene duplications have subsequently given rise to one of the largest and most complex of multigene families. Some CYP450 genes are highly polymorphic, resulting in enzyme variants with differing metabolic capacities [45–48]. CYP450s catalyze

Table 3 Drug Metabolism and Transport Genes [43]

Cytochrome P450 Genes		Transporters		Others
CYP11A1	CYP2E1	ABCB1	SLC22A5	UGT1A1
CYP11B1	CYP2A4	ABCC2	SLC10A1	UGT1A4
CYP11B2	CYP2F1	ABCC3	SLC10A2	UGT2B7
CYP19A1	CYP2J2	ABCC4	SLC15A1	POR
CYP1A1	CYP2S1	ABCC5	SLC15A2	SLC28A1
CYP26A1	CYP2W1	ABCC6	SLC16A1	SLC38A1
CYP27B1	CYP3A43	ABCG2	SLC22A11	SLC47A1
CYP2A13	CYP3A5		SLC22A12	SLC47A2
CYP2A6	CYP3A7		SLC22A2	SLCO1A2
CYP2B6	CYP4A11		SLC22A4	SLCO1B1
CYP2C18	CYP4B1		SLC22A6	SLCO2B1
CYP2C19	CYP4F2		SLC22A7	SLCO4C1
CYP2C8	CYP4F22		SLC22A8	OSTalpha
CYP2C9	CYP7A1			OSTbeta
CYP2D6	CYP7B1			
CYP8B1				

oxygenation of lipophilic drugs and other xenobiotics to give rise to more hydrophilic metabolites suitable for their elimination in the urine or for further metabolism by other enzymes [45,46]. Frequently, a CYP450 gene polymorphism is a major factor affecting drug plasma concentration, drug detoxification, and drug activation. The CYP450 group consists of 57 CYP genes, which are categorized into 18 families and 44 subfamilies coding more than 50 isoenzymes located primarily in liver [5]. The *CYP1* to *CYP3* families are involved in the first phase of drug metabolism, whereas *CYP4* to *CYP51* are associated with metabolism of endogenous compounds. The Human Cytochrome P450 Allele Nomenclature website is a database of more than 660 alleles of a total of 30 genes that include 29 CYPs as well as the cytochrome P450 oxidoreductase (*POR*) gene [5].

CYP2D6, CYP2C9, CYP2C19, and *CYP3A4* are the most functionally important genes in pharmacogenomics [43]. The majority of the clinically important variants are the effects of stop codons, nonsynonymous mutations, CNVs, and splice defects [43]. Both single-nucleotide polymorphisms (SNPs) and CNVs can affect CYP450 activity by altering gene expression. CNV detection is challenging for some CYP450 genes owing to the presence of homologous gene family members, pseudogenes (humans possess 19 pseudogenes in the CYP450 complex), and complex rearrangements including chimeras and gene conversions [47,48]. Pseudogenes, being relics of gene duplications, are deficient in the essential regulatory elements for transcription.

Martis et al. examined the role of CYP450 CNVs in selected CYP450 pharmacogenes using multiplex ligation-dependent probe amplification and quantitative polymerase chain reaction [48]. Samples from 542 individuals from various ethnic groups were tested. The *CYP2A6, CYP2B6,* and *CYP2E1* combined deletion/duplication allele frequencies ranged from 2 to 10%. High-resolution microarray-based comparative genomic hybridization and DNA sequencing localized *CYP2A6, CYP2B6,* and *CYP2E1* break points to directly oriented low-copy repeats resulting in the *CYP2B6*29* partial deletion allele and novel *CYP2B6/2B7P1* duplicated fusion allele (*CYP2B6*30*) [48]. These authors also identified novel CYP450 CNV alleles (*CYP2B6*30* and *CYP2E1*1Cx2*) [48]. Detection of these CNVs should be considered when interrogating these genes to predict the drug response [48].

The CYP alleles known so far have generally not been identified through large genomic sequencing projects. Most frequently the variant alleles have been identified based on an altered phenotype within an individual. Moreover targeted sequence screening projects have been performed to search for nonsynonymous mutations. Results from such studies are expected to yield a high number of novel variants. It is evident that thousands of mutations are localized in introns and gene-flanking regions that are not present in the databases [49]. Owing to current NGS-based projects, new knowledge regarding the intron sequence variability among the CYP genes should be revealed and new alleles could be recognized explaining yet unidentified genetic backgrounds of individual differences in drug metabolism [49].

The CYP genes are highly polymorphic in humans, with hundreds of SNPs, insertions and deletions, and CNVs described to date [43]. The intronic polymorphisms in the CYP genes account for only a small number of the important variant alleles. Among the P450 genes, only 15 different alleles with intronic mutations causing functional alterations have been identified so far [43,49]. Among these mutations, all but three abrogate enzyme activity as a result of erroneous splicing [49]. Their discoveries have been based on phenotypic changes after a long process of identification [49]. As a result of the NGS-based projects, it is anticipated that new information regarding the intron sequence variability among the CYP genes will be discovered and that further alleles will be identified that can partly fill the gap in our knowledge of the genetic basis of interpersonal differences in CYP-mediated drug metabolism [49].

CYP2D6 is one of the most polymorphic CYP genes in humans among the CYPs, accounting for around 80 different allelic variants and 130 genetic variations described [50]. The *CYP2D6*4* allele was the first defective *CYP2D6* variant allele to be identified (in 1990) and constitutes the main explanation for the poor metabolizer (PM) phenotype among Caucasians [49]. *CYP2D6* variants have been shown to affect mRNA transcripts and alter proteins and catalytic activity, all affecting drug metabolism [50]. *CYP2D6* is located within a large *CYP2* gene cluster on the long arm of chromosome 19 [5]. *CYP2D6* represents less than 5% of the CYP liver content, but is responsible for the metabolism of up to 25% of common drugs, mainly antidepressants, antipsychotics, beta-blockers, antiretroviral agents, antiarrhythmics, morphine derivatives, and tamoxifen, many of which have a narrow therapeutic window [51,52]. Marked interethnic variations in the frequency of various alleles have been reported [53] and are available in various online databases (dbSNP [54], ALFRED [55], 1000 Genomes [25,50]). For example, the *CYP2D6*10* allele is the most common *CYP2D6* variant in many Asian populations, and *CYP2D6*4* is among the most common variants in Caucasians [49,56]. Subjects who possess certain allelic variants will show normal, decreased, or no CYP2D6 function depending on the allele. The CYP2D6 function may be described as one of the following [43]:

- little or no CYP2D6 function—poor metabolizers;
- a rate of metabolism between the poor and the extensive metabolizers—intermediate metabolizers;
- normal CYP2D6 function—extensive metabolizers;
- high CYP2D6 function—ultrarapid metabolizers, subjects with multiple copies of the *CYP2D6* gene expressed.

The extensive metabolizers have two normal alleles and normal metabolism; intermediate metabolizers have one defective allele and may have slower drug metabolism; the ultrafast metabolizers have gene duplications and have increased drug metabolism. The poor metabolizers are carrying two defective alleles, resulting in substantially decreased drug metabolism and, in particular situations, higher levels of drugs and increased risk for adverse drug reactions [43].

Other genes that are highly polymorphic in this gene family are *CYP2C9* and *CYP2C19*, while other genes with important functional polymorphisms are *CYP1A2, CYP1B1, CYP2A6, CYP2A13, CYP2B6, CYP2C8, CYP2J2, CYP2R1, CYP2W1, CYP3A4, CYP3A5, CYP3A7, CYP4A22, CYP4B1, CYP4F2, CYP5A1, CYP8A1, CYP19A1, CYP21A2,* and *CYP26A1* [43].

CYP2C19 catalyzes the metabolism of many commonly used drugs, including phenytoin, omeprazole, and benzodiazepines. More than 20 polymorphisms of *CYP2C19* have been reported [43].

CYP2C9 is involved in the metabolism of many clinically important drugs, including tolbutamide, glipizide, phenytoin, warfarin, and certain nonsteroidal anti-inflammatory drugs. More than 30 variants of *CYP2C9* have been identified [43].

CYP3A4 is the most abundant P450 enzyme in human liver and is responsible for the metabolism of more than 50% of clinically important drugs. More than 20 *CYP3A4* variants have been identified; however, important functional alterations have not been found [43,49].

The *CYP3A5*3* allele (6986A > G) is the most frequently occurring allele of *CYP3A5* that results in a splicing defect that abolishes enzyme activity. Individuals with at least one allele of 6986A, designated as *CYP3A5*1,* are classified as CYP3A5 expressers [43]. The *CYP3A5*3* and *CYP3A5*5* alleles are much less common in African populations [49].

The highest number of variant alleles among the cytochromes P450 is seen in *CYP21A2*, which encodes the steroid 21 hydroxylase, for which 119 rare variants have been identified. In addition to the CYPs, NADPH cytochrome P450 reductase, the electron donor for CYP enzymes, has been shown to have important polymorphic alterations, and the second electron donor, cytochrome b_5, has also been shown to exhibit functionally actionable polymorphisms, although functionally variant alleles are rare [5,43,49].

NON-P450 DRUG-METABOLIZING ENZYMES

Additionally several non-P450 drug-metabolizing enzymes also play critical roles in the metabolism of a variety of drugs. Polymorphisms of these enzymes influence the metabolism and therapeutic effects of drugs, some of which are clinically significant [43].

Thiopurine methyltransferase (TPMT) catalyzes the methylation of 6-mercaptopurine, azathioprine, and thioguanine to inactive metabolites. More than 20 variant alleles of the *TPMT* gene have been identified, among which *TPMT*2*, *TPMT*3A*, and *TPMT*3C* are defective alleles that encode a protein with low enzymatic activity [43]. Approximately 90% of Caucasians inherit high enzyme activity, 10% inherit intermediate activity, and 0.3% inherit low or no activity [43]. The widespread use of pretreatment *TPMT* testing has not been widely accepted. The FDA agreed that the evidence was sufficient to mention testing for TPMT deficiency, thus allowing the establishment of safe doses of mercaptopurine without compromising efficacy [58]. There is ample evidence supporting the use of *TPMT* testing in patients receiving mercaptopurine to prevent serious myelosuppression [58].

UDP-glucuronosyltransferase 1A1 catalyzes the glucuronidation of many commonly used drugs or metabolites [43].

Serum butyrylcholinesterase hydrolyzes succinylcholine and thereby determines the serum concentration of the drug and the duration of muscle relaxation [43].

N-acetyltransferases catalyze the acetylation of aromatic amines and hydrazines [43].

DRUG TRANSPORTERS

Drug transporters modulate the absorption, distribution, and elimination of drugs by controlling the influx and efflux of drugs in cells. Increasing evidence indicates that genetic polymorphisms of transporters can have a profound impact on drug efficacy, and safety [43].

The *ABCB1* gene encodes the P-glycoprotein that transports many important drugs out of cells and can be responsible for multidrug resistance [43].

The breast cancer resistance protein is an ABC transporter (*ABCG2*) important in the intestinal absorption and biliary excretion of drugs, their metabolites, and some toxic xenobiotics [43,59].

The organic anion-transporting polypeptides (OATPs) are a large family of membrane transporter proteins for the transport of organic anions, including drugs and metabolites, across the cell membrane. The *SLC21A6* gene encodes OATPC, a liver-specific transporter important for hepatic uptake of a variety of compounds [46,60]. OATP1B1, encoded by *SLCO1B1,* is critical for hepatic uptake of simvastatin acid, the active metabolite of simvastatin. A polymorphism of *SLCO1B1* (c. 521T > C) that is associated with reduced activity of OATP1B1, increases the blood concentration of simvastatin acid, thus increasing toxicity and reducing efficacy of the drug [45,61].

Table 3 provides a list of drug metabolism and transport genes.

CLINICAL APPLICATIONS OF PHARMACOGENOMICS
ANTICOAGULANTS

Warfarin represents the most advanced application of pharmacogenetics clinical medicine. Initiation and continuation of warfarin therapy are challenging, as the drug has a narrow therapeutic window. The safety and efficacy of warfarin therapy requires monitoring of the international normalized ratio (INR), which has to be maintained within the target range for the underlying clinical condition. The drug response to warfarin is determined by variants in a few genes, including *VKORC1, GGCX, CYP2C9, CALU,* and *CYP4F2*. VKORC1 (vitamin K epoxide reductase) can assist in the selection of the starting dose of warfarin [62]. Polymorphisms in *CYP2C9* and *VKORC1* account for approximately 40% of the variance in warfarin dose. In particular, warfarin interferes with the VKORC1 enzyme, and variants in *CYP2C9* are related to differences in the drug's clearance [63]. The *VKORC1* G1639A polymorphism is associated with lower dose requirements and increased bleeding risk [63,64]. Moreover, decreased initial warfarin doses have been associated with the *CYP2C9*2* and *CYP2C9*3* variants [63,64]. Algorithms for genetics-based dosing have been developed and are available online (online calculator for initiation of warfarin dosing based on pharmacogenetic algorithms [65,66]). Among several algorithms identified from the medical literature, the most accurate equations were established by the International Warfarin Pharmacogenetics Consortium [64,67].

To date, despite many studies including large-scale prospective observational studies and comparative effectiveness clinical trials, the identified clinical and genetic factors account for only approximately 50% of the interindividual variability in warfarin dose requirement—for review see [68]. The implementation of pharmacogenetic-guided warfarin dosing is limited by the lack of evidence supporting a clinical benefit over conventional dosing. The routine use of genotyping to set up the dose for patients newly started on warfarin is still an unresolved issue despite several clinical trials that have been carried out [68]. The CoumaGen ($n=206$) and Marshfield trials ($n=230$) randomized patients to pharmacogenetic-guided or standard dosing. Although pharmacogenetic-guided dosing more accurately predicted the therapeutic dose than standard dosing, a reduction in out-of-range INRs or prothrombin time ratio was not achieved [68]. The EU-PACT trial ($n=455$) randomized patients to pharmacogenetic-guided dosing or a fixed dosing. Pharmacogenetic-guided dosing resulted in significantly fewer patients with INR ≥ 4 and less time to achieve therapeutic INR levels; however, bleeding event rates did not differ and thromboembolic events were too rare to compare [68]. The COAG trial ($n=1015$) randomized patients to pharmacogenetic-guided dosing or clinical algorithm dosing. In contrast to EU-PACT, no significant differences were detected in prothrombin time ratio, nor in time to first therapeutic INR, nor in adverse event rate. COAG ultimately did not support incorporation of genetic information into clinical dosing algorithms for initiation of warfarin therapy [68].

Genotype-guided warfarin therapy is the focus of ongoing prospective randomized clinical trials evaluating the utility of dosing algorithms that incorporate genetic polymorphisms in *CYP2C9* and *VKORC1* to determine dosages. The largest is the Genetics Informatics Trial—a randomized controlled trial ($n=1600$) assessing the safety and effectiveness of pharmacogenetic-guided warfarin dosing compared with clinical algorithm dosing [69]. In contrast to previous trials, the dosing algorithms will guide therapy for the first 11 days of treatment and the primary end point is venous thromboembolism, major bleeding, INR ≥4, or death.

Clopidogrel is a prodrug that metabolizes to the antiplatelet agent. The important influence of *CYP2C19* polymorphisms on clopidogrel response was first reported in 2006 [43]. *CYP2C19*2* is the most common loss-of-function polymorphism, with a frequency of 30% in European and African populations and 70% in Asian populations; PMs homozygous for the *CYP2C19*2* allele make up approximately 3–4% and 15%, respectively, resulting in clopidogrel resistance [43]. In patients with acute coronary syndromes, particularly among those who are undergoing percutaneous coronary intervention, *CYP2C19*2* is strongly associated with an increased risk of stent thrombosis [43].

PSYCHIATRY

In contemporary psychiatry there is an increasing emphasis on testing genes that may affect the patient's response to antidepressant and antipsychotic drugs, the medications used in attention-deficit hyperactivity disorder (ADHD) and in schizophrenia [70]. These tests include pharmacokinetic genes from the cytochrome P450 superfamily and pharmacodynamic genes related to the activity of neurotransmitters (e.g., *CYP2D6* (tricyclics, thioridazine, perphenazine, haloperidol, risperidone, selective serotonin recapture inhibitors, nonselective recapture inhibitors), *CYP2D19, CYP1A2*, the serotonin transporter gene *SLC6A4*, and the serotonin 2A receptor gene *HTR2A*) [71,72]. There are tests analyzing genes affecting patient response to both stimulant and nonstimulant medications in ADHD. NGS technologies are expected to aid in the identification of new clinically significant variants for those complex phenotypes. The Pharmacogenetics Working Group of the Royal Dutch Pharmacists Association established dose recommendations for several tricyclic antidepressants (TCAs), selective serotonin re-uptake inhibitors (SSRIs) and norepinephrine serotonin re-uptake inhibitors (NSRIs) on the basis of *CYP2D6* genotype [50,73].

ANALGESICS

Several pharmacokinetic and pharmacodynamic genes mediate the effects of common analgesics, including opioids. Opioids, including codeine, tramadol, and oxycodone, are among the pain medications metabolized by CYP2D6. The two extreme phenotypes (ultrarapid and poor metabolizers) are related to pain response and/or adverse effects. There is also direct evidence of an association between *OPRM1, COMT, ABCB1, CYP3A, UGT2B7,* and *MDR1* variants and response to analgesics [74].

TRANSPLANTOLOGY

NGS can also be used as a cutting edge technology to analyze clinically important genomic variants to adjust the dose of immunosuppressive agents in transplant recipients (i.e., tacrolimus), to balance the risk of adverse drug reactions and the risk of graft rejection [43].

ANTI-HUMAN IMMUNODEFICIENCY VIRUS DRUGS

The human leukocyte antigen (HLA) is associated with toxicity and hypersensitivity to several drugs such as abacavir, carbamazepine, and flucloxacillin. Several clinical studies proved that the *HLA-B*5701* allele reduces abacavir hypersensitivity [43]. Moreover, pre-prescription genotyping for *HLA-B*5701* was cost-effective [43,75].

MULTIGENE PHARMACOGENETIC TESTS ASSESSING PHARMACOKINETICS AND PHARMACODYNAMICS RESPONSE

Despite extensive research, only a few multigene pharmacogenetic tests are currently used in clinical practice. Much of the research remains in the discovery phase, with researchers struggling to demonstrate clinical utility and validity.

High-throughput microarray testing (AutoGenomics INFINITI CYP4502D6-I assay) of *CYP2D6* polymorphisms offers higher effectiveness in the identification of patient responders/nonresponders compared to single-allele analysis [76].

The FDA has approved the AmpliChip™ CYP450 test (Roche, Basel, Switzerland) based on the Affymetrix (Santa Clara, CA, USA) microarray technology for genotyping 27 alleles in *CYP2D6* and three alleles in *CYP2C19* associated with various metabolizing phenotypes [43]. The test is recommended for the assessment of patients who are either ultrarapid metabolizers or PMs [43].

The Drug Metabolizing Enzymes and Transporters Plus Panel (Affymetrix) genotyping platform interrogates 1936 genetic variations (CNVs, insertions/deletions, and biallelic and triallelic SNPs) in 225 genes involved in the ADME of a very wide range of therapeutics, as well as a number of genes that regulate intracellular processes that facilitate drug metabolism [43].

Another test currently in use is the Luminex xTAG® CYP2D6 Kit version 3, analyzing 20 cytochrome P450 2D6 variants [43].

The NGS technologies will inevitably accelerate the discovery rate of novel P450 genes; however, as of this writing only a few diagnostic tests based on NGS, adapted for routine clinical practice (excluding oncology), have been described (Table 4). Apart from targeted DNA sequencing options, NGS of the transcriptome (RNA sequencing) has opened up a new avenue. Such an approach may overcome some of the limitations of next generation DNA sequencing of complicated genes such as CYPs. Illumina launched the TruSeq targeted RNA expression cytochrome P450 panel as a gene expression profiling assay for studying cytochrome P450 genes [57]. The test targets 28 CYP genes important in pharmacogenomics. The panel is validated to ensure a strong correlation with traditional RNA sequencing. The kit is applicable for the benchtop MiSeq system [57].

Finally pharmacogenetics has moved from investigating a few candidate genes to examining complex genome-wide approaches, including a large panel of pharmacogenes [57]. Whole-genome sequencing is particularly useful to determine significant SNPs associated with a phenotype among a large set of polymorphisms and for identification of new coding and noncoding DNA variations modulating drug response. As whole-genome sequencing has emerged, studies of single variants (pharmacogenetics) have moved toward evaluating the entire genome for associations with pharmacologic phenotypes (pharmacogenomics). With the rapid advances in sequencing technologies, decrease in costs, and short turnaround times, large-scale genomic information will become available in the clinical setting, facilitating the implementation of pharmacogenomics. Moreover, one of the major impacts of whole-genome sequencing is that it has become possible to discover variants outside the genes already known to be involved in drug response and/or toxicity, increasing the patient's chance for effective and safe drug treatment. Developed by Illumina, the TruSight One panel targets more than 4800 clinically relevant genes (mostly exonic regions sheltering disease-causing mutations), including 45 CYPs and a plethora of other pharmacogenes (selected examples of included genes are shown in Table 4) [57]. The panel was designed to cover the most commonly ordered genetic tests, enabling labs to perform all these within one assay. Laboratories can analyze all of the genes in the panel or focus on a specific

Table 4 Multiplex Pharmacogenetic Tests Based on NGS for Nononcological Applications [57,77]

Company	Panel	Target Genes
AIBiotech	Personalized medicine panel (PMP)	10 genes including the cytochrome P450 genes *CYP2C19*, *CYP2C9* (with *VKORC1*), *CYP2D6*, *CYP3A4*, and *CYP3A5*, as well as prothrombin, factor V Leiden, *APOE*, and *MTHFR*.
myGenomics	myRxAct	Unspecified *CYP450* exons
Illumina	TruSight One	*CYP11A1, CYP11B1, CYP11B2, CYP17A1, CYP19A1, CYP1A1, CYP1A2, CYP1B1, CYP21A2, CYP24A1, CYP26A1, CYP26B1, CYP27A1, CYP27B1, CYP2A13, CYP2A6, CYP2B6, CYP2C18, CYP2C19, CYP2C8, CYP2C9, CYP2D6, CYP2D7P1, CYP2E1, CYP2F1, CYP2G1P, CYP2J2, CYP2R1, CYP2W1, CYP3A4, CYP3A43, CYP3A5, CYP3A5P1, CYP3A7, CYP46A1, CYP4A11, CYP4A22, CYP4B1, CYP4F12, CYP4F2, CYP4F22, CYP4F3, CYP4V2, CYP7A1, CYP7B1, POR, ABCB1, ABCG2, BCHE, UGT1A1, SLCO1B1*
Illumina	TruSeq targeted RNA expression P450	*CYP11A1, CYP19A1, CYP1A1, CYP1A2, CYP24A1, CYP26B1, CYP27A1, CYP27B1, CYP27C1, CYP2B6, CYP2C19, CYP2C8, CYP2C9, CYP2J2, CYP2R1, CYP3A4, CYP3A43, CYP4A11, CYP4F12, CYP4F2, CYP4F3, CYP4F8, CYP4X1, CYP4Z1, CYP51A1, CYP7A1, CYP7B1, PTGIS*

subset—that is, pharmacogenes. Genomic targets were identified based on information in the Human Gene Mutation Database (HGMD Professional), the Online Mendelian Inheritance in Man catalog, and GeneTests.org [57]. Combining data from these sources ensures that the TruSight One panel covers most of genes currently reviewed in clinical settings. Moreover, the provided bioinformatics tool (VariantStudio software) enables quick and accurate analysis, classification, and reporting of relevant genomic variants [57].

The PMP panel (Table 4) is being used as a diagnostic tool in the clinical study of genotyping frequencies on all markers for over 10,000 clinical specimens. Comparisons will be made between published frequencies from smaller studies in four ethnic populations (African-American, Caucasian, Hispanic, and Asian). The data generated from this study will generate new insight into the ethnic frequency of mutations that affect drug dosing and management [77].

QUALITY REQUIREMENTS FOR NGS-BASED PHARMACOGENOMIC TESTS

A routine test based on NGS technology needs to fulfill analytical quality requirements for clinical laboratory tests according to the Clinical Laboratory Improvement Amendments (CLIA) in the United States or other national certifying organizations ensuring analytical quality and patient safety [78]. NGS sequencing has to be done in certified facilities by authorized personnel. Clinical laboratories may also develop and validate tests in-house and market them as a laboratory service, although such tests must meet the general regulatory standards of a certifying agency. The laboratory offering the service must be licensed by CLIA (or another organization) for high-complexity testing. NGS tests require a standardized protocol for preanalytic, analytic, and postanalytic processes, securing high-quality laboratory testing including adequate precision and accuracy and referring to the clinically targeted NGS platform [79].

All preanalytical, analytical, and postanalytical laboratory procedures should be consistently implemented to ensure adequate quality of the obtained results. The postanalytic variables (i.e., entry and transfer of data, validation and interpretation of results, reporting of the results, and data storage) relevant to NGS need to be taken into consideration [15,80–82].

The postanalytic process includes the alignment of the raw sequence data using a reference human genome followed by variant identification with corresponding clinical interpretations and clinical decision-making [82]. External quality assurance programs for the postanalytical phase have to be introduced. The high quality of interpretative comments has to be ensured. The interpretation of data needs to be standardized, including algorithms for variant calling, especially when evaluating large numbers of variants, including uncommon alterations. Moreover, the analytic and clinical validity of those tests may be strongly dependent on the sequencing platforms used to generate genomic data [82,83]. When implementing new testing panels in the routine clinical practice, it is important to ensure that all important genetic variants are evaluated using a single-testing strategy. In oncology testing the same issues with preanalytic sample quality, such as tumor content of specimens, DNA integrity, and yield, will still apply and must be emphasized to avoid prelaboratory errors. Moreover, several potentially important mechanisms of gene disruption, such as exon loss, triplet repeat expansion, and epigenetic changes, cannot as of this writing be reliably assessed using NGS [83].

One helpful strategy to increase the postanalytic yield of potentially important genetic variants is to simultaneously test normal tissue to more easily and reliably identify somatic mutations [84]. In cancer testing care should be taken to not ignore the possibility that germ-line variants may also be informative in disease pathogenesis. While developing and validating testing strategies in individual laboratories, particular attention should be paid to detection limits, accuracy, and precision [84]. Moreover universal criteria for sample rejection when quality-control standards have not been met should be adhered to rigorously. Special attention must be paid to the amount of sequencing data that is produced at individual gene regions (i.e., the depth of coverage) to avoid making false positive or false negative calls [84]. It is also prudent to perform the testing in duplicates to confirm the presence of positive results until the laboratory has confidence in its analysis pipelines. A major challenge that remains to be resolved is to accurately prioritize variants of clinical significance and to depreciate variants that can safely be ignored as not contributing to disease and to filter out variants that appear in the general population at a certain minimum frequency [84]. In addition, pathogenically important variants may nevertheless appear in such repositories at low frequencies [84]. It would also be helpful to comprehensively catalog human genetic variation. Several efforts have been initiated to reach this aim, such as the National Center for Biotechnology Information-sponsored data repository ClinVar. Other large pharmacogenomic data repositories such as The Pharmacogenomics Knowledge Base (PharmGKB) developed by Stanford University [88] and the OpenPGx Consortium [85] and other data sets, including those available from the National Institutes of Health (NIH) genome-wide association studies (GWAS) collection [86], have made it possible to computationally analyze personal genomes for potential translation of pharmacogenomics into clinical practice.

GUIDELINES FOR CLINICAL APPLICATION OF PHARMACOGENOMICS

Although a large number of biomarkers have been validated, guidelines that link the results of a pharmacogenomic test to specific therapeutic dose recommendations are scarce. Development of simple, easily accessible, reliable clinical algorithms and guidelines must support clinicians in the interpretation of genetic data and guide the treatment decision process. These point-of-care tools will be embedded in electronic health records systems and it will be critical to accelerate individualized medicine [87]. PharmGKB shelters detailed primary data of gene variants and their effects on drug responses, which has facilitated the formation of several data-sharing consortia for the pharmacogenomics of warfarin, tamoxifen, selective serotonin reuptake inhibitors, clopidogrel, antihypertensives, and statins, to detect complex genotype–phenotype associations with greater statistical power [88]. PharmGKB has moved toward the clinical implementation of pharmacogenomic tests into clinical practice. PharmGKB and the Pharmacogenomics Research Network (PGRN) have provided guidelines to help clinicians understand how available pharmacogenomic test results should be used to optimize drug therapy. Large multinational consortia have been already established, such as the NIH-funded PGRN and NIH-funded Electronic Medical Records and Genomics Network, which are cataloged in the Pharmacogenomics Knowledge Base [89]. As part of the PGRN, The Genome Institute has improved the technology to target selected genomic regions and to help understand the roles these variants play in clinical medicine [90]. The NIH's PGRN is a union of research groups funded "… to lead discovery and advance translation in genomics in order to enable safer and more effective drug therapies", with the final goal of implementing personalized medicine in routine clinical practice [91]. PGRN's accomplishments and future projects aim to provide peer-reviewed, freely accessible guidelines for gene–drug pairs, thus facilitating the translation and interpretation of pharmacogenomic tests [91,92].

Another implementation project, the DPWG, supported by the Royal Dutch Pharmacist's Association, aims to develop pharmacogenetics-based therapeutic guidelines through systematic review of the literature [93]. Both projects try to provide peer-reviewed recommendations to guide clinicians and pharmacists by translating laboratory test results into actionable treatment decisions [91,92]. Actual drug-dosing guidelines based on individual genotypes are being published. These guidelines are updated periodically by PharmGKB and are available on the PharmGKB website [88].

Another pharmacogene database enhanced by the 1000 Genomes Project was created to immediately evaluate and utilize sequencing data every time they are released [94]. Particularly, this database can be used to access SNP genotypic calls of 39 pharmacogenetic candidate genes, maintained by the Very Important Pharmacogenes (VIP) project of the Pharmacogenetics Knowledge Base [88]. The VIP project is an initiative to provide valuable genomic information of particular relevance for pharmacogenetics.

Pharmacogenomic information has drawn attention from the official regulatory agencies and medical societies in many countries, such as the U.S. FDA and European Medicines Agency. The FDA issued their "Guidance for Industry: Pharmacogenomic Data Submission" to facilitate scientific progress in pharmacogenomics and to help the pharmaceutical industry integrate genomics into development plans [95]. To date, about 10% of FDA-approved drugs contain pharmacogenomics information in their labels [96.97]. Many biomarkers related to drug responses have been identified; however, the translation of this knowledge into clinical practice remains slow and limited [98]. Pharmacogenomics is gradually being applied from the bench to the bedside, which becomes a key component of personalized medicine [99].

NGS IN PHARMACOGENOMICS—OTHER POSSIBLE APPLICATIONS

Drug repositioning means finding novel indications for currently marketed drugs. This strategy may reduce the costs of new drug development and advance the delivery of new therapeutics to patients with incurable diseases. Genomics can be a major contributor to defining novel indications for drugs in development or after approval. The Institute of Medicine's Roundtable on Translating Genomic-Based Research for Health has explored state-of-the-art drug repositioning [100].

Human genetic information is used to support the finding of new targets for both novel and known drugs. High-throughput genomic and bioinformatics technologies allow repurposing to advance drug discovery and development. Comprehensive databases of integrated genomic information and phenotypic and clinical data are being developed [101].

LIMITATIONS OF NGS IN PHARMACOGENOMICS

The promise of NGS for personalized therapy is obvious; however, several critical issues must be addressed to ensure the appropriate and efficient use of this new technology for pharmacogenomics. It is evident that all methods to determine human genome variation contain errors. Many of the pharmacogenes, including the CYP (especially the *CYP2D6* superfamily) and HLA genes, are not ideally suited to NGS as their complexity is the source of errors in this technology [102]. The disadvantages of the second generation of NGS lie primarily in the short lengths of reads. Potentially this problem can be partly resolved by novel technologies improving the quality of assemblies for large and complex genes such as CYPs [103]. Moreover, assemblers that surpass long-range continuity in contigs are prone to local errors such as insertions/deletions (indels) [103]. To prevent fragmentation of contigs the appropriate assembly algorithms should be implemented to achieve longer continuity and to avoid misassembly errors. These technologies are widely applied for research purpose, but can potentially be used for constructing diagnostic tests. Now available third generation sequencing technology has similar throughput, longer read lengths, but also new read types. Still challenging is to overcome the problem of assembly, as well as correcting for errors in reads and taking heterozygosity into account [102]. Currently some tradeoffs are necessary as longer read length often causes more errors. All currently used benchtop sequencers manifest the problems associated with short-read technology, including extensive processing of tissue samples and complex data analysis. Moreover, prevailing short-read sequencing technologies cannot accurately resolve allelic variants in those pharmacogenes, as the process requires either 60- to 100-fold coverage for a single individual or low-coverage whole-genome sequence data from a large population. All short-read NGS methods rely on the use of a reference genome as a background, whereas different reference genomes can include unusual variants [104]. Widely used alignment methods can detect SNPs and short indel variants. Such variants have been found to account for only up to 50% of known differences in drug responses; in addition, most variants that influence individual drug responses have not yet been identified [105]. Common SNPs, with a frequency of more than 0.5%, have yielded modest effects in GWAS for determination of complex traits [106]. Early results from pharmacogenomic GWAS appear to indicate a greater ability to discover SNPs with substantial effect size, but are limited in power by small cohort sizes [107]. Furthermore, SNP variants do not explain the full repertoire of human genome variation in relation to the treatment outcome. SNPs have been the easiest genomic variant to detect, but other variants, such as CNVs, may be more important determinants of drug efficacy and toxicity. CNVs and their significance in pharmacogenomics need to be fully explored in the future. It is known that CNVs involve some known metabolizing enzymes, such as CYP2D6 and GSTM1 and potential drug targets such as CCL3L1 and can influence the phenotype through alterations in gene dosage, structure, and expression [108,109]. Martis et al. showed that a common CYP450 CNV formation is probably mediated by nonallelic homologous recombination resulting in both full gene and gene-fusion copy number imbalances [53]. Although the identification of CNVs using NGS data causes serious problems, particularly for large insertions, it can be feasible as the technology moves forward. Taking into consideration all the above drawbacks, the rapid introduction of NGS tests required for deep pharmacogenomics analysis into the clinical practice is extremely demanding.

CONCLUSIONS

Pharmacogenomic research aimed at finding new biomarkers for drug safety and efficacy and their translation to the clinic is a great endeavor. However, as the proof of clinical validity and utility is lacking, only a few pharmacogenetic biomarkers have reached the clinic. It is important that not only Caucasians but also diverse ethnic populations are included in the process. The development of NGS, including through whole-genome, whole-exome, and whole-transcriptome analysis, allows fast, cheap, reliable production of large volumes of DNA or RNA sequence data. Nevertheless, the implementation of NGS and related technologies covering different variants in different pharmacogenes is a major challenge in the clinical setting. The majority of published pharmacogenomic research is still hypothesis driven. This may in part reflect higher costs of genome-wide studies and the necessity to recruit larger patient cohorts, which often implies large consortium efforts [47].

The numerous barriers need be overcome before pharmacogenomics can be implemented into clinical medicine, meanwhile offering the promise of identifying new treatment modalities. An important obstacle is the opinion that the costs of pharmacogenotyping outweigh its potential benefits. However, some studies have documented that genotype-guided personalized therapy can be more cost-effective than the classical approach [110].

Major advances will be needed to transfer pharmacogenomic information to patient care and it is too early to know whether the use of a personalized profile of the patient will improve treatment outcomes. Another limitation of targeted pharmacogenomics is the lack of good sequence data for untranslated regions, making this approach useful only for receiving a genetic profile for those drugs that have relevant variants exclusively in coding regions. This can be overcome by a genome-wide NGS approach. Other issues include the need for genotyping accuracy and validation of the obtained results. One of the major obstacles in clinical practice is the lack of facilities available for genetic testing. Inadequate pharmacogenomics knowledge and awareness of clinicians is another important barrier. The majority of health care workers do not know when to order pharmacogenetic testing, how to communicate the results to patients, or how to select a drug and adapt a dose accordingly. Currently, prescription genetic testing is limited to specialized laboratories and university hospitals [43]. On the other hand, advances in pharmacogenomics are already visible on labels of many drugs, including information about related pharmacogenes [111]. In terms of research dedicated to advancing pharmacogenomics, numerous consortia have been established [25,27,58,67,88,112]. Sophisticated databases covering pharmacogenomic information are widely available online. Nevertheless, with a few examples proven by large-scale clinical trials, the effects of the vast number of genetic/genomic variations on the metabolism and drug response remain to be elucidated. For most diseases, sufficient clinical data are not available to support the use of genetic testing to monitor treatment decisions. As of this writing, identifying and validating tests identifying large panels of actionable pharmacogenes can be approached by the use of an NGS platform applying various methods, including DNA, RNA, and chromatin immunoprecipitation sequencing to identify DNA-interacting proteins [113]. The analysis of the data produced from exome and genome sequencing is enormous. Even more challenging is the interpretation of these data. Improvements in bioinformatics will be needed to assess the functional significance of rare variants identified in genes of potential pharmacogenetic impact. Oncology will probably be the most promising field in promoting personalized medicine. Furthermore, the contributions of epigenomics, transcriptomics, and proteomics to interindividual variability in drug response are best studied in cancer patients.

REFERENCES

[1] Monte AA, Heard KJ, Vasiliou V. Prediction of drug response and safety in clinical practice. J Med Toxicol 2012;8:43–51.

[2] Crews KR, Hicks JK, Pui CH, Relling MV, Evans WE. Pharmacogenomics and individualized medicine: translating science into practice. Clin Pharmacol Ther 2012;92:467–75.

[3] Metzker ML. Sequencing technologies – the next generation. Nat Rev Genet 2010;11:31–46.

[4] DePristo MA, Banks E, Poplin R, Garimella KV, Maguire JR, et al. A framework for variation discovery and genotyping using next-generation DNA sequencing data. Nat Genet 2011;43:491–8.

[5] http://www.cypalleles.ki.se/.

[6] Filipski KK, Mechanic LE, Long R, Freedman AN. Pharmacogenomics in oncology care. Front Genet 2014;8(5):73. eCollection.

[7] Gillis N, Patel J, Innocenti F. Clinical implementation of germ line cancer pharmacogenetic variants during the next-generation sequencing era. Clin Pharmacol Ther 2014;95:269–80.

[8] Hertz DL, McLeod HL. Use of pharmacogenetics for predicting cancer prognosis and treatment exposure, response and toxicity. J Hum Genet 2013;58:346–52.

[9] Ou SH, Bartlett CH, Mino-Kenudson M, Cui J, Iafrate AJ. Crizotinib for the treatment of ALK-rearranged non-small cell lung cancer: a success story to usher in the second decade of molecular targeted therapy in oncology. Oncologist 2012;17:1351–75.

[10] Ong FS, Das K, Wang J, Vakil H, Kuo JZ, Blackwell WL, et al. Personalized medicine and pharmacogenetic biomarkers: progress in molecular oncology testing. Expert Rev Mol Diagn 2012;12:593–602.

[11] Balabanov S, Braig M, Brümmendorf TH. Current aspects in resistance against tyrosine kinase inhibitors in chronic myelogenous leukemia. Drug Discov Today Technol 2014;11:89–99.

[12] Paugh SW, Stocco G, McCorkle JR, Diouf B, Crews KR, Evans WE. Cancer pharmacogenomics. Clin Pharmacol Ther 2011;90:461–6.

[13] Cheng EH, Sawyers CL. In cancer drug resistance, germline matters too. Nat Med 2012;18:494–6.

[14] Ng KP, Hillmer AM, Chuah CT, Juan WC, Ko TK, Teo AS, et al. A common BIM deletion polymorphism mediates intrinsic resistance and inferior responses to tyrosine kinase inhibitors in cancer. Nat Med 2012;18:521–8.

[15] Patel JN. Application of genotype-guided cancer therapy in solid tumors. Pharmacogenomics 2014;1:79–93.

[16] Marrone M, Filipski KK, Gillanders EM, Schully SD, Freedman AN. Multi-marker solid tumor panels using next-generation sequencing to direct molecularly targeted therapies. PLoS Curr Evid Genomic Tests 2014. Edition 1.

[17] MacConaill LE, Van Hummelen P, Meyerson M, Hahn WC. Clinical implementation of comprehensive strategies to characterize cancer genomes: opportunities and challenges. Cancer Discov 2011;1:297–311.

[18] Garraway LA. Genomics-driven oncology: framework for an emerging paradigm. J Clin Oncol Off J Am Soc Clin Oncol 2013;31(15):1806–14.

[19] Burrell RA, McGranahan N, Bartek J, Swanton C. The causes and consequences of genetic heterogeneity in cancer evolution. Nature 2013;501:338–45.

[20] Arup Laboratories. Solid tumor mutation panel by next generation sequencing. http://ltd.aruplab.com/Tests/Pub/2007991.

[21] FoundationOne. http://foundationone.com/docs/FoundationOne_tech-info-and-overview.pdf. March 2014.

[22] SuraSeq. http://asuragen.com/wp-content/uploads/2013/03/brochure-NGS-0116132.pdf. March 2014.

[23] Frampton GM, Fichtenholtz A, Otto GA, et al. Development and validation of a clinical cancer genomic profiling test based on massively parallel DNA sequencing. Nat Biotech 2013;31:1023–31.

[24] Dreyer NA, Tunis SR, Berger M, Ollendorf D, Mattox P, Gliklich R. Why observational studies should be among the tools used in comparative effectiveness research. Health Aff (Millwood) 2010;29:1818–25.

[25] Abecasis GR, Altshuler D, Auton A, Brooks LD, 1000 Genomes Project Consortium, et al. A map of human genome variation from population-scale sequencing. Nature 2010;467:1061–73.

[26] Gamazon ER, Zhang W, Huang RS, Dolan ME, Cox NJ. A pharmacogene database enhanced by the 1000 Genomes Project. Pharmacogenet Genomics 2009;19:829–32.

[27] Technology Evaluation Center Assessment Program. Special report: multiple molecular testing of cancers to identify targeted therapies. Exec Summ 2013;28(1):1–2.

[28] Dienstmann R, Serpico D, Rodon J, et al. Molecular profiling of patients with colorectal cancer and matched targeted therapy in phase I clinical trials. Mol cancer Ther 2012;(9):2062–71.

[29] Tsimberidou AM, Iskander NG, Hong DS, et al. Personalized medicine in a phase I clinical trials program: the MD Anderson Cancer Center initiative. Clin Cancer Res Off J Am Assoc Cancer Res 2012;18:6373–83.

[30] Von Hoff DD, Stephenson Jr JJ, Rosen P, et al. Pilot study using molecular profiling of patients' tumors to find potential targets and select treatments for their refractory cancers. J Clin Oncol 2010;28(33):4877–83.

[31] https://asco.org/CancerLinQ.

[32] Marrone M, Filipski KK, Gillanders EM, et al. IMAGE study: personalized molecular profiling in cancer treatment at Johns Hopkins, ClinicalTrials.gov NCT01939847.

[33] Comprehensive gene sequencing in guiding treatment recommendations patients with metastatic or recurrent solid tumors. ClinicalTrials.gov NCT01987726.

[34] A study to select rational therapeutics based on the analysis of matched tumor and normal biopsies in subjects with advanced malignancies (WINTHER). ClinicalTrials.gov NCT01856296.

[35] https://clinicaltrials.gov.

[36] Ciardiello F, Normanno N, Maiello E, Martinelli E, Troiani T, Pisconti S, et al. Clinical activity of FOLFIRI plus cetuximab according to extended gene mutation status by next generation sequencing: findings from the CAPRI-GOIM trial. Ann Oncol 2014. pii: mdu230.

[37] Tran B, Brown AM, Bedard PL, Winquist E, Goss GD, Hotte SJ, et al. Feasibility of real time next generation sequencing of cancer genes linked to drug response: results from a clinical trial. Int J Cancer 2013;132(7):1547–55.

[38] Guchelaar HJ, Gelderblom H, van der Straaten T, Schellens JH, Swen JJ. Pharmacogenetics in the cancer clinic: from candidate gene studies to next-generation sequencing. Clin Pharmacol Ther 2014;95:383–5.

[39] Rieder MJ, Carleton B. Pharmacogenomics and adverse drug reactions in children. Front Genet 2014;16(5):78. http://dx.doi.org/10.3389/fgene.00078. eCollection 2014.

[40] Brothers KB. Ethical issues in pediatric pharmacogenomics. J Pediatr Pharmacol Ther 2013;18:192–8.

[41] Chalmers D, Nicol D, Otlowski M, Critchley C. Personalised medicine in the genome era. J Law Med 2013;20:577–94.

[42] http://am.asco.org/consensus-standards-multiplex-cancer-genomic-testing-joint-roundtable-fosters-awareness-expectations.

[43] Sanoudou D. Clinical applications of pharmacogenetics. Rijeka, Croatia: InTech; 2012. www.intechopen.com.

[44] Sim SC, Ingelman-Sundberg M. The human cytochrome P450 Allele Nomenclature Committee Web site: submission criteria, procedures, and objectives. Methods Mol Biol 2006;320:183–91.

[45] Ma Q, Lu AYH. Pharmacogenetics, pharmacogenomics, and individualized medicine. Pharmacol Rev 2011;63:437–59.

[46] Ma Q, Lu AY. The challenges of dealing with promiscuous drug-metabolizing enzymes, receptors and transporters. Curr Drug Metab 2008;9:374–83.

[47] Gurwitz D, Howard L, McLeod HL. Genome-wide studies in pharmacogenomics: harnessing the power of extreme phenotypes. Pharmacogenomics 2013;14:337–9.

[48] Martis S, Mei H, Vijzelaar R, Edelmann L, Desnick RJ, Scott SA. Multi-ethnic cytochrome-P450 copy number profiling: novel pharmacogenetic alleles and mechanism of copy number variation formation. Pharmacogenomics J 2013;13:558–66.

[49] Ingelman-Sundberg M, Sim SC. Intronic polymorphisms of cytochromes P450. Hum Genomics 2010;4:402–5.

[50] Samer CF, Ing Lorenzini K, Rollason V, Daali Y, Desmeules JA. Applications of CYP450 testing in the clinical setting. Mol Diagn Ther 2013;17:165–84.

[51] Eichelbaum M, Ingelman-Sundberg M, Evans WE. Pharmacogenomics and individualized drug therapy. Annu Rev Med 2006;57:119–37.

[52] Zanger UM, Raimundo S, Eichelbaum M. Cytochrome P450 2D6: overview and update on pharmacology, genetics, biochemistry. Naunyn Schmiedeb Arch Pharmacol 2004;369:23–37.

[53] McGraw J, Waller D. Cytochrome P450 variations in different ethnic populations. Expert Opin Drug Metab Toxicol 2012;8:371–82.

[54] http://www.ncbi.nlm.nih.gov/SNP/.

[55] http://alfred.med.yale.edu/.

[56] Zhou SF, Di YM, Chan E, Du YM, Chow VD, Xue CC, et al. Clinical pharmacogenetics and potential application in personalized medicine. Curr Drug Metab 2008;9:738–84.

[57] http://www.illumina.com/technology/next-generation-sequencing.ilmn.

[58] Relling MV, Gardner EE, Sandborn WJ, Schmiegelow K, Pui CH, Yee SW, et al. Clinical Pharmacogenetics Implementation Consortium guidelines for thiopurine methyltransferase genotype and thiopurine dosing. Clin Pharmacol Ther 2011;89:387–91.

[59] Noguchi K, Katayama K. Human ABC transporter ABCG2/BCRP expression in chemoresistance: basic and clinical perspectives for molecular cancer therapeutics. Pharmgenomics Pers Med 2014;7:53–64.

[60] Nishizato Y, Ieiri I, Suzuki H, Kimura M, Kawabata K, Hirota T, et al. Polymorphisms of OATP-C (SLC21A6) and OAT3 (SLC22A8) genes: consequences for pravastatin pharmacokinetics. Clin Pharmacol Ther 2003;73:554–65.

[61] SEARCH Collaborative Group, Link E, Parish S, Armitage J, Bowman L, Heath S, Matsuda F, et al. SLCO1B1 variants and statin-induced myopathy—a genomewide study. N Engl J Med 2008;359:789–99.

[62] Rieder MJ, Reiner AP, Gage BF, Nickerson DA, Eby CS, et al. Effect of VKORC1 haplotypes on transcriptional regulation and warfarin dose. N Engl J Med 2005;352:2285–93.

[63] Wadelius M, Chen LY, Downes K, Ghori J, Hunt S, et al. Common VKORC1 and GGCX polymorphisms associated with warfarin dose. Pharmacogenomics J 2005;5:262–70.

[64] Sanderson S, Emery J, Higgins J. CYP2C9 gene variants, drug dose, and bleeding risk in warfarin-treated patients: a HuGEnet systematic review and meta-analysis. Genet Med 2005;7:97–104.

[65] www.WarfarinDosing.org.

[66] Klein TE, Chang JT, Cho MK, Easton KL, Fergerson R, et al. Integrating genotype and phenotype information: an overview of the PharmGKB project. Pharmacogenetics Research Network and Knowledge Base. Pharmacogenomics J 2001;1:167–70.

[67] The International Warfarin Pharmacogenetics Consortium. Estimation of the warfarin dose with clinical and pharmacogenetic data. NEJM 2009;360(8):753–64.

[68] Scott SA, Lubit SA. Warfarin pharmacogenetic trials: is there a future for pharmacogenetic-guided dosing Pharmacogenomics 2014;15:719–22.

[69] Do EJ, Lenzini P, Eby CS, et al. Genetics informatics trial (GIFT) of warfarin to prevent deep vein thrombosis (DVT): rationale and study design. Pharmacogenomics J 2012;12:417–24.

[70] Drögemöller BI, Wright GE, Niehaus DJ, Emsley R, Warnich L. Next-generation sequencing of pharmacogenes: a critical analysis focusing on schizophrenia treatment. Pharmacogenet Genomics 2013;23:666–74.

[71] Chou WH, et al. Extension of a pilot study: impact from the cytochrome P450 2D6 polymorphism on outcome and costs associated with severe mental illness. J Clin Psychopharmacol 2000;20:246–51.

[72] Kirchheiner J, et al. Pharmacogenetics of antidepressants and antipsychotics: the contribution of allelic variations to the phenotype of drug response. Mol Psychiatry 2004;9:442–73.

[73] Swen JJ, et al. Pharmacogenetics: from bench to byte–an update of guidelines. Clin Pharmacol Ther 2011;89:662–73.

[74] Branford R, Droney J, Ross JR. Opioid genetics: the key to personalized pain control? Clin Genet 2012;82:301–10.

[75] Martin MA, Hoffman JM, Freimuth RR, Klein TE, Dong BJ, Pirmohamed M, et al. Clinical pharmacogenetics implementation consortium (CPIC) guidelines for HLA-b genotype and abacavir dosing: 2014 update. Clin Pharmacol Ther 2014;95:499–500.

[76] Savino M, Seripa D, Gallo AP, Garrubba M, D'Onofrio G, Bizzarro A, et al. Effectiveness of a high-throughput genetic analysis in the identification of responders/non-responders to CYP2D6-metabolized drugs. Clin Lab 2011;57:887–93.

[77] www.albiotech.com.

[78] http://wwwn.cdc.gov/CLIA/Default.aspx.

[79] Hadfield J, Eldridge MD. Multi-genome alignment for quality control and contamination screening of next-generation sequencing data. Front Genet 2014. http://dx.doi.org/10.3389/fgene.2014.00031.

[80] MacConaill LE. Existing and emerging technologies for tumor genomic profiling. J Clin Oncol 2013;31:1815–24.

[81] Simon R, Polley E. Clinical trials for precision oncology using next-generation sequencing. Precis Med 2013;10:485–95.

[82] Van Allen EM, Wagle N, Levy MA. Clinical analysis and interpretation of cancer genome data. J Clin Oncol Off J Am Soc Clin Oncol 2013;31:1825–33.

[83] Daber R, Sukhadia S, Morrissette JJ. Understanding the limitations of next generation sequencing informatics, an approach to clinical pipeline validation using artificial data sets. Cancer Genet 2013;206:441–8.

[84] http://am.asco.org/consensus-standards-multiplex-cancer-genomic-testing-joint-roundtable-fosters-awareness-expectations.

[85] www.openpgx.org.

[86] Sherry ST, Ward MH, Kholodov M, Baker J, Phan L, et al. dbSNP: the NCBI database of genetic variation. Nucleic Acids Res 2001;29:308–11.

[87] Fackler JL, McGuire AL. Paving the way to personalized genomic medicine: steps to successful implementation. Curr Pharmacogenomics Pers Med 2009;7(2):125.

[88] http://www.pharmgkb.org/.

[89] McCarty CA, Wilke RA. Biobanking and pharmacogenomics. Pharmacogenomics 2010;11:637–41.

[90] http://genome.wustl.edu/projects/detail/pharmacogenomics-research-network-project-pgrn/.

[91] Long RM, Berg JM. What to expect from the Pharmacogenomics Research Network. Clin Pharmacol Ther 2011;89:339–41.

[92] Roden DM, Tyndale RF. Genomic medicine, precision medicine, personalized medicine: what's in a name? Clin Pharmacol Ther 2013;94:169–72.

[93] http://www.pharmgkb.org/guideline/PA166104937.

[94] http://genemed1.bsd.uchicago.edu/pharmacodb/thougen/main.php.

[95] http://www.fda.gov/Drugs/GuidanceComplianceRegulatoryInformation/Guidances/default.htm.

[96] Weng L, Zhang L, Peng Y, Huang RS. Pharmacogenetics and pharmacogenomics: a bridge to individualized cancer therapy. Pharmacogenomics 2013;14:315–24.

[97] http://wwwfdagov/drugs/scienceresearch/researchareas/pharmacogenetics/ucm083378htm.

[98] http://www.fda.gov/downloads/RegulatoryInformation/Guidances/UCM126957.pdf.

[99] Cheng Y, He Yi-J, Tang J, Zhou H-H. Translating pharmacogenetics and pharmacogenomics: the last 60 years and the rise of collective innovation as a force multiplier for personalized medicine. Curr Pharmacogenomics Pers Med 2014;12:15–31.

[100] Drug Repurposing and Repositioning: Workshop Summary. Roundtable on Translating Genomic-Based Research for Health; Board on Health Sciences Policy; Institute of Medicine (US). The National Academies Collection: Reports funded by National Institutes of Health.

[101] Power A, Berger AC, Ginsburg GS. Genomics-enabled drug repositioning and repurposing insights from an IOM roundtable activity. JAMA 2014;3:11–20.

[102] Gamazon ER, Skol AD, Perera MA. The limits of genome-wide methods for pharmacogenomic testing. Pharmacogenet Genomics 2012;22(4):261–72.

[103] Henson J, Tischler G, Ning Z. Next-generation sequencing and large genome assemblies. Pharmacogenomics 2012;13:901–15.

[104] Nielsen R, Paul JS, Albrechtsen A, Son YS. Genotype and SNP calling from next-generation sequencing data. Nat Rev Gen 2011;12:443–51.

[105] Cooper GM, et al. A genome-wide scan for common genetic variants with a large influence on warfarin maintenance dose. Blood 2008;112:1022–7.

[106] Guessous I, Gwinn M, Khoury MJ. Genome-wide association studies in pharmacogenomics: untapped potential for translation. Genome Med 2009;1:46.

[107] Sato Y, et al. A new statistical screening approach for finding pharmacokinetics related genes in genome-wide studies. Pharmacogenomics J 2009;9:137–46.

[108] Group SC, et al. *SLCO1B1* variants and statin-induced myopathy—a genome wide study. N Engl J Med 2008;359:789–99.

[109] Takeuchi F, et al. A genome-wide association study confirms VKORC1, CYP2C9, and CYP4F2 as principal genetic determinants of warfarin dose. PLoS Genet 2009;5:e1000433.

[110] Kazi DS, Garber AM, Shah RU, Dudley RA, Mell MW, Rhee C, et al. Cost-effectiveness of genotype-guided and dual antiplatelet therapies in acute coronary syndrome. Ann Intern Med 2014;160:221–32.

[111] http://www.fda.gov/drugs/scienceresearch/researchareas/pharmacogenetics/ucm08337.

[112] Clancy JP, Johnson SG, Yee SW, McDonagh EM, Caudle KE, Klein TE, et al. Clinical pharmacogenetics implementation consortium (CPIC) guidelines for ivacaftor therapy in the context of CFTR genotype. Clin Pharmacol Ther 2014;95:592–7.

[113] Bottillo I, Morrone A, Grammatico P. Pharmacogenetics in the era of next generation sequencing. J Pharmacovigil 2013;1:109. http://dx.doi.org/10.4172/2329-6887.1000109.

THE ROLE OF NEXT GENERATION SEQUENCING IN GENETIC COUNSELING

12

Asude Durmaz, Burak Durmaz

Department of Medical Genetics, Ege University Faculty of Medicine, Izmir, Turkey

CHAPTER OUTLINE

INTRODUCTION

Genetic counseling is defined as "a process that gives information regarding the risk of developing a genetic condition, or transmitting a genetic condition to the next generation, as well as management advice, and treatment options, for the genetic ailment." Genetic counseling sessions should be given to individuals or families by a trained person such as a clinical/medical geneticist, genetic counselor, or other appropriately trained professional. Genetic specialists may recommend additional genetic testing. In addition to targeted analysis of genes known to be linked to certain syndromes, multiplex genetic testing, using high-throughput technologies such as next generation sequencing (NGS), have strongly influenced the practice of genetic counseling. NGS is an efficient method to search for alleles underlying rare Mendelian disorders and for detection of an enormous number of variants helpful in assessing the risk of genetic disease. An important element is the discovery of variants implicated in adult-onset conditions such as cancer predisposition genes, neurodegenerative disease genes, or variants of unknown significance (VUS). Pretest genetic counseling is important, as VUS and incidental results may be confusing for the patient. The genetic counselor, during the informed consent process, should discuss with patients alternatives regarding the return of results. The complex nature and sheer volume of reported results require professional interpretation to provide information and advise patients and their families. Patients should be aware of the benefits, limitations, and challenges in the interpretation of huge amounts of genomic data obtained via NGS technology. Genetic counselors are suited to providing the interpretation of germ-line results and make recommendations regarding their disclosure. There is an urgent need to establish guidelines on the disclosure of genetic information and to balance privacy issues with the potential advantages and drawbacks of sharing genetic data with patients and their relatives. The patient would benefit from understanding the strengths and limitations of NGS during genetic counseling. Thus, it is imperative that genetic counselors contribute to multidisciplinary care teams implementing current models. Making effective and efficient genetic testing available to patients and their health care providers is the primary aim. Importantly, a strong partnership with patients is critical for understanding how to maximize the translation of genetic information for the future benefit of all patients.

GENETIC COUNSELING

Genetic counseling is defined as the process of dealing with the risk and occurrence of a known, or suspected, genetic disorder present in a particular family. It is a process for assisting the "right" informed decision, while also coping with any medical or nonmedical issues associated with the genetic condition [1]. Basically, the process involves helping patients, or other family members, to understand the genetic and clinical facts of a specific disorder, how heredity contributes to this condition, and the risk of recurrence and then presenting options to deal with the condition [1,2]. The term "genetic counseling" was originally coined in 1947 by Sheldon Reed, the director of the Dight Institute of Human Genetics in Minnesota, an institution established only in 1941 [3]. In Reed's opinion, a genetic counselor was a health care professional, not necessarily with a medical background, with a specialty in genetics and a primary role of dispensing advice to the larger medical community. Reed also described genetic counseling as a form of social work, rather than a purely medical service [4]. Significant advances occurred in medical genetics during the 1960s and 1970s, perhaps far beyond the scope of what even Sheldon Reed foresaw; these included the introduction of prenatal genetic testing, improved cell culturing techniques, and newborn screening tests, as well as better banding techniques for identification of chromosomal diseases in greater detail [5].

In 1973 the National Genetics Foundation convened a workshop on genetic counseling and medical genetics services. Participating geneticists included members of the Committee on Genetic Counseling of the American Society of Human Genetics [5]. The Society described the process of genetic counseling as a communication process dealing with human problems relating to the occurrence, or the risk of occurrence, of a genetic disorder within a family [6]. The aims of the process were concerned with helping the patient or family to understand: (1) the medical facts, including diagnosis, probable course of the disorder, and available management; (2) how heredity contributes to the disorder and the risk of recurrence in specified relatives, (3) the alternatives for dealing with the risk of recurrence. Once this information had been successfully disseminated, the counselor could then assist in the selection of an appropriate course of action in view of risk, family goals, ethical concerns, and religious standards. Following the decision-making process, the counselor would then remain actively involved to assist with adjustment to the decision and offer advice on the potential for recurrence. These services were generally provided by trained health care professionals such as genetic counselors, nurses, medical geneticists, and, occasionally, other persons appropriately trained for the task. The skill of the counselor was in assisting patients to make "the right choice" and for the patients to be satisfied with their decision afterward. This was achieved by answering all the questions directed to the counselor and trying to understand the person behind the patient he or she was dealing with. The genetic counselor was required to be nondirective; but nevertheless, nondirective counseling was, and is, very difficult, sometimes impossible. It is the guiding principle; and the counselor has to provide the family all the possible options free of personal biases or preconceived ideas. It is generally considered that knowledge itself is not sufficient if it cannot be appropriately understood by the patient. Convincing a patient of the known facts, therefore, has always been paramount and should never be considered as persuasion [7]. The National Society of Genetic Counselors (NSGC) composed a committee to discuss and renew the very definition of genetic counseling [8]. The new definition approved by the NSGC Board of Directors in 2005 is as follows: Genetic counseling is the process of helping people understand and adapt to the medical, psychological, and familial implications of genetic contributions to disease. This process integrates the following:

1. Interpretation of family and medical histories to assess the chance of disease occurrence or recurrence,
2. Education regarding inheritance, testing procedures, management, prevention, resources available, and research,
3. Counseling to promote informed choices, understanding of potential risk, and adaptation to the condition [8].

Additionally, it is obvious that the demand for genetic counseling will increase in unison with identification of rare syndromes and genetic conditions. Implementation of new genetic testing procedures and utilization of cutting edge technologies are making all this a reality [9]. Genetic services are increasingly being offered by primary health care providers, meaning the sheer volume of tests performed is also rising significantly. The onus is on genetic counselors to be informed of all potential risks and to then accurately present this information. Although genetic testing is not compulsory, genetic counseling should be an accepted part of the general screening process and should, therefore, be recommended. In pretest counseling, the patient is informed about the test, the accuracy of the test in relation to the genetic condition, the diagnosis and treatment options, the limitations of the test, and the benefits associated with performing the test on other family members, as well as other privacy and confidentiality clauses where appropriate. Informed consent is a very important step in genetic counseling and should always be obtained prior to any testing. Written information materials, in conjunction with other informative sources, including a summary of the counseling process, should be available upon request. Immediately following the test results, the counselor must focus on the emotional impact on the patient. The results should be discussed in detail to a point at which the counselor is sure that the patient fully understands. The counselor may recommend a further consultation in regard to the context of the disease and the test results; this is completely within the context of posttest genetic counseling. In short, the committee states that genetic counseling must be an integral part of genetic testing and anyone being offering genetic testing must be provided with appropriate genetic counseling prior to and after testing [10]. With the development of novel methods in genetic testing many serious questions have been raised in terms of the significance of the results. As yet, neither physicians nor counselors are informed well enough to deal with the complexities of some results. Novel technologies, such as NGS, have brought new insights into the research and clinical use of genetic tests and, by association, into the field of genetic counseling.

GENETIC COUNSELING IN THE NGS ERA

The Human Genome Project (HGP) has enabled the sequencing of the human genome. From the development of high-throughput technologies, such as NGS, both time and cost concerns have been reduced considerably. During the HGP it took more than a decade to sequence the complete human genome, and it was, finally, accomplished only through a $3 billion publicly funded project fund and the combined input of more than 200 leading scientists. Following the launch of NGS onto the free market, sequencing costs have dropped precipitously, more steeply even than predicted by Moore's law [11]. In 2014 one of the world's largest NGS companies announced that it could sequence a human genome for US$1000 in less than a day [12]. This tremendous decline, both in cost and in turnaround time, will make NGS far more accessible to clinicians and, from this, a central component of future routine health care management [13,14]. Because the human genome comprises 3.2 billion pairs of nucleotides and is estimated to have 4 million DNA sequence variants, analyzing and interpreting this huge amount of data is sure to complicate the clinical applications of NGS technologies [15]. Before the NGS era, there was only a limited number of patients, and genetic tests, requiring a counseling session to facilitate the genetic test. Basic requirements included directing patients to the proper genetic tests and then reporting and interpreting the results. Following the introduction of NGS into routine clinical practice, as well as the research area, the number of potential patients needing genetic counseling sessions will increase exponentially. A patient suspected of having a single gene disorder could be analyzed for the responsible disease-carrying gene to reveal the suspected mutation. The result of the DNA sequencing analysis will be conclusive as to the presence of the suspected mutation. However, assessment of the sequencing of a single gene is less complicated compared with NGS technologies; it is estimated that whole-genome sequencing (WGS) will reveal 3–4 million variants, with whole-exome sequencing (WES) revealing approximately 20,000 of them [16]. Just handling such a massive amount of sequence data generates some serious concerns. In addition to the new inherent problems beset by genetic counselors, many researchers and clinicians also have serious misgivings in relation to interpreting data and returning accurate results based purely on WGS/WES. Thus, the roles and responsibilities of genetic counselors need to be broadened to satisfy the demands of both the patients and the clinicians/researchers. In many studies it has been indicated that genetic counselors should be actively involved in every step of WES/WGS both clinically and in the research area [17,18]. Genetic counseling should, therefore, include pretest and posttest genetic counseling sessions, which could be broken down further into multiple sessions if the situation so requires (Figure 1). Genetic counselors are routinely facing patients demanding specific information regarding their genetic diseases or associated risks, and through association with all steps of the WES/WGS process counselors will be in a unique position to accurately dispense this often complex information.

PRETEST GENETIC COUNSELING

It is important for a clinician or counselor, during the initial meeting with the patient, to determine whether NGS is an appropriate alternative to cytogenetics or traditional molecular technologies. Good candidates, for example, might be:

1. a patient carrying a suspected single-gene disorder with an unknown underlying genetic defect that other tests had failed to identify or
2. a patient presenting with clinical features consistent with a specific genetic disease showing locus heterogeneity [14].

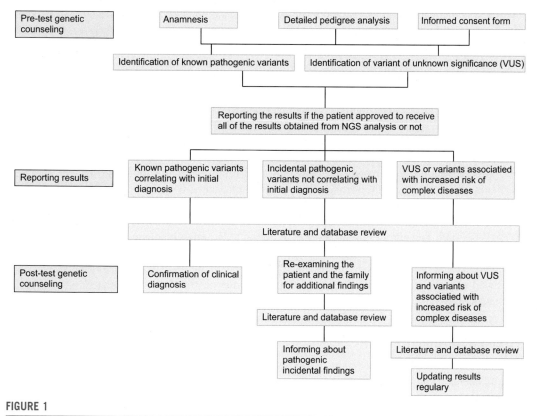

FIGURE 1

Brief summary of genetic counseling in NGS.

The American College of Medical Genetics and Genomics (ACMG) has suggested guidelines for the implementation of NGS technologies on a clinical basis. In an ACMG report, it was recommended that NGS be offered to patients having a phenotype not corresponding to a specific disorder, having a defined genetic disorder with a genetic heterogeneity, or having a likely genetic disorder undiagnosed after other specific testing procedures including targeting sequencing. This would also apply to pregnant women when a genetic disorder in the fetus could not be confirmed by other means [19]. Understanding the benefits and pitfalls of NGS will help the clinician to choose the appropriate course of testing for the patient. Where phenotypically single-gene disorders are suspected, but all known single-gene mutations have been eliminated, or for a disease for which multigene testing is extremely expensive, NGS presents as a very reasonable option. Currently, NGS is frequently applied to reveal unknown genetic etiology of many diseases, raising the expectations of a clinician or a patient. Therefore, a genetic counseling session should be performed prior to offering the test, to inform and reassure the patient and his or her doctor, reducing their anxiety level even when such expectations would not be met. The ACMG suggests that pretest counseling should be performed by a medical geneticist or an affiliated genetic counselor and should include a formal consent process [19]. Prior to performing any NGS test detailed information, including family history and phenotype, needs to be gathered and evaluated systematically. This systematic evaluation of the patient should be performed by an expert medical geneticist to determine whether the patient fulfills the diagnostic criteria for a known genetic disorder, thus narrowing the options of the genetic tests to be performed. Where NGS is considered the most appropriate test for a patient the "informed consent process" should follow.

An extremely thorough genetic counseling session, estimated to last from 6 to 8 h, is required prior to any WGS/WES analysis. During this process, participants will be informed regarding the expected outcomes and benefits of the test and the incidental findings (IFs) and what they could mean, and they will discuss which results the patient does NOT want disclosed [16]. It is during this process that counselors need to use the allocated time efficiently, 6 to 8 h not being a long time to complete the difficult task of ensuring that patients fully understand the advantages and challenges of the technology and how results gleaned from it can ultimately affect them or their relatives. Even with the advantages of WES/WGS being well understood, the views and experiences of genetic counselors still differ greatly. In a 2014 survey completed by genetic counselors from various primary areas of practice (such as pediatrics, cancer, and prenatal counseling) several factors were identified explaining the variation in the use of WES/WGS [20]. The aim of this survey was to highlight the roles, challenges, and needs of genetic counselors during the initial implementation of WES/WGS. Interestingly, billing and insurance issues, duration and content of the consent process, disclosure of IFs, and the significance of unknown variants, as well as complete result interpretation, all presented as central challenges related to the offering of this technology [20].

INFORMED CONSENT FORM

Informed consent has been described as "a process of communication between a patient and physician that results in the patient's authorization or agreement to undergo a specific medical intervention" [21]. In general, before undergoing a clinical genetic test or participating in a research project, the prospective patient should be fully informed regarding the purpose of the test, medical implications, economic considerations, confidentiality, alternative options, potential risks, and perceived benefits.

In 2012, the ACMG Board of Directors published recommendations for "Points to Consider for Informed Consent for Genome/Exome Sequencing." They focused on the necessity and content of an appropriate informed consent that should be mandatorily obtained prior to any clinical or research applications of genome sequencing and exome sequencing (GS/ES) for germ-line testing (Box 1) [22]. It must be noted that this is only an educational tool, available to clinical geneticists and genetic counselors, and it should be considered within the process of adopting their own professional clinical procedures.

The benefits and limitations of WES/WGS need to be clearly defined during the informed consent process. The major advantage of NGS is that it is a powerful tool in identifying novel genes underlying single-gene disorders and complex diseases. Previously used techniques in typical strategies have usually depended on the presence of more than one affected individual sharing similar phenotypes in multiple families, for example, in linkage analysis, in which candidate genomic regions shared in affected family members were systematically sequenced to reveal disease-causing mutations. NGS, however, is frequently well suited to discovering disease-causing mutations within a limited number of study groups and even with single-case mutations following an unknown diagnosis [23,24].

The success of NGS has improved following the usage of targeted panels of disease genes for clinical diagnosis of solid tumors, myeloid malignancies, inherited diseases, cardiomyopathies, hearing

---

BOX 1 ACMG RECOMMENDATIONS FOR INFORMED CONSENT FORM [22].

1. Before initiating GS/ES, counseling should be performed by a medical geneticist or an affiliated genetic counselor and should include written documentation of consent from the patient.
2. Incidental/secondary findings revealed in either children or adults may have high clinical significance for which interventions exist to prevent or ameliorate disease severity. Patients should be informed of this possibility as a part of the informed consent process.
3. Pretest counseling should include a discussion of the expected outcomes of testing, the likelihood and type of incidental results that may be generated, and the types of results that will or will not be returned. Patients should know if and what type of IFs may be returned to their referring physician by the laboratory performing the test.
4. Patients should be counseled regarding the potential benefits and risks of GS/ES, the limitations of such testing, potential implications for family members, and alternatives to such testing.
5. GS/ES is not recommended before the legal age of majority except for:
   a. Phenotype-driven clinical diagnostic uses,
   b. Circumstances in which early monitoring or interventions are available and effective, or
   c. Institutional review board-approved research.
6. As part of the pretest counseling, a clear distinction should be made between clinical and research-based testing.
7. Patients should be informed as to whether individually identifiable results may be provided to databases, and they should be permitted to opt out of such disclosure.

---

loss, and autism [25–27]. By applying WGS/WES to cancer, numerous variants have been found in association with clinical diagnosis and potential therapeutic targets, the first step of an effective treatment plan [28]. Compared with WES/WGS, the targeted sequencing of selected genes offers a cost-effective way of identifying disease-causing mutations. In addition to having a lower cost, panels are also more user-friendly at a clinical level because of their shorter turnaround time and more flexible protocols. The conclusions of several studies have raised concerns regarding current counseling models and the necessity for new genetic counseling models specifically designed for patients undergoing targeted panel testing [29,30]. For instance, high-penetrance genes, those associated with autosomal dominant cancer syndromes, can be easily evaluated via a pedigree analysis. The patient is informed about the genes that are responsible for the suspected cancer syndromes and is not tested for the genes that are unlikely to be mutated. Thus, the counselor and the patient deal only with the suspected genes prior to the testing procedure. Following the use of WES/WGS, moderate-penetrance genes associated with lower risks of a genetic condition could be found, and new genetic counseling models should be considered to cope with this issue [31]. The management of carriers of moderate-penetrance genes and the clinical response to a mutation in a moderate-penetrance gene are not yet well enough defined. For example, the lifetime breast cancer risk for a person (without family history) having a truncating mutation in the moderate-penetrance gene *CHEK2* is 20% but the risk increases to 44% with a family history of breast cancer [32]. There is no consensus on the standard management of patients with *CHEK2* mutations detected in cancer panels. All management protocols need to be clearly defined during the informed consent process in pretest genetic counseling sessions, including those for mutations that are detected in a moderate-penetrance gene and other such anomalies. The amount and quality of a DNA sample used in a genetic test can be a first-step limitation. Effective analysis is dependent on a specific quantity of DNA. The amount of DNA required for Sanger sequencing is a few micrograms, whereas NGS analysis can be conducted with 50 ng of DNA [33]. Improved DNA preparation techniques will bring great advantages in the usage of NGS in both clinical and research areas.

Clinical and research-based sequencing objectives should be clearly defined prior to performing any WES/WGS analysis. Research projects, using WES/WGS in particular, are very likely to create unprecedented amounts of data not directly associated with the condition [34]. The participants need to reach an informed decision as to whether they want results obtained from the research project disclosed and to what extent [35]. A systematic evaluation of studies, concerning the application of NGS in complex disorders, revealed that the use of NGS creates the necessity of genetic counseling in research projects [18]. It should be noted that participants may not be willing to receive results unrelated to the study in which are they enrolled [36]. In this case every participant should be considered as an individual person. In addition to receiving results from WES/WGS analysis, data management and sharing is another important aspect that needs to be fully explained during the initial genetic counseling session. The patient should then also be informed regarding the maintenance, storage, and security protocols associated with the handling of sensitive genomic data.

RETURN OF RESULTS

There is a growing argument regarding the handling of results: which results should be returned to clinical investigations, whether sensitive results should be reported in a research project, and to what extent the results should be reported to participants [37,38]. Previously, it had been recommended not to return genetic results to the research participants; and this was later amended to the opposite practice in cases "where the associated risk for the disease is significant; the disease has important health implications such as premature death or substantial morbidity or has significant reproductive implications; and where proven therapeutic or preventive interventions are available" [39]. Although the researchers have no obligation to search for genetic results unrelated to the primary goal of the study, the participant has a right to know all individual health-related genetic information obtained through the sequencing. Although, theoretically, every patient has a right to know every result, the importance of the consent of the participant has been emphasized regarding the return of research results, as well as IFs [40,41]. During the consent process the participant may consider the potential role of the results in the future treatment or prevention of a disease [18,42]. It has been reported that, when fully informed about the benefits and pitfalls of a test, research participants who are provided with the option of receiving all results obtained from a specific research project usually show high interest in receiving all clinically significant data [43]. As the technology and the data obtained from NGS are rapidly evolving, new insights and consensus need to be considered. A new model for returning results, termed "self-guided management," has been proposed. It allows individuals "to determine whether and when they receive results, in a personalized and time-sensitive context, that is responsive to their value system and their perception of possible benefits and risks, at a given time point" [44]. It is noted that the main problem is not which results should be returned, but rather how the access can translate into improvements in the individual's health. According to this model a person may wish to access no information, receive all genetic results at once, or alternatively receive results serially over time. The patient also retains the option of changing his or her preferences, meaning they can change their decision at any stage of the genetic counseling process. This makes it a continuous, dynamic process rather than a single informed session. There is still only limited clinically relevant evidence for most genomic changes. A variant known as a VUS, or an insignificant variant at the time of sequence data interpretation, may be reclassified over time. This presents the problem of when/if the WES/WGS data of a patient should be reevaluated, how often it should be reevaluated, and who is financially responsible for these subsequent reevaluations. Is it reasonable that the burden should fall to the department of social security?

The content of the WES/WGS report is also a challenging point to be considered. The Clinical Sequencing Exploratory Research (CSER) consortium has developed a protocol refining the structure and content of clinical genomic reports; it provides detailed, systematic results and suggests methods for interpretation (Table 1) [45]. It concludes that extensive findings in a patient tested for a particular disease, including IFs in other genes, could be confusing for clinicians. Because of this, dividing the process into two reports is recommended. One focuses on the results concerning genes of interest (i.e., colorectal cancer/polyposis susceptibility genes) and another separate report deals with IFs and pharmacogenomics variants, which should be returned during a separate clinic visit. The first page of the clinical test report should include a brief table describing the gene name, variant, clinical significance, and recommendation. The second page of the clinical test report should include an interpretation of each variant listed within a table showing the gene, the variant name with a detailed genomic location, the frequency of the variant, in silico predictions for VUS, functional evidence, and a review from the literature describing the association with the disease. Another, separate report for IFs and pharmacogenomics results obtained from WES/WGS should also be prepared giving the same information on each variant.

Interpretation of results

It is challenging to analyze and interpret the clinical significance of DNA variants identified through WES/WGS. Interpretation of a routine single-gene sequencing result comprises a few hundred to a few thousand bases, thus generating only a few variants to analyze. A WES or WGS test may produce results classified as VUS or likely to be pathogenic that require a detailed database search. NGS brings with it an increased likelihood of producing a huge amount of results within the data, including VUS or IFs. At this instance, an information session dedicated to variant interpretation would be of great benefit. The significance of the variants varies greatly; thus, each variant has to be analyzed carefully. In a study using targeted cancer panels, 10% of panels detected positive results, whereas 30% of panels detected VUS that required careful evaluation [30]. The assessment of data included evaluation or prediction of the effect on the protein, the frequency of the variant within the population, and the predisposition of the variant to cause disease. This kind of assessment is often achieved through detailed database searches, such as UCSC, ClinVar, and HGMD (Table 2).

For many genetic counselors, the interpretation of genomic data using web-based tools can be time-consuming and challenging. For example, an average of 20 min of review per variant is spent for analysis of genes proposed by the ACMG, with some variants requiring hours of review and discussion [46]. A detailed comprehensive database needs to be completed, in conjunction with a literature review, to assist in the interpretation of data. Thus, it is very important to work in collaboration with genetic counselors and researchers. In 2014, a group of cancer genetic counselors in Colorado published their results of a collaboration on a cross-institutional project to create a resource that could be consulted for the interpretation of abnormal results. It also provided an educational platform for genetic counselors to gain a better understanding of the mutations in hereditary cancer syndrome genes included in clinical NGS panels [47]. This collaboration saved counselors from a time-consuming data interpretation task. The development of collaborative groups in different disease areas such as cancer, metabolic disease, and neurologic disease is essential for future NGS implementations. Because the variant found in a patient may have been previously investigated and uploaded to the database, this means there is the potential for counselors to spend far less time doing individual investigation and review. Various medical centers and research programs such as Coriell Personalized Medicine Collaborative, Clinical

Table 1 **Exome/Genome Report Content Recommended by the CSER Consortium [45]**

Laboratory/Patient/Sample Identifiers

Name and address of reporting laboratory (optional: phone, FAX, e-mail, Website)

Patient's name (first and last with middle initial or middle name)

ID No. (medical record number) and/or date of birth

Date of specimen collection

Date of receipt or accession in laboratory, with accession number

Specimen source and how tissue was received (fresh, frozen, paraffin-embedded, etc.)

Ordering physician

Indication

Reason for referral with pertinent clinical history

Results

For each identified variant, list the variant-containing standardized gene name (HUGO gene name), variant ID (HGVS convention) with zygosity, clinical significance, and recommendation (if applicable)

List genes examined associated with indication for testing

Interpretation

Interpretation involves synthesizing analytic and clinical information to describe what the result means for the patient. Results should be interpreted in a clinically relevant manner and explain how technical limitations might affect the use of test results. When appropriate, test results may be explained with reference to family members and whether they possess the identified variant.

Should include gene and chromosome names, variant position (with respect to genome reference used), and nucleotide and protein variant designations according to HGVS nomenclature. The interpretation also includes text describing the current state of knowledge regarding the detected sequence variant and its disease-causing classification according to current recommendations for sequence interpretation.

Comments/recommendations

Significance of the result in general or in relation to this patient

Correlate with prior results (if applicable)

Recommend additional measures (additional testing, genetic counseling, etc.)

Condition of specimen that may limit accuracy of testing (e.g., partially degraded DNA)

Pertinent assay performance characteristics or interfering substances

Residual risk of disease (or carrier status) by Bayesian analysis

Document interdepartmental consultation

Incorporate information specifically requested on the requisition (e.g., ethnicity)

Answer specific questions posed by the requesting clinician (e.g., rule out)

Reason specimen rejected or not processed to completion

Disposition of residual sample (e.g., sample repository)

Chain of custody documentation, if needed

If the report is an amended or addendum report, describe the changes or updates

Describe discrepancies between preliminary and final reports

Name of testing laboratory, if transmitting or summarizing a referral laboratory's results

References

Cite medical literature supporting interpretation

Procedure

Type of procedure (e.g., exome sequencing, genome sequencing, etc.)

Defined target (e.g., exome reference)

Pertinent details of procedure, for example, analyte-specific reagent or kit version and manufacturer, instrument type

Disclaimer on non-FDA-approved test

Limitations of assay (e.g., does not detect large deletions and duplications)

Signature of laboratory director or designee when interpretation is performed (reports may be signed electronically)

Date of report

Table 1 Exome/Genome Report Content Recommended by the CSER Consortium [45]—cont'd

Demographic information

Accession number and specimen number from referring laboratory
Genetic counselor, when appropriate
Clinic/inpatient location or name/address/phone of outside facility

Research statement

If data were produced for a research project requiring return of results to the study participant, a brief statement describing the assay's use and limitations should be provided prior to the indication section

---

Table 2 Examples of Databases Used for Interpretation of Genomic Sequence Data

Database	URL	Definition
UCSC genome browser	https://genome.ucsc.edu	Genome bioinformatics database
Ensembl genome browser	http://www.ensembl.org/index.html	A detailed genomic information browser
ENCODE Encyclopedia of DNA Elements	https://genome.ucsc.edu/ENCODE/	Identifying functional elements in the human genome
dbSNP	http://www.ncbi.nlm.nih.gov/SNP/	Information about sequence variations
ClinVar	http://www.ncbi.nlm.nih.gov/clinvar/	Information about sequence variation and its relationship to human health
Human Genome Mutation Database	http://www.hgmd.org/	Database of known (published) mutations responsible for human inherited diseases
Human Genome Variation Society	http://www.hgvs.org/dblist/dblist.html	Database of genomic variations including population distribution and phenotypic associations
PolyPhen2	http://genetics.bwh.harvard.edu/pph2/	Predicts possible effects of human nsSNPs on the structure and function of a human protein
SIFT	http://sift.jcvi.org/	Predicts possible effects of amino acid substitutions based on the degree of conservation
Mutation Taster	http://www.mutationtaster.org/	Predicts possible effects of human variations

---

Genome Resource, Electronic Medical Records and Genomics Network, International Collaboration for Clinical Genomics, and Return of Results Working Group are constantly evaluating these databases and establishing new approaches to identifying the clinical significance of variants and novel methods for adopting these variants into clinical practice [48].

Developing publicly available databases to enhance the understanding of rare variants is crucial for further scientific advancements; but it also raises some serious concerns regarding patient confidentiality and privacy. It is, therefore, vital to identify any potential risks associated with both data sharing in public and restricted access databases. Even if the participants have supported the idea of data sharing, several reports indicate that they would feel deceived if their genomic data were shared without their express permission [49–51].

Incidental findings

Genomic IFs are described as unanticipated information discovered in the course of genetic testing or medical care, much like the "incidentalomas" that are often discovered in radiological studies [52]. Other terms have also been used to describe IFs, such as "unanticipated findings" or "off-target results" [53,54]. Considering whole-genome analysis, a person is assumed to carry 4 million variants. Among this vast number, more than 100,000 variants with the potential to cause inherited disorders have currently been identified in the Human Genome Mutation Database [55–57]. All these variants should be interpreted carefully before returning any results to the patient.

In 2013 the ACMG published a policy statement based on the reporting of IFs in clinical exome and genome sequencing [58]. The ACMG working group suggested the assessment and reporting of findings in 56 genes associated with 24 inherited conditions. It was considered that all 56 genes directly related to perceived medical benefit for either the patient or the relatives of the patient (Table 3). These beneficial conditions included preventable and treatable disorders, as well as disorders in which pathogenic mutations might lie dormant, being asymptomatic for long periods of time. While these variants have been restricted to only known pathogenic variants, VUS, or likely pathogenic variants, there is still no consensus on a protocol concerning whether these 56 genes, or other potentially beneficial genes, should even be reported [59].

The CSER network, funded by the National Human Genome Research Institute and the National Cancer Institute, has also reported medically actionable genes that, if present, should be reported as IFs [60]. In this report, management protocols of IFs related to some very challenging cases are also discussed. Such findings as neurofibromatosis type 1, homozygous Factor V Leiden mutation, hemochromatosis, Gaucher, maturity-onset diabetes of the young, and long-QT syndrome all present their own unique dilemmas. It can be especially confusing when many of the variants detected by WGS/WES have a discernible clinical significance. The implementation of genome-scale sequencing currently requires further evidence and guidance. A binning system to analyze and report on IFs has been proposed to deal with such enormous quantities of ambiguous data [61]. This system, which categorizes genes into three groups according to their clinical utility, outlines which genes, and what kinds of variants, need to be reported if detected incidentally. Studies revealing the attitudes of genetic counselors and patients toward the reporting of IFs may help to clarify this uncertainty. In a research study documenting the views and perspectives of genetic professionals on reporting IFs from clinical genome-wide sequencing in Canada, geneticists and genetic counselors indicated that variants associated with serious and treatable diseases, as well as pharmacogenetic information of the patient, should be disclosed regardless of patient preference [62]. Geneticists and genetic counselors did not differ in their responses regarding which types of IFs should be included. Significant differences were present, however, in relation to returning IFs to an adult or a child patient. There are many arguments for and against disclosure of results for adult-onset genetic disorders in children [63].

Before defining a variant as an IF, the genetic counselor and the consulting physician should clearly review all relevant patient information, including a detailed physical examination and three generations of family history. Any variant, even disease causing, obtained from WGS/WES analysis

Table 3 ACMG Recommendations on Minimum List of Genes that Should be Sought for a Mutation and Reported if an IF is Detected in Clinical Sequencing [58]

Gene Name	Disease	Inheritance	Age of Onset	Gene Name	Disease	Inheritance	Age of Onset
BRCA1	Hereditary breast and ovarian cancer	AD	Adult	MYBPC3	Hypertrophic cardiomyopathy, Dilated cardiomyopathy	AD	Child/adult
BRCA2				MYH7			
TP53	Li-Fraumeni syndrome	AD	Child/adult	TNNT2			
STK11	Peutz-Jeghers syndrome	AD	Child/adult	TNNI3			
MLH1	Lynch syndrome	AD	Adult	TPM1			
MSH2				MYL3			
MSH6				ACTC1			
PMS2				PRKAG2			
APC	Familial adenomatous polyposis	AD	Child	MYL2			
MUTYH	MYH-associated polyposis; Adenomas, multiple colorectal, FAP type 2; colorectal adenomatous polyposis, autosomal recessive, with pilomatricomas	AR	Adult	LMNA			
VHL	Von Hippel Lindau syndrome	AD	Child/adult	GLA		X-L	
MEN1	Multiple Endocrine Neoplasia type 1	AD	Child/adult	RYR2	Catecholaminergic polymorphic ventricular tachycardia	AD	
RET	Multiple Endocrine Neoplasia type 2	AD	Child/adult	PKP2	Arrhythmogenic right-ventricular cardiomyopathy	AD	Child/adult
RET	Familial Medullary Thyroid cancer (FMTC)	AD	Child/adult	DSP			
NTRK1		AD?		DSC2			
PTEN	PTEN Hamartoma tumor syndrome	AD	Child	TMEM43			
RB1	Retinoblastoma	AD	Child	DSG2			

Continued

Table 3 ACMG Recommendations on Minimum List of Genes that Should be Sought for a Mutation and Reported if an IF is Detected in Clinical Sequencing [58]—cont'd

Gene Name	Disease	Inheritance	Age of Onset	Gene Name	Disease	Inheritance	Age of Onset
SDHD *SDHAF2* *SDHC*	Hereditary paraganglioma-pheochromocytoma syndrome	AD	Child/adult	*KCNQ1* *KCNH2* *SCN5A*	Romano-Ward long QT syndromes types 1, 2, and 3, Brugada syndrome	AD	Child/adult
				LDLR	Familial hypercholesterolemia	SD	Child
TSC1 *TSC2*	Tuberous Sclerosis complex	AD	Child	*APOB* *PCSK9*		SD AD	
WT1	WT1-related Wilms tumor	AD	Child	*RYR1* *CACNA1S*	Malignant hyperthermia susceptibility	AD	Child/adult
NF2	Neurofibromatosis type 2	AD	Child/adult				
COL3A1	EDS, vascular type	AD	Child/adult				
FBN1 *TGFBR1* *TGFBR2* *SMAD3* *ACTA2* *MYLK* *MYH11*	Marfan syndrome, Loeys-Dietz syndromes, and familial Thoracic Aortic Aneurysms and Dissections	AD	Child/adult				

and pursued without a proper indication or differential diagnosis can be considered as an IF [64]. For example, if a detailed pedigree analysis could not be obtained for a patient who had undergone WGS/WES for a diagnosis of a genetic disorder, a mutation detected in the huntingtin (*HTT*) gene may be considered as an IF. If, however, the patient was extensively questioned during a pretest counseling session and it was learned that his or her grandparent, even without a clinical diagnosis, exhibited symptoms resembling Huntington disease, the situation is different. In this case a variant previously considered as an IF may be reevaluated to an expected variant in the test results. There is also the option for this kind of family history exploration during posttest counseling once information has been brought to light. Either way, it highlights the value of precise information concerning both the patient and the family members when attempting to reveal the significance of variants detected in WGS/WES. This posttest evaluation, including additional literature and database searches in conjunction with a detailed follow-up examination of the patient, requires an advanced level of expertise in both clinical and molecular genetics.

There is currently no valid consensus concerning the disclosure of IFs, to whom they should be disclosed, how often the results should be reevaluated, and by whom, or how further medical follow-up procedures should be arranged. A serious lack of guidance in relation to all these questions has resulted in many of the concerns that genetic counselors and the patients now have. ACMG recommendations lead to an opportunity to standardize and optimize genomic testing. The gathering of more data and evidence regarding the experiences of the counselors, as well as patient preferences, is also crucial for a smooth integration of genomics into clinical practice.

FUTURE PERSPECTIVES

In recent years the traditional term "genetic counseling" has tended to change to "genomic counseling" to reflect the use of whole-genome-based techniques [65]. The original model for traditional genetic counseling was based on finding the etiology of a well-defined genetic condition using single-gene-based techniques. This model is not applicable for genome-based testing, therefore invoking a necessity for "genomic counselors" [66]. In a genomic counseling session the information given differs greatly from that given in traditional genetic counseling and covers such things as number and/or type of diseases for which testing is available, purpose of testing, intervention, clinical utility, and accessibility to resources. In addition to disease-causing mutations, WGS/WES also reveal variants associated with an increase of risk for certain complex disorders. Thus, the role of a genetic counselor needs to be broadened not only to explain the results of a WGS/WES, but also to suggest and encourage preventive behavior. As a direct result of this new disease prevention role, additional knowledge of "risk-reducing behaviors for complex diseases" will be required. Following the recognition of the genome-informed preventive medicine area, a group of genetic counselors have established a new special interest group focused on personalized medicine within the NSGC.

The potential role of genetic counselors, to define the most appropriate way of evaluating and dispensing information derived from the results of WGS/WES, necessitates both educational and professional organizations forming a consensus that clearly outlines the definition of genomic counseling in the new era. In conjunction with this, the role and scope of genetic counseling also must evolve to include management of information obtained from genome-based techniques.

LIST OF ACRONYMS AND ABBREVIATIONS

ACMG American College of Medical Genetics and Genomics
ASHG American Society of Human Genetics
HGP Human Genome Project
HGVS Human Genome Variation Society
HUGO Human Genome Organization
IFs Incidental findings
NSGC National Society of Genetic Counselors
VUS Variants of unknown significance
WGS whole-genome sequencing
WES whole-exome sequencing

REFERENCES

[1] Rantanen E, Hietala M, Kristoffersson U, Nipper I, Schmidtke J, Sequeiros J, et al. What is ideal genetic counseling? A survey of current international guidelines. Eur J Hum Genet 2008;16(4):445–52.

[2] Pilnick A, Dingwall R. Research directions in genetic counseling: a review of the literature. Patient Educ Couns 2001;44(2):95–105.

[3] Reed SC. A short history of genetic counseling. Soc Biol 1974;21(4):332–9.

[4] Resta RG. Defining and redefining the scope and goals of genetic counseling. Am J Med Genet C Semin Med Genet 2006;142C(4):269–75.

[5] Fraser FC. Genetic counseling. Am J Hum Genet 1974;26(5):636–59.

[6] Epstein CJ. Who should do genetic counseling, and under what circumstances? Birth Defects Orig Artic Ser 1973;9(4):39–48.

[7] Pennacchini M, Pensieri C. Is non-directive communication in genetic counseling possible? Clin Ter 2011;162(5):e141–4.

[8] Resta R, Biesecker BB, Bennett RL, Blum S, Hahn SE, Strecker MN, et al. A new definition of genetic counseling: National Society of Genetic Counselors' task force report. J Genet Couns 2006;15(2):77–83.

[9] Austin J, Semaka A, Hadjipavlou G. Conceptualizing genetic counseling as psychotherapy in the era of genomic medicine. J Genet Couns 2014;23(6):903–9.

[10] Revel M, editor. Proceedings of the third session (Genetic counseling). Paris: International Bioethics Committee of UNESCO; 1995.

[11] Hayden EC. Technology: the $1000 genome. Nature 2014;507(7492):294–5.

[12] http://www.technologyreview.com/news/523601/does-illumina-have-the-first-1000-genome/.

[13] Ong FS, Lin JC, Das K, Grosu DS, Fan JB. Translational utility of next-generation sequencing. Genomics 2013;102(3):137–9.

[14] Biesecker LG, Green RC. Diagnostic clinical genome and exome sequencing. N Engl J Med 2014;370(25):2418–25.

[15] Ng PC, Levy S, Huang J, Stockwell TB, Walenz BP, Li K, et al. Genetic variation in an individual human exome. PLoS Genet 2008;4(8):e1000160.

[16] Johansen Taber KA, Dickinson BD, Wilson M. The promise and challenges of next-generation genome sequencing for clinical care. JAMA Intern Med 2014;174(2):275–80.

[17] Ropers HH. On the future of genetic risk assessment. J Community Genet 2012;3(3):229–36.

[18] Egalite N, Groisman IJ, Godard B. Genetic counseling practice in next generation sequencing research: implications for the ethical oversight of the informed consent process. J Genet Couns 2014;23(4):661–70.

[19] ACMG Board of Directors. Points to consider in the clinical application of genomic sequencing. Genet Med 2012;14(8):759–61.

[20] Machini K, Douglas J, Braxton A, Tsipis J, Kramer K. Genetic counselors' views and experiences with the clinical integration of genome sequencing. J Genet Couns 2014;23(4):496–505.

[21] American medical association encyclopedia of medicine. New York: Harcourt and Brace; 1994.

[22] ACMG Board of Directors. Points to consider for informed consent for genome/exome sequencing. Genet Med 2013;15(9):748–9.

[23] Vozzi D, Licastro D, Martelossi S, Athanasakis E, Gasparini P, Fabretto A. Alagille syndrome: a new missense mutation detected by whole-exome sequencing in a case previously found to be negative by DHPLC and MLPA. Mol Syndromol 2013;4(4):207–10.

[24] Kalay E, Yigit G, Aslan Y, Brown KE, Pohl E, Bicknell LS, et al. CEP152 is a genome maintenance protein disrupted in Seckel syndrome. Nat Genet 2011;43(1):23–6.

[25] Deeb KK, Sram JP, Gao H, Fakih MG. Multigene assays in metastatic colorectal cancer. J Natl Compr Canc Netw 2013;11(Suppl. 4):S9–17.

[26] Kammermeier J, Drury S, James CT, Dziubak R, Ocaka L, Elawad M, et al. Targeted gene panel sequencing in children with very early onset inflammatory bowel disease-evaluation and prospective analysis. J Med Genet 2014;51(11):748–55.

[27] Gu X, Guo L, Ji H, Sun S, Chai R, Wang L, et al. Genetic testing for sporadic hearing loss using targeted massively parallel sequencing identifies 10 novel mutations. Clin Genet 2014;87(6):588–93.

[28] Tran B, Dancey JE, Kamel-Reid S, McPherson JD, Bedard PL, Brown AM, et al. Cancer genomics: technology, discovery, and translation. J Clin Oncol 2012;30(6):647–60.

[29] Domchek SM, Bradbury A, Garber JE, Offit K, Robson ME. Multiplex genetic testing for cancer susceptibility: out on the high wire without a net? J Clin Oncol 2013;31(10):1267–70.

[30] Mauer CB, Pirzadeh-Miller SM, Robinson LD, Euhus DM. The integration of next-generation sequencing panels in the clinical cancer genetics practice: an institutional experience. Genet Med 2014;16(5):407–12.

[31] Rainville IR, Rana HQ. Next-generation sequencing for inherited breast cancer risk: counseling through the complexity. Curr Oncol Rep 2014;16(3):371.

[32] Cybulski C, Wokołorczyk D, Jakubowska A, Huzarski T, Byrski T, Gronwald J, et al. Risk of breast cancer in women with a CHEK2 mutation with and without a family history of breast cancer. J Clin Oncol 2011;29(28):3747–52.

[33] Simbolo M, Gottardi M, Corbo V, Fassan M, Mafficini A, Malpeli G, et al. DNA qualification workflow for next generation sequencing of histopathological samples. PLoS One 2013;8(6):e62692.

[34] Lohn Z, Adam S, Birch PH, Friedman JM. Incidental findings from clinical genome-wide sequencing: a review. J Genet Couns 2014;23(4):463–73.

[35] Wendler D. What should be disclosed to research participants? Am J Bioeth 2013;13(12):3–8.

[36] Sijmons RH, Van Langen IM, Sijmons JG. A clinical perspective on ethical issues in genetic testing. Acc Res 2011;18(3):148–62.

[37] Jarvik GP, Amendola LM, Berg JS, Brothers K, Clayton EW, Chung W, et al. Return of genomic results to research participants: the floor, the ceiling, and the choices in between. Am J Hum Genet 2014;94(6):818–26.

[38] Evans JP, Rothschild BB. Return of results: not that complicated? Genet Med 2012;14(4):358–60.

[39] Bookman EB, Langehorne AA, Eckfeldt JH, Glass KC, Jarvik GP, Klag M, et al. Reporting genetic results in research studies: summary and recommendations of an NHLBI working group. Am J Med Genet A 2006;140(10):1033–40.

[40] Wolf SM, Crock BN, Van Ness B, Lawrenz F, Kahn JP, Beskow LM, et al. Managing incidental findings and research results in genomic research involving biobanks and archived data sets. Genet Med 2012;14(4):361–84.

[41] Issues PCftSoB. Privacy and progress in whole genome sequencing. 2012. [October 2012]. Available from: http://bioethics.gov/sites/default/files/PrivacyProgress508.pdf.

[42] Wendler D, Pentz R. How does the collection of genetic test results affect research participants? Am J Med Genet A 2007;143A(15):1733–8.

[43] Haga SB, Zhao JQ. Stakeholder views on returning research results. Adv Genet 2013;84:41–81.

[44] Yu JH, Jamal SM, Tabor HK, Bamshad MJ. Self-guided management of exome and whole-genome sequencing results: changing the results return model. Genet Med 2013;15(9):684–90.

[45] Dorschner MO, Amendola LM, Shirts BH, Kiedrowski L, Salama J, Gordon AS, et al. Refining the structure and content of clinical genomic reports. Am J Med Genet C Semin Med Genet 2014;166C(1):85–92.

[46] Dorschner MO, Amendola LM, Turner EH, Robertson PD, Shirts BH, Gallego CJ, et al. Actionable, pathogenic incidental findings in 1000 participants' exomes. Am J Hum Genet 2013;93(4):631–40.

[47] Wolfe Schneider K, Anguiano A, Axell L, Barth C, Crow K, Gilstrap M, et al. Collaboration of colorado cancer genetic counselors to integrate next generation sequencing panels into clinical practice. J Genet Couns 2014;23(4):640–6.

[48] Ramos EM, Din-Lovinescu C, Berg JS, Brooks LD, Duncanson A, Dunn M, et al. Characterizing genetic variants for clinical action. Am J Med Genet C Semin Med Genet 2014;166C(1):93–104.

[49] Trinidad SB, Fullerton SM, Bares JM, Jarvik GP, Larson EB, Burke W. Genomic research and wide data sharing: views of prospective participants. Genet Med 2010;12(8):486–95.

[50] Ludman EJ, Fullerton SM, Spangler L, Trinidad SB, Fujii MM, Jarvik GP, et al. Glad you asked: participants' opinions of re-consent for dbGap data submission. J Empir Res Hum Res Ethics 2010;5(3):9–16.

[51] Trinidad SB, Fullerton SM, Ludman EJ, Jarvik GP, Larson EB, Burke W. Research ethics. Research practice and participant preferences: the growing gulf. Science 2011;331(6015):287–8.

[52] Kohane IS, Masys DR, Altman RB. The incidentalome: a threat to genomic medicine. JAMA 2006;296(2):212–5.

[53] Mayer AN, Dimmock DP, Arca MJ, Bick DP, Verbsky JW, Worthey EA, et al. A timely arrival for genomic medicine. Genet Med 2011;13(3):195–6.

[54] Parker LS. The future of incidental findings: should they be viewed as benefits? J Law Med Ethics 2008;36(2):341–51. 213.

[55] Wheeler DA, Srinivasan M, Egholm M, Shen Y, Chen L, McGuire A, et al. The complete genome of an individual by assively parallel DNA sequencing. Nature 2008;452(7189):872–6.

[56] McKernan KJ, Peckham HE, Costa GL, McLaughlin SF, Fu Y, Tsung EF, et al. Sequence and structural variation in a human genome uncovered by short-read, massively parallel ligation sequencing using two-base encoding. Genome Res 2009;19(9):1527–41.

[57] Berg JS, Adams M, Nassar N, Bizon C, Lee K, Schmitt CP, et al. An informatics approach to analyzing the incidentalome. Genet Med 2013;15(1):36–44.

[58] Green RC, Berg JS, Grody WW, Kalia SS, Korf BR, Martin CL, et al. ACMG recommendations for reporting of incidental findings in clinical exome and genome sequencing. Genet Med 2013;15(7):565–74.

[59] Burke W, Matheny Antommaria AH, Bennett R, Botkin J, Clayton EW, Henderson GE, et al. Recommendations for returning genomic incidental findings? We need to talk! Genet Med 2013;15(11):854–9.

[60] Berg JS, Amendola LM, Eng C, Van Allen E, Gray SW, Wagle N, et al. Processes and preliminary outputs for identification of actionable genes as incidental findings in genomic sequence data in the Clinical Sequencing Exploratory Research Consortium. Genet Med 2013;15(11):860–7.

[61] Berg JS, Khoury MJ, Evans JP. Deploying whole genome sequencing in clinical practice and public health: meeting the challenge one bin at a time. Genet Med 2011;13(6):499–504.

[62] Lohn Z, Adam S, Birch P, Townsend A, Friedman J. Genetics professionals' perspectives on reporting incidental findings from clinical genome-wide sequencing. Am J Med Genet A 2013;161A(3):542–9.

[63] Anderson JA, Hayeems R, Shuman C, Szego MJ, Monfared N, Bowdin S, et al. Predictive genetic testing for adult-onset disorders in minors: a critical analysis of the arguments for and against the 2013 ACMG guidelines. Clin Genet 2014;87(4):301–10.

[64] Tennessen JA, Bigham AW, O'Connor TD, Fu W, Kenny EE, Gravel S, et al. Evolution and functional impact of rare coding variation from deep sequencing of human exomes. Science 2012;337(6090):64–9.

[65] Sturm AC, Manickam K. Direct-to-consumer personal genomic testing: a case study and practical recommendations for "genomic counseling". J Genet Couns 2012;21(3):402–12.

[66] Mills R, Haga SB. Genomic counseling: next generation counseling. J Genet Couns 2014;23(4):689–92.

NEXT GENERATION SEQUENCING IN UNDIAGNOSED DISEASES

13

Urszula Demkow

Department of Laboratory Diagnostics and Clinical Immunology of Developmental Age,
Medical University of Warsaw, Warsaw, Poland

CHAPTER OUTLINE

OVERVIEW

Undiagnosed diseases comprise rare and ultrarare conditions, new disease entities, or even common diseases with very unusual, misleading phenotypes. By definition rare or ultrarare disorders have respectively low (below 1:2000) or very low (below 1:200,000) prevalence [1]. The European Organization for Rare Diseases estimates that approximately 7000 distinct rare diseases exist [2]. Some examples include ribose-5-phosphate isomerase deficiency, pantothenate kinase-associated neurodegeneration, Noonan syndrome, epidermolytic hyperkeratosis, ataxia–telangiectasia, Kabuki syndrome, DiGeorge syndrome, dihydropteridine reductase deficiency, glutathione synthetase deficiency, and Joubert syndrome [3]. The most common undiagnosed disorders involve neurologic phenotypes, rheumatologic diseases, fibromyalgia, gastrointestinal diseases, chronic pain, and skin and cardiovascular diseases [4]. According to the Rare Genomics Institute, rare diseases usually have a genetic background [5].

Undiagnosed diseases may serve as a model for the clinical application of emerging genomic technologies. When a patient presents with very nonspecific symptoms, genetic testing of single genes is usually uninformative. The majority of patients with previously unexplained conditions may be diagnosed only by means of genome-wide approaches, staying at the border between research and clinical testing [6,7]. Unfortunately, integration of these research tools into clinical practice for ultrarare disorders is challenging, because only a very few laboratories may be involved in studying the potentially causative rare variant. Furthermore, rare, orphan diseases do not generate interest from the industrial sector providing diagnostic tests.

State-of-the-art next generation sequencing (NGS) technology, eventually combined with other genetic tests, is an optimal strategy to address an unmet need of clinical medicine by attaining a diagnosis for patients with infrequent disorders in whom a disease cannot be identified despite exhaustive effort. The described NGS-based approach has already been proven successful in a number of cases, and the National Institutes of Health (NIH) have launched the Undiagnosed Diseases Program to identify candidate genes and establish diagnoses for unexplained conditions or to define new disorders [8].

THE OVERALL GENETIC TESTING STRATEGY IN UNDIAGNOSED DISEASES—LOOKING FOR THE NEEDLE IN A HAYSTACK

A majority of rare disorders result from a large spectrum of pathogenic variants, including a variety of mutations, copy number variations, large gene deletions or duplications, or other gross genomic alterations. Clinical and family history defines the priorities for the analysis of various regions of the genome, depending on the phenotype or the suspected inheritance pattern [9]. An important initial decision in genetic testing for a rare disorder is the sequencing protocol.

Current genome-scale sequencing in rare diseases utilizes one of the two strategies:

- *targeted*—custom capture, gene panel or whole-exome sequencing (WES) with flanking regions or
- *nontargeted*—whole-genome sequencing (WGS).

> The optimal approach in undiagnosed diseases is to combine NGS with either dense single-nucleotide polymorphism (SNP) arrays or array comparative genomic hybridization (aCGH), which detect copy number variants (CNVs) and pinpoint regions of homozygosity [10]. Such strategy is recommended by the American College of Clinical Genetics [8,11].

The majority of available clinical studies have used the targeted strategy, usually WES. The selection of a targeted approach is usually justified by lower cost compared to WGS, more straightforward interpretation of the obtained data, prioritization of better signal-to-noise ratio over uniformity of coverage, and the assumption that the clinically most important variants are located in protein-coding regions [11].

Target capture NGS may include a defined panel of preselected genes. A custom panel can be constructed to concentrate on regions where exhaustive genotype calling is requested. As an illustration, targeted NGS was used to find variants previously associated with Noonan syndrome while also detecting new variants in the region of interest [12]. As the simplest and cheapest strategy, such testing can be sufficient to diagnose a number of unexplained conditions. Nevertheless, the awareness of missing data is critical for the interpretation of the results of such an approach [11].

WES has emerged as a successful diagnostic tool in rare genetic diseases, being particularly effective in identifying deleterious variants that are refractory to linkage analysis. Recently, Zhu et al. applied trio WES to a cohort of 119 patients with undiagnosed conditions to identify both known and novel genes. A genetic diagnosis was obtained for 29 (24%) of the examined patients. Thirteen of 29 cases (45%) were due to *de novo* mutation, seven (24%) resulted from a homozygous genotype, five

(17%) were attributed to a hemizygous genotype, and four (14%) were due to a compound heterozygous genotype [4]. Fifteen (13%) of the 119 interpreted trios had a variant found in databases such as OMIM [13], ClinVar [14], and HGMD [15] and previously reported in patients with similar phenotypes. Furthermore, using WES, Yang et al. established a diagnosis in 25% of 250 patients with unknown conditions [16], whereas other studies have reported variable, sometimes even higher, diagnostic rates [17–19]. However, a definite diagnosis was attainable only if the causal gene had been previously implicated in a similar condition. Exome capture technology has biases, such as poor capture of some regions of the genome, which may be underrepresented and thus not accessible for further analysis [8,11]. The huge potential of WES in identifying novel pathogenic variants is still largely unexplored. The critical factor for the WES approach to undiagnosed patients is the coverage of protein-coding regions that should allow accurate detection of genomic alterations within all those areas [4,8,11].

WGS has significant advantages over WES in a non-hypothesis-driven approach, as is often needed for undiagnosed conditions, because it covers a larger area, including noncoding regions, and is less susceptible to allele skewing and dropout (but at the cost of the overall signal-to-noise relation) [11]. Some studies such as ENCODE [20] and Roadmap Epigenomics [21] have pinpointed an important role for noncoding variants in regulating gene expression and function of the genome, suggesting that some noncoding variants may also cause diseases with major effects [22,23]. Novel disease-causing variants have been identified in a noncoding region of the genome.

Some examples of noncoding mutations influencing the risk of a disease include [22] the following:

- *Splicing regulatory variants* are implicated in several diseases, that is, skipping of exon 7 of the *SMN* gene is associated with spinal muscular atrophy [24].
- *Enhancers* [25], that is, Hirschprung disease, a complex disorder involving around 10 genes, including the tyrosine kinase receptor *RET* and the gene *GDNF*, which encodes its ligand. The most common variant in the main susceptibility gene *RET* is an SNP in an enhancer [26].
- *Sequences regulating translation, stability, and localization*, that is, loss-of-function mutations in the 5′-UTR of *CDKN2A* predispose to melanoma [27]; further, a mutation that creates a binding site for the miRNA hs-miR-189 in the transcript of the gene *SLITRK1* is associated with Tourette syndrome [28]. Both rare and common mutations in the gene *RMRP* encoding an RNA component of the mitochondrial RNA processing ribonuclease have been associated with cartilage–hair hypoplasia [29].
- *Promoters—APOE* promoter mutations are associated with Alzheimer disease [30].
- *Synonymous mutations within protein-coding sequences*. A synonymous variant in *DRD2* associated with psychiatric disorders has been demonstrated to modulate dopamine receptor production through changes in mRNA folding and stability [31].

WGS can be more easily used to differentiate hemizygosity from homozygosity, as in contrast to WES, the distribution of read depths for single-copy (hemizygous) states and double-copy (dizygous) states in the course of WGS is not overlapping. With WGS short-read coverage is more uniform than that associated with WES; moreover, edge-detection techniques allow one to reliably differentiate copy number variations [23]. On the other hand, the cost of WGS at the time of this writing is a few times higher than that of WES, and the analysis of whole-genome data is much more problematic (see below).

In the clinical setting, karyotyping, fluorescence in situ hybridization, and microarrays may be used as a screening step preceding genome-scale sequencing. Microarray data can be used to determine structural variations that are difficult to detect using WES. For instance, SNP microarray data can be used to find copy number variations, recombination segments, mosaicism, and regions of homozygosity [8].

Scanning for very large genes may be useful prior to sequencing although the dropping price of NGS makes these approaches less and less attractive. Most popular scanning technologies include denaturing gradient gel electrophoresis, protein truncation testing, single-strand conformational polymorphism, heteroduplex analysis, and denaturing high-performance liquid chromatography [8]. These methods will find point mutations including small deletions/insertions; nonsense, missense, and splicing mutations; SNPs; and unclassified variants, but they will not detect large gene deletions or duplications or other gross genomic rearrangements [8]. Several techniques for the detection of intragenic deletion and duplication exist, including real-time quantitative PCR, multiplex ligation probe amplification, aCGH, and Southern blot. A combination of technologies may be useful as confirmatory testing of variants identified by NGS. These techniques can also be used along with sequencing or mutation scanning to provide more comprehensive variant detection [8].

THE ANALYSIS OF NGS TESTING RESULTS FOR RARE DISEASES

To prioritize variants generated by sequencing the results of NGS testing need to be filtered using databases of population frequency such as the 1000 Genomes database, NHLBI-ESP5400, CG46 [32], or the recently available powerful resource, the ExAC database, including >60,000 subjects [33,11]. For each proband, variants that are found in these databases with allele frequency >1% are usually removed, although this threshold may miss some frequent recessive mutations. In numerous cases, novel variants will be uncovered. It is also important that these new variants are well defined and reported back to the databases for the benefit of the scientific community [8].

Established custom algorithms select variants according to criteria such as quality of variant, inheritance pattern, and relation between a variant and a protein structure and function. The sequence analysis software must support the assembly of contigs for analysis, including both forward and reverse directions of sequenced and overlapping fragments, the cDNA reference sequence, and the genomic reference sequence [34]. Zhou et al. proved that the application of appropriate bioinformatic signatures can point toward novel genes and implies broadened phenotypes for known deleterious variants [4]. Petrovski et al. introduced a "hot zone" bioinformatic signature to prioritize *de novo* mutations among other variants [35]. According to some authors, comparing the *de novo* hot-zone mutations in cases versus controls represents a novel tool to understand the genetics of the rare diseases [4,35]. Among the genes with a *de novo* hot-zone mutation, some are pathogenic, but others are not disease-associated. Such approach can help to find a genetic background of severe rare diseases, as the collection of hot-zone *de novo* mutations will hide some pathogenic variants. Zhou et al. demonstrated that healthy subjects seldom have a hot-zone *de novo* mutation within essential genes (1.9% of individuals), in contrast to rare disease patients (15.5% of patients) [4]. Furthermore, Zhou found that five (31%) of the 16 hot-zone mutations in essential genes were already considered to be causal (*KCNQ2*, *NOTCH2*, *GNAO1*, and two variants of *DYNC1H1 de novo* mutation). In addition, seven could be found in OMIM or PubMed as a disease associated with various clinical phenotypes (*DCTN1*, *BIN1*, *GRIN1*, *GRIN2B*, *COL4A1*, *MYO5A*, and *HUWE1*). Finally, four mutations occurred in genes (*SLC9A1*, *HNRNPU*, *CELSR3*, and *EWSR1*) not known for an association with a disease [4].

The analysis of genomic data from patients with undiagnosed conditions may further benefit from the acquisition of sequences from family members to filter variants that do not segregate in a manner consistent with the expected Mendelian model [11].

The data analysis for noncoding variants is more challenging, as functional prediction algorithms for noncoding variants are far less developed than for coding variants [23]. As more than 99% of the variants in WGS data are noncoding, the number of nonexomic rare variants is still higher than that of the whole-exome variants without any prior filtering [11,23]. Therefore, it is critical to apply well-defined filtering criteria and bioinformatic prioritization tools to select variants that most likely are relevant. Furthermore, functional prediction approaches should be applied to prioritize noncoding variants. Despite the increasing clinical use of NGS, many of the prerequisites for a successful diagnosis are best realized in the research setting [4,11]. All these highlight the important role of research geneticists in implementing diagnostic WES and WGS.

SNP chips and aCGH provide additional genome-wide data that can be used to complement and improve the analysis of NGS data in undiagnosed conditions [11]. In particular, if mosaicism or uniparental disomy is suspected, SNP data are preferable to WES data. The NIH Undiagnosed Diseases Program obtains SNP chip data for all patients qualified for genome-scale sequencing. Currently, adding an SNP chip to exome sequencing is an alternative and less expensive strategy compared to WGS [8,11]. A further strategy to discover potentially relevant variants associated with a rare disorder is the integration of NGS, SNP, and CNV analysis results using appropriate bioinformatic methodologies. SNP data can be filtered using public databases such as dbSNP 130, dbSNP 131, or the 1000 Genomes project [32,36]. If a single-nucleotide variant is discovered in a patient, the laboratory must determine if it is a recognized SNP of no clinical significance; this can be done by searching the public SNP databases [37,38], by consulting researchers studying that gene, and/or by examining a significant number of normal sequences [8].

Gross alterations (deletions, duplications, or rearrangements) can be detected in some sequencing assays. Nevertheless, the information concerning such variants in available databases is scarce, as they are not routinely tested for and reported. Standard nomenclature is provided on the Human Variation Society Web site [39].

An important issue in genetic testing for undiagnosed diseases is the situation in which a normal result of genetic testing is obtained. The undetected mutation could result from: (1) the nature of the mutation, for example, deletions or duplications; (2) the presence of mutations outside of the tested regions, for example, in the promoter; (3) the presence of unexpected variation in a particular patient at a primer binding site, leading to allele dropout or preferential amplification of a competing allele; or (4) mosaicism [8].

PATHOGENICITY

Pathogenicity filtration is the final stage of the genomic analysis and should be done with caution, as it can be a source of errors, and the available methods and programs to evaluate the pathogenicity of variants are not fully reliable [11]. Lack of clear criteria for the pathogenicity classification of variants is a critical issue in genomic medicine. A software to determine the probability of whether a variant is likely to have a functional effect on protein can be applied; however, the results of such analysis can be discordant or inconclusive. Pathogenicity of individual variants can also be assessed using programs such as CDPred [40]. CDPred assigns a numeric score to each variation that can be aligned to a residue in the NCBI Conserved Domain database [41]. Nevertheless, it is worth emphasizing that such assessment is also prone to misinterpretation.

Finally, the findings should be validated by functional studies to support the evidence that the mutated gene product may play a role in the pathogenesis of a disease and, if possible, by demonstrating the disease association in independent patients [11]. The ultimate test for any of these unclassified variants would be the development of a functional assay to test whether the variation affects protein function [42]. There are very few examples of such a test in a clinical laboratory. As an illustration, Bouwman et al. described, and further validated, a high-throughput assay for the functional classification of *BRCA1* sequence variants of unknown significance [43]. This cDNA-based functional test to classify *BRCA1* relies on an ability to functionally complement *BRCA1*-deficient mouse embryonic stem cells in the context of the full-length protein. Using this assay, 74 unclassified *BRCA1* missense mutants were successfully analyzed, for which all pathogenic variants were confined to the BRCA1 RING and BRCT domains.

It is necessary to assess whether a variant truly associates with the rare disease and the conclusions need to go beyond a very limited set of affected individuals. Ideally, at least two peer-reviewed publications should provide very strong evidence of the involvement of the particular gene in the development of the disease [8]. Particular caution should be taken when considering disorders in which: (1) the phenotype may be due to the interaction of multiple loci; (2) mutations in more than one gene can lead, independently, to the phenotype; (3) there are phenocopies of the disorder due to other genes than the one being tested; (4) the clinical presentation associated with a mutation is poorly defined [8].

To interpret the meaning of an observed variant, it is crucial to determine if there is significant cosegregation of the variant with the phenotype and absence of the mutation in clinically unaffected family members. The interpretation depends on the mode of inheritance—autosomal dominant, recessive, or X-linked. In autosomal dominant disorders, if one asymptomatic parent has the mutation or CNV, it is unlikely that the alteration is meaningful. However, if the variant has arisen *de novo*, it may be more likely to be causative. Variants of unknown significance in autosomal recessive disorders can be more difficult to interpret [8].

Custom mutation analysis (testing of any gene for families with previously identified mutations) applies to those genetic disorders in which only rare private mutations have been described (the majority of ultrarare genetic disorders). GeneTests lists laboratories that provide a custom mutation analysis service [44].

To validate the sequencing findings and to enhance the understanding of rare disorders, basic research scientists and physician scientists need to cooperate. Additionally, metabolites of selected body fluids, including cerebral spinal fluid, are being analyzed. Furthermore, pluripotent stem cells induced from fibroblasts (iPSCs) can also be used to investigate disorders of inaccessible tissues such as brain or retina [8,11]. Tucker et al., by exome sequencing, identified 21 retinitis pigmentosa patients with a causative homozygous ALU insertion in exon 9 of male germ cell-associated kinase (*MAK*) [45]. To functionally validate this finding by investigating the mechanism through which the mutation causes disease, these authors used skin-derived iPSCs from the proband and a retinitis pigmentosa patient with another mutation (control). The retinal cell lineage from the control sample exhibited only the *MAK* transcript bearing exon 9, whereas the sample from the proband with the ALU insertion in *MAK* exhibited only the transcript lacking exon 9, while the developmental switch to the *MAK* transcript bearing exons 9 and 12 did not occur. A similar analysis of RNA extracted from the neural retina of a normal adult human eye revealed only the exon 9-containing *MAK* transcript. Thus, the ALU insertion seems to disrupt the correct splicing of exon 9, thereby preventing retinal cells from expressing the correct MAK protein. Western blotting revealed very little MAK protein in undifferentiated iPSCs from the proband; in contrast, the iPSCs derived from the retinitis pigmentosa patient with the *MAK* ALU insertion expressed a protein that was about 40 kDa smaller.

GOOD LABORATORY PRACTICE IN GENETIC TESTING FOR UNDIAGNOSED DISEASES

The American College of Medical Genetics proposed and further revised *Standards and Guidelines for Clinical Genetics Laboratories Concerning Molecular Genetic Testing for Ultra-Rare Disorders* [8]. This document was developed by the Ultra-Rare Disorders Working Group under the auspices of the Molecular Subcommittee of the Laboratory Quality Assurance Committee as an educational resource helping to provide optimal quality laboratory services for ultrarare disorders. As the results of research studies are now translated to the clinical setting, it has become evident that there is a need for elaboration of quality assurance standards, proficiency testing, and interpretation and reporting of genetic tests for ultrarare conditions [46]. Final recommendations included adapted CLIA requirements for research laboratories ensuring compliance with quality assurance and quality control guidelines that are not required of research laboratories [47]. Moreover, an assessment of the clinical validity and clinical utility of the test is a challenge for rare disorders because of the paucity of data. Most of these technologies require in-house development of the assay; thus it is recommended that only highly experienced laboratories undertake this challenge. The laboratory is responsible for validation of all analytical components of testing according to standards proposed by the Centers for Disease Control [48], the College of American Pathology, and the American College of Medical Genetics and Genomics (ACMG). The ACMG has published a list of genes/variants that it recommends be reported for exome sequencing performed for clinical diagnostics [49]. There is an ongoing debate whether variants found during research sequencing fall under similar or different standards [8,46,49].

TEST VALIDATION

The accepted guidelines for determining clinical validity of tests for common genetic disorders are not directly applicable for rare diseases owing to the paucity of data and private mutations specific only for single families. However, all laboratories are required to validate their particular assay on a number of samples with known sequence information and then on a series of blinded samples containing a diverse array of mutations. The ACMG has developed a position statement entitled *Recommendations for Standards for Interpretation of Sequence Variations* to help clinical laboratories with interpretation and reporting of genomic variants [50]. It is critical to assess whether a variant is deleterious or benign and whether it has been reported as pathogenic, neutral, unreported, or unknown.

Critical points to be considered in test design include the existence of any pseudogenes or other homologous genes. Pseudogenes are genomic DNA sequences characterized by high similarity to their corresponding coding genes. A pseudogene with a high proportion of coding sequence identity makes it impossible to rely on NGS for a gene screening, and a Sanger sequencing protocol excluding the pseudogene is recommended [51]. Pseudogenes arise from duplication of DNA sequence or retrotransposition and subsequent reintegration of the cDNA into the genome. Owing to the high sequence similarity to their paralogs, pseudogenes are responsible for errors in the NGS process (PCR/target enrichment) and mistakes in the database (read mapping) causing false positive signals and diagnostic pitfalls. This problem may also result from a short read length produced by some NGS platforms (SOLiD and Illumina) and ambivalent mapping during alignment. To prevent such misinterpretations, sequence alignment programs (i.e., FASTA or BLAST) could be applied to detect matching regions in the DNA sequence. Such tools allow clustering of genes from annotated genomes into paralog families, further used to screen the whole genome for copies or homologs [52]. Subsequently every matching sequence is further validated as a pseudogene by searching for repeated elements, overlap with other homologs, and cross-references with exon assignments from genome annotations [52]. The identified pseudogenes are further assigned to the paralog family of the most homologous gene (or assigned to a singleton gene if no obvious paralog was found) [52]. Numerous examples of the presence of a pseudogene impeding diagnostics have been described. As an example, the identification of a *STRC* deletion responsible for hearing loss was disturbed by the presence of a nonprocessed pseudogene with 98.9% genomic and 99.6% coding sequence identity [51]. This structure was located in a region encompassing a duplication with four genes: *CATSPER2*, *HISPPD2A*, *STRC*, and *CKMT1A* [51]. Vona et al. developed a Sanger sequencing method for *STRC* pseudogene exclusion [51]. Furthermore, Knies et al. recognized this problem in Fanconi anemia (FA). These authors showed that the c.2314G $\rightarrow$ T mutation attributed to *FANCD2* was due to incorrect mapping of the variant-containing reads, corresponding to the pseudogene, *FANCD2-P2* [53]. Another reported missense mutation (c.2204G $\rightarrow$ A) also represented the *FANCD2-P2* pseudogene sequence. In the same exon these authors identified two more base substitutions representing pseudogene sequence, but the corresponding reads were misleadingly mapped to *FANCD2*. Only gene-specific resequencing resolved the correct sequence [53]. These authors conclude that in terms of FA genes, special attention needs to be paid to *FANCD2*, for which *FANCD2-P1* LOC100421239 is listed in the NCBI database but not the other reported pseudogene, *FANCD2-P2* [53].

THE IMPORTANCE OF GENETIC DIAGNOSIS IN RARE DISEASES

Patients and their families usually benefit from a definitive genetic diagnosis, as it ends a period of uncertainty even though it brings no new therapy [11]. The valid diagnosis helps to prepare for the future and may have an impact on the reproductive plans of the family. However, in some cases, genotype-related patient management is possible, including new candidate drugs, certain diets, and lifestyle changes. New therapeutic strategies could be attempted based on the genomic profile of the patient (genome-forward medicine). The selection of a therapeutic strategy is based on the potential biological effects of the pathogenic mutation and known mechanisms of action of a candidate drug [4,54]. Similarly, proper dietary interventions in certain patients with metabolic disorders could be introduced after the specific mutation was found. Zhou et al. found four examples in which the genetic diagnosis caused adjustment of the therapy. For two patients, the genetic diagnosis was critical for the introduction of a new, effective drug. One of the patients had a *de novo* missense mutation in *KCNQ2* and has been receiving retigabine [55], and another patient had a *de novo* missense mutation in *KCNT1* and has been treated with quinidine [54]. Quinidine was considered as a possible therapy for refractory seizures in

KCNQ2-associated epileptic encephalopathy only after the *KCNT1* genotype was revealed. In two further patients with the metabolic disorder, the genetic diagnoses opened new possibilities for specific dietary interventions that substantially improved the patients' health.

CONCLUSIONS

NGS is the most powerful diagnostic tool in unresolved cases; however, the collaborative efforts of researchers, clinical pathologists, and treating clinicians are necessary to fully integrate NGS technology into patient solutions, as the value of NGS can be exploited only if an extensive review of patient records is integrated with sequencing data [56]. This strategy has shed light on the path toward personalized medicine. Furthermore, to increase the diagnostic potential of NGS, a deeper knowledge about population frequencies and pathogenicity of rare variants is required. Genome-wide testing of rare-disease patients encompasses the existing single-gene tests; therefore the translation of molecular signatures into clinical tests is also cost-effective [4,56]. NGS technology enabled the discovery of genetic complexities of numerous unexplained conditions in the search for pathogenic variants scattered across the genome, including both common and rare mutations, hence improving the chance of finding disease-causal variants, given the potential ability to perform functional annotation on the identified variants. Infrequent disorders, in addition to being a diagnostic dilemma, also serve as a rich resource for knowledge about unknown disease mechanisms.

REFERENCES

[1] www.orpha.net.
[2] www.eurordis.org.
[3] http://globalgenes.org/rarelist/.
[4] Zhu X, Petrovski S, Xie P. Whole-exome sequencing in undiagnosed genetic diseases: interpreting 119 trios. Genet Med 2015 Jan 15. [Epub ahead of print]. http://dx.doi.org/10.1038/gim.2014.191.
[5] www.raregenomics.org.
[6] Ali-Khan SE, Daar AS, Shuman C, et al. Whole genome scanning: resolving clinical diagnosis and management amidst complex data. Pediatr Res 2009;66:357–63.
[7] Vento JM, Schmidt JL. Genetic testing in child neurology. Semin Pediatr Neurol 2012;19:167–72.
[8] www.acmg.net/StaticContent/SGs/UltraRareDisorders.pdf.
[9] Rehm HL, Bale SJ, Bayrak-Toydemir P, et al. ACMG clinical laboratory standards for next-generation sequencing. Genet Med 2013;15:733–47. http://dx.doi.org/10.1038/gim.2013.92.
[10] Vona B, Müller T, Nanda I, et al. Targeted next-generation sequencing of deafness genes in hearing-impaired individuals uncovers informative mutations. Genet Med 2014;16:945–53. http://dx.doi.org/10.1038/gim.2014.65.
[11] Markello TC, Adams DR. Genome-scale sequencing to identify genes involved in mendelian disorders. Curr Protoc Hum Genet. Jonathan L. Haines, et al., editorial board. 2013;79:6.13.1–6.13.19. http://dx.doi.org/10.1002/0471142905.hg0613s79.
[12] Lepri FR, Scavelli R, Digilio MC, Gnazzo M, Grotta S, Dentici ML, et al. Diagnosis of Noonan syndrome and related disorders using target next generation sequencing. BMC Med Genet 2014;15:14.
[13] http://www.omim.org/.
[14] www.ncbi.nlm.nih.gov/clinvar/.

[15] Human Gene Mutation Database (HGMD): 2003 update Stenson PD, Ball EV, Mort M, Phillips AD, Shiel JA, Thomas NS, et al. Hum Mutat 2003;21:577–81.

[16] Yang Y, Muzny DM, Reid JG, et al. Clinical whole-exome sequencing for the diagnosis of mendelian disorders. N Engl J Med 2013;369:1502–11.

[17] Need AC, Shashi V, Hitomi Y, et al. Clinical application of exome sequencing in undiagnosed genetic conditions. J Med Genet 2012;49:353–61.

[18] Dixon-Salazar TJ, Silhavy JL, Udpa N, et al. Exome sequencing can improve diagnosis and alter patient management. Sci Transl Med 2012;4:138–78.

[19] de Ligt J, Willemsen MH, van Bon BW, et al. Diagnostic exome sequencing in persons with severe intellectual disability. N Engl J Med 2012;367:1921–9.

[20] Dunham I, Kundaje A, Aldred SF, Collins PJ, Davis CA, Doyle F, ENCODE Project Consortium, et al. An integrated encyclopedia of DNA elements in the human genome. Nature 2012;489:57–74. http://dx.doi.org/10.1038/nature11247.

[21] Bernstein BE, Stamatoyannopoulos JA, Costello JF, Ren B, Milosavljevic A, Meissner A, et al. The NIH roadmap epigenomics mapping consortium. Nat Biotechnol 2010;28:1045–8. http://dx.doi.org/10.1038/nbt1010-1045.

[22] Ward LD, Kellis M. Interpreting noncoding genetic variation in complex traits and human disease. Nat Biotechnol 2012;30:1095–106. http://dx.doi.org/10.1038/nbt.2422.

[23] Shi L, Zhang X, Golhar R, et al. Whole-genome sequencing in an autism multiplex family. Mol Autism 2013;4:8. http://dx.doi.org/10.1186/2040-2392-4-8.

[24] Lorson CL, Hahnen E, Androphy EJ, Wirth B. A single nucleotide in the SMN gene regulates splicing and is responsible for spinal muscular atrophy. PNAS 1999;96:6307–11.

[25] Sakabe NJ, Savic D, Nobrega MA. Transcriptional enhancers in development and disease. Genome Biol 2012;13:238.

[26] Amiel J, et al. Hirschsprung disease, associated syndromes and genetics: a review. J Med Genet 2008;45:1–14.

[27] Bisio A, et al. Functional analysis of CDKN2A/p16INK4a 5′-UTR variants predisposing to melanoma. Hum Mol Genet 2010;19:1479–91.

[28] Abelson JF, et al. Sequence variants in SLITRK1 are associated with Tourette's syndrome. Science 2005;310:317–20.

[29] Bonafé L, et al. Evolutionary comparison provides evidence for pathogenicity of RMRP mutations. PLoS Genet 2005;1:e47.

[30] Bray NJ, et al. Allelic expression of APOE in human brain: effects of epsilon status and promoter haplotypes. Hum Mol Genet 2004;13:2885–92.

[31] Duan J, et al. Synonymous mutations in the human dopamine receptor D2 (DRD2) affect mRNA stability and synthesis of the receptor. Hum Mol Genet 2003;12:205–16.

[32] http://www.1000genomes.org/.

[33] http://exac.broadinstitute.org.

[34] Crowgey EL, Stabley DL, Chen C, et al. An integrated approach for analyzing clinical genomic variant data from next-generation sequencing. J Biomol Tech 2015;26(1):19–28. http://dx.doi.org/10.7171/jbt.15-2601-002. jbt.15-2601-002.

[35] Petrovski S, Wang Q, Heinzen EL, Allen AS, Goldstein DB. Genic intolerance to functional variation and the interpretation of personal genomes. PLoS Genet 2013;9:e1003709.

[36] Sherry ST, Ward MH, Kholodov M, et al. dbSNP: the NCBI database of genetic variation. Nucleic Acids Res 2001;29:308–11.

[37] http://snp.cshl.org.

[38] www.ncbi.nlm.nih.gov/entrez/query.fcgi?db=snp.

[39] http://www.genomic.unimelb.edu.au/mdi/dblist/dblist.html.

[40] Gahl WA, Markello TC, Toro C, et al. The NIH undiagnosed diseases program: insights into rare diseases. Genet Med 2012;14(1):51–9. http://dx.doi.org/10.1038/gim.0b013e318232a005.

[41] Marchler-Bauer A, Lu S, Anderson JB, et al. CDD: a Conserved Domain Database for the functional annotation of proteins. Nucleic Acids Res 2011;39:D225–9.

[42] www.acmg.net.

[43] Bouwman P, van der Gulden H, van der Heijden I, et al. A high-throughput functional complementation assay for classification of *BRCA1* missense variants. Cancer Discov 2013;3:1142–55.

[44] http://www.genetests.org.

[45] Tucker BA, Scheetz TE, Mullins RF, et al. Exome sequencing and analysis of induced pluripotent stem cells identify the cilia-related gene *male germ cell-associated kinase (MAK)* as a cause of retinitis pigmentosa. Proc Natl Acad Sci USA 2011;108(34):E569–76. http://dx.doi.org/10.1073/pnas.1108918108.

[46] Richards CS, Bale S, Bellissimo DB, Das S, Grody WW, Hegde M, Molecular Subcommittee of ACMG, Quality Assurance Committee, et al. ACMG recommendations for standards for interpretation of sequence variations: Revisions 2007. Genet Med 2008;10:294–300.

[47] http://www.genome.gov/10001733.

[48] Chen B, Gagnon M, Shahangian S, Anderson NL, Howerton DA, Boone DJ. Good laboratory practices for molecular genetic testing for heritable diseases and conditions. Division of Laboratory Systems. National Center for Preparedness, Detection, and Control of Infectious Diseases, Coordinating Center for Infectious Diseases MMWR 2009;58:1–43.

[49] Green RC, Berg JS, Grody WW, Kalia SS, Korf BR, Martin CL, et al. ACMG recommendations for reporting of incidental findings in clinical exome and genome sequencing. Genet Med 2013;15:565–74. http://dx.doi.org/10.1038/gim.2013.73.

[50] https://www.acmg.net/StaticContent/StaticPages/Sequence_Variations.pdf.

[51] Vona B, Hofrichter MAH, Neuner C, et al. DFNB16 is a frequent cause of congenital hearing impairment: implementation of STRC mutation analysis in routine diagnostics. Clin Genet 2014;87(1):49–55. http://dx.doi.org/10.1111/cge.12332.

[52] www.pseudogene.org.

[53] Knies K, Schuster B, Ameziane N, et al. Schuelke M, editor. Genotyping of fanconi anemia patients by whole exome sequencing: advantages and challenges. PLoS ONE 2012;7(12):e52648. http://dx.doi.org/10.1371/journal.pone.0052648.

[54] Milligan CJ, Li M, Gazina EV, et al. KCNT1 gain of function in 2 epilepsy phenotypes is reversed by quinidine. Ann Neurol 2014;75:581–90.

[55] Orhan G, Bock M, Schepers D, et al. Dominant-negative effects of KCNQ2 mutations are associated with epileptic encephalopathy. Ann Neurol 2014;75:382–94.

[56] Hennekam RCM, Biesecker LG. Next-generation sequencing demands next-generation phenotyping. Hum Mutat 2012;33:884–6.

ORGANIZATIONAL AND FINANCING CHALLENGES

14

Jacub Owoc

Lubuski College of Public Health, Zielona Góra, Poland

CHAPTER OUTLINE

The rapidly dropping costs of next generation sequencing (NGS), around $100,000,000 per genome in 2001 to around $4000 in 2014 [1], make it an increasingly affordable tool that opens up a unique opportunity to revolutionize the way we predict, prevent, and treat diseases. However, before we see this happen there are numerous complex issues to be addressed not only from a clinical point of view but also regarding the readiness of health care systems around the world to implement it into routine clinical practice. If some of the exhilarating applications of NGS described in other chapters of this book are to become universal instruments available equally and universally to patients, then the crucial prerequisite is the public reimbursement and acceptance by third-party payers [2,3].

A pathway for new and innovative medical technologies, medicines, and treatments to reimbursement and coverage by third-party payers, be they public or private, seems rather clear in the world of evidence-based medicine (EBM) and health technology assessment (HTA). Health care systems have limited resources and require decisions about what and how is to be covered to gain the maximum health effect with available resources. The difficulty in making the best and most informed choices has led to the emergence of EBM—the conscientious, explicit, and judicious use of current best evidence in making decisions about the care of individual patients [4]—calling for the information used by policy makers to be based as much as possible on rigorous research [5,6]. EBM is closely linked to HTA (Figure 1).

The expansion of HTA is associated with the rapid development of medical equipment and technologies in 1970s [7]. Some of them, such as contraceptives, organ transplantation, or life-sustaining equipment, raised social, ethical, legal, and political concerns and challenged codes and norms of human life concerning its essential aspects, such as birth or death [8]. There was a need for their comprehensive and objective assessment also in terms of clinical and cost effectiveness. In this respect HTA is a tool well suited to being used in assessing the application of genomic research, as there are some evident parallels between current challenges related to NGS and what we already have experienced in the past. The discipline of bioethics emerged in the 1960s as a result of some of the issues associated currently with genomic research, such as ethics in general or informed consent. The debate that accompanied the relevant concerns and questions has led over the years to the emergence of guidelines, legal regulations, and institutions commonly accepted around the developed world.

HTA has been defined as "a form of policy research that systematically examines the short- and long-term consequences, in terms of health and resource use, of the application of a health technology, a set of related technologies or a technology related issue" [9]. The declared purpose of HTA is to support the process of decision-making in health care at the policy level by providing reliable information [6]. It has also been compared to a bridge between the world of research and the world of decision-making [10].

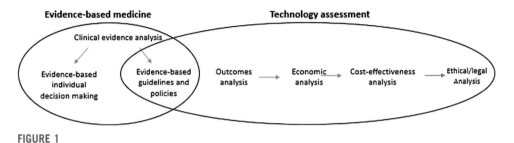

FIGURE 1

The link between evidence-based medicine and health technology assessment.

Nowadays, HTA covers a broad range of medical areas:

- pharmaceuticals,
- devices and equipment,
- medical procedures, and
- other related areas (such as patient record systems, telemedicine).

And HTA analysis may include [8]:

- technical properties,
- safety,
- efficacy,
- cost effectiveness,
- feasibility, and
- impact on:
  - social issues,
  - legal issues,
  - ethical issues,
  - political issues.

However, in most cases the two most important issues for HTA are [11]:

- Clinical effectiveness—how do the health outcomes of the technology compare with the available treatment alternatives?
- Cost effectiveness—are these improvements in health outcomes commensurate with the additional costs of the technology?

If third-party payers see cost and clinical effectiveness of more advanced NGS tests, then reimbursement decisions will be a matter of time. This approach already proves to be correct in various countries, such as the United States [12], Great Britain [13], and Holland. Payers pay for procedures and testing that are necessary for the treatment of specific diseases and there is a very strict limit as to what they are willing to pay. They do not view their role in the health care as supporting science or clinical research and focus on providing payment for testing, services, or procedures that directly affect patient care that is medically necessary as opposed to experimental/investigational. This is the position taken by the largest American health insurance companies interviewed about gene panels tests by the Center for Business Models in Healthcare [14]. Most of them indicated that testing numerous genes combined clinical care and research thus did not meet reimbursement requirements. Thus, evidence for all targeted genes in a panel is necessary for positive coverage decisions, although the companies believe at the same time that targeted sequencing has the potential to transform care [14].

There are already numerous examples of how NGS-based testing can save costs by speeding up some genetic diagnoses [15]. Researchers from Duke University have calculated that in the case of patients who had to return to their clinic for subsequent visits to obtain a genetic diagnosis the estimated average cost was US$25,000. Based on the results of an earlier study of whole-exome sequencing with a 50% diagnostic yield, they concluded that using NGS during the first visit could have saved about US$5000 in each case [16]. Numerous single-gene tests are the commonly reimbursed procedures, so if payers learn that testing many genes simultaneously can save them money, they will eventually be willing to reimburse for them. Also, the pharmacogenomics described in more detail in another chapter is not only about personalized, more effective care but also about saving costs spent otherwise on unnecessary and ineffective treatment.

However, cost and clinical effectiveness are not the only issues needed to be addressed before genomic testing becomes a universal tool. In fact, they may be the easiest to address, as they are commonly based on objective results of research. As NGS techniques and genomic research quickly expand, there are other far more complex questions and concerns being raised. Answers to those questions and relevant solutions are a prerequisite for the development of NGS in terms of public and insurance reimbursement. This is especially relevant for public payers such as Medicare or the British National Health Service (NHS) and sickness funds across European countries that spend billions of dollars in public funds (budgetary or mandatory public health insurance premiums) to provide universal and equal access to health care to an entire population and require consistent and clear health policy.

The following issues, some of them related to one another, are being raised by researchers, clinicians, and policy makers:

1. Interpretation and significance of certain variants in genes
   The capacity to generate data greatly outpaces the ability to analyze and interpret it [17]. There is a need for collecting clinical data that would unequivocally support certain diagnoses and clear guidelines in this context. As long as there is room for uncertainty in some areas, a common application in clinical setting and payers' acceptance is unlikely. This is less of a concern with targeted NGS than with genome or exome sequencing. However, with the dropping costs of the latter we may eventually reach the point at which routine genome or exome sequencing for a future use could be more cost effective than many targeted ones [18]. As technological reality exceeds most of expectations so far, this futuristic concept may not be so distant. And this vision brings us to another related issue.
2. Incidental findings
   The positions of various stakeholders vary significantly in this context. On one hand, we have a not-at-all reporting of incident findings recommended by the United Kingdom's NHS, which

maintains that analyzing genomes beyond what is strictly called for by the clinical context represents "opportunistic screening" that has no scientific basis [19]. A less strict position is presented by the European Commission, which suggests being careful with incidental findings and warns against unnecessary medicalization. It also recommends learning from the experience of radiology, which has been dealing with incidental findings for years and has worked out protocols and guidelines that are both accurate and flexible. On the other hand, the American College of Medical Genetics and Genomics calls for "reporting all pathogenic variants irrespective of patient age" [20]. Clearly, there is a need for better consensus, as it is the patient's well-being at stake and approaches cannot vary so significantly.

Furthermore, additional knowledge may be a burden for a physician who may not wish to take the additional responsibility associated with knowing that a patient carries a specific gene variant [30]. In this context targeted sequencing seems much more suitable.

3. Education and knowledge among physicians

    As there are numerous medical specialties that can benefit from NGS, there is a need for education and knowledge dissemination among physicians so that they are not only aware of tests but also know how to use them and interpret their results. The rapidly increasing amount of genetics-related information is difficult to follow for regular physicians and this has already been confirmed by various studies [21,22]. One of the barriers is a lack of easy-to-use analysis tools for physicians to facilitate routine NGS data interpretation at hospitals [2]. The same concerns structured training programs. The need to address the issue will be more and more pressing with more NGS tests available, also directly to consumers, and their dropping costs. Low-cost, direct-to-consumers tests without any interpretation (already available from some companies) will inevitably build up expectations for even family physicians to advise on genetic issues. One of the suggested solutions is to create a strategic collaboration framework between disciplines to deal with genomics [23].

4. Coding is another important issue associated with reimbursement for services by third-party payers. Codes such as Current Procedural Terminology maintained by the American Medical Association (AMA) are a basic tool to describe medical services provided to patients. They are important to payers for administrative, financial, and analytical purposes. If a code does not precisely describe a service being provided to a patient, insurance companies have trouble with classifying and paying for them. This is the case with most NGS tests and it will remain a challenge as new tests emerge at a rapid pace, while a process of coding may take years. There are attempts from various stakeholders to address this issue. The Association for Molecular Pathology (AMP) developed an NGS coding proposal in March 2013 and submitted it to the AMA for review. In this proposal, the AMP coined the term "genomic sequencing procedures" to describe both current and future diagnostic technologies that analyze the human genome in complex and diverse ways, such as NGS [24]. The new codes were implemented in January 2015 and include specific codes for numerous diseases (such as hereditary colon cancer syndrome, hearing loss). They do not impose any methods of testing but rather define a minimum number of genes to be analyzed and their symbols. Some of the new codes also look into the future by providing for the possibility of exome or genome analysis in cases of unexplained disorders, a need for comparative analysis, and even a need for reevaluation of previously obtained data.

    Nonspecific coding that usually describes processes rather than a particular test used makes it much more difficult or even impossible for payers to analyze outcomes and assess the value or efficacy of specific diagnostic tools. The problem with coding new NGS test has even led to halting their reimbursement in Germany (described further in the chapter).

5. Data storage, management, sharing, and ownership
    Although this is more a technical than a medical issue in part concerning data storage and management, one may assume that it will have to be addressed by payers as they are the most relevant entities to gather medical information about their clients and share it with medical service providers. With the amount of information related to genome or even exome sequencing, this will certainly be a very serious and costly challenge. Some estimates say that every dollar spent on sequencing hardware must be matched by a similar amount spent on informatics [25]. There is also a need for user-friendly software that would interpret vast amounts of data in a standardized and accepted way.
    There are already, however, a number of compression tools that are dedicated strictly to genomic data, so the eventual emergence of a standard that would combine quality with efficiency (high compression ratio) may be achieved in the foreseeable future.

Numerous concerns lead to some situations in which not only payers but also clinicians make some of their choices based on their prejudice for or against technology rather than on individual tests, although some NGS tests are accurate enough to be used in clinical settings [26]. However, the European Commission has already recommended that national health care systems should put the systems in place for the reimbursement of "gene panels" with clinical utility [27].

Some countries are making attempts to benefit from the development of new NGS technologies.

UNITED STATES

In addition to the process of new CPT codes implementation, Medicare has started its gap-filling process that aims to determine levels of payment for each of 21 new genetic testing codes. The final reimbursement amounts are to become effective in 2016 [28]. Progress is also being made in terms of reimbursement of tests in specific fields, especially noninvasive prenatal testing for high-risk pregnancy, owing to recommendations from professional organizations.

GREAT BRITAIN

"UK to become world number one in DNA testing with plan to revolutionize fight against cancer and rare diseases"—this modest statement is the driving force behind the 100,000 Genome Project initiated in 2013 by the U.K. Department of Health. The project is carried out by Genomic England—a company owned by the Department of Health—together with the NHS—a public entity with an annual budget exceeding £100 billion spent on health care provided to all citizens in the United Kingdom. The cost of sequencing 100,000 genomes of NHS patients is estimated to be £300,000,000 by 2017, and the main objectives cited at the company's website are:

- to bring benefit to patients,
- to create an ethical and transparent program based on consent,
- to enable new scientific discovery and medical insights,

- to kick-start the development of a U.K. genomics industry,
- to achieve the ultimate goal of making genomic testing a routine part of clinical practice [29].

The company would like to eventually work out a data service that would link a whole genome sequence with an individual's medical records. Genomics England will also try to address some the issues associated with genome and exome sequencing by cooperating closely with their ethics committee, working out a model for informed consent and feedback for patients from sequencing. However, all of this will be taken step by step as the project develops or be tested as a part of pilot projects. Access to all the data will be allowed for university scientists and pharmaceutical companies for research purposes related to health only. All research proposals will have to be approved under conditions worked out by the company's Ethics Advisory Committee.

The partner in this project is Illumina—the leading producer of sequencing instruments, including their most recent one, a $10,000,000 Hiseq X Ten, which reportedly will be able to sequence genomes for $1000 each, in cases of large population studies [30]. The company has already managed to sell 13 units all over the world in the first two quarters of 2014, which suggests that many larger studies are only a matter of time. The company will be paid £78,000,000 for all the sequencing services but will in turn invest £162,000,000 to create new jobs in the field of genome sequencing [31].

HOLLAND

The major Dutch health insurance companies providing statutory public health services to all Dutch citizens (the Dutch health care system is based on a public–private partnership—public financing and private health insurance companies offering open enrollment and comprehensive care provision) agreed in 2011 to cofinance a 500-person pilot program led by Radboud University Nijmegen Medical Center and designed to demonstrate that reimbursing for exome sequencing is more cost effective than single-gene tests, particularly in cases in which the diagnosis is uncertain [32]. The reimbursement amount was set at €1500—twice the usual amount for genetic testing. Genetic counseling is reimbursed by the system if it is medically justified.

GERMANY

The German statutory health care system halted reimbursement of NGS-based tests in 2013, forcing laboratories to revert to traditional single-gene Sanger technology. This followed a decision made by the Joint Federal Committee (G-BA), which determines what medical services are provided by the public health care system. The reasoning behind this was that there has never been a dedicated code for NGS tests, and laboratories had used a code covering tests designed to detect single-gene mutations. Codes are the basis for reimbursement and its level. A working group was set up to work out new codes for NGS tests by June 2014; however, the deadline has been kept so far.

Another decision made by the G-BA will bring reimbursement for noninvasive prenatal testing in 2015 for a trial period of one year. The only accepted test is from the LifeCodexx company using Illumina equipment.

It appears that publically reimbursed genome or exome screening is highly unlikely in the foreseeable future. Even if the cost of sequencing a single genome drops below $1000 the amount of resources necessary to store, manage, and interpret the data at the population level as well as the eventual analysis as a part of routine care is difficult even to estimate. Some experts and officials in various countries have already indicated that. The Center for Medicare and Medicaid Services (CMS) Coverage and Analysis Officer commented in October 2011 "I hope people realize that whole genome sequencing itself is probably something that CMS would never cover" [2]. The reasoning is that it is considered experimental and investigational, not medically necessary, and thus does not meet criteria that payers take into account. Similarly, British experts concluded that there is no point in the NHS offering individual genome sequencing. They were, however, less definite, by excluding situations in which it would be justified by specific, evidence-based medical purposes and noting that in cases of privately sequenced and analyzed genomes the NHS should provide follow-up and care for clinically significant findings [33].

Some precedents have, however, already been set, with reimbursement for tests based on NGS technology in the United States. The key to success is testing panels of selected genes that are already reimbursed as single-gene tests. Some laboratories launched cancer panels that include 40 or more genes and plan to expand them. The reimbursement rate for such tests is 80–90% [34]. Other panels include cardiomyopathy and, as the approach proved successful, there are more being developed such as one for renal disease. Representatives of American health insurance companies officially confirm that and also emphasize that coverage policy for sequencing does not concern only the validity of included targets, but also potential outcomes from therapies informed by it [14].

CONCLUSIONS

On one hand, the major challenge for NGS technology seems clear: to demonstrate that it can meaningfully contribute to patient care through diagnoses or improved therapeutic decisions that cannot currently be made and/or savings compared to existing test strategies to include the retirement of other tests no longer needed [2]. On the other hand, the number and complexity of issues and questions that need to be addressed complicate this straightforward message. What is more, most of them still seem far from being answered. This is very much confirmed by the 100,000 Genome project, which is receiving substantial publicity (admittedly, encouraged to some extent by the government officials involved) and has very ambitious goals while not addressing all the accompanying challenges—informed consent; data storage, management, and interpretation; or incidental findings policy. The company itself admits that it needs to be flexible and thus has a commercial status rather than a public one, which would require a dedicated legislation and addressing a priori the troubling issues.

It seems that the only way forward with implementing NGS technologies into routine medical care is a step-by-step approach. The ultimate goal of improving or perhaps even revolutionizing patient care is shared by all the stakeholders in health care and it is their points of view and priorities that vary. One shall also remember that there will be members of many disciplines involved in pursuing that goal: physicians, biologists, lawyers, mathematicians, bioinformaticians, computer scientists, statisticians, health economists, ethicians, and last, but not the least, government officials. They will definitely need to give much thought to establishing decision-making policies and future strategy for the regulatory, ethical, and financial framework.

REFERENCES

[1] Wetterstrand K.A. DNA Sequencing Costs: data from the NHGRI Genome Sequencing Program (GSP), Available at: www.genome.gov/sequencingcosts [accessed in September 2014].

[2] Gullapalli R, Desai K, Santana-Santos L, Kant J, Becich M. Next generation sequencing in clinical medicine: challenges and lessons for pathology and biomedical informatics. J Pathol Inform 2012;3:40.

[3] Workshop Summary. Reimbursement models to promote evidence generation and innovation for genomic test. Bethesda (Maryland): National Institutes of Health; October 24, 2012.

[4] Sackett DL, Rosenberg WM, Gray JA, Haynes RB, Richardson WS. Evidence based medicine: what it is and what it isn't. BMJ 1996;312(7023):71–2.

[5] Ham C, Hunter DJ, Robinson R. Evidence based policymaking. Br Med J 1995;310:71–2.

[6] Busse R, Velasco-Garrido M. Health technology assessment. An introduction to objectives, role of evidence, and structure in Europe. Eur Obs Health Syst Policies 2005.

[7] Jonsson E, Banta HD. Management of health technologies: an international view. Br Med J 1999;319:1293.

[8] Goodman CS. HTA 101—Introduction on health technology assessment. National Information Center on Health Services research and Health Care Technology (NICHSR); 2004.

[9] Henshall C, et al. Priority setting for health technology assessment: theoretical considerations and practical approaches. Int J Technol Assess Health Care 1997;13:144–85.

[10] Battista RN. Towards a paradigm for technology assessment. In: Peckham M, Smith R, editors. The scientific basis of health services. London: BMJ Publishing Group; 1996.

[11] Taylor R, Taylor R. What is health technology assessment. Hayward Med Commun 2009.

[12] Bredemeyer A. Clinical Next-Generation sequencing to guide cancer treatment decisions. Published online: on June 26, 2013 at www.onclive.com.

[13] Next-Generation Screening Goes National in UK. Cancer Discov June 2013. 3; OF6.

[14] Trosman J., Weldon C., Kelley R., Phillips K. Barriers to insurance coverage of next-generation tumor sequencing by U.S. payers. Center for Business Models in Healthcare.

[15] Shashi V, McConkie-Rosell A, Rosell B, Schoch K, Vellore K, McDonald M, et al. The utility of the traditional medical genetics diagnostic evaluation in the context of next-generation sequencing for undiagnosed genetic disorders. Genet Med 2013. August 8.

[16] Levenson D. Next-generation sequencing may reduce cost and wait time for some genetic diagnoses: experts argue that clinical evaluation remains crucial. Am J Med Genet Part A December 2013;161(12):vii–i.

[17] Nekrutenko A, Taylor J. Next-generation sequencing data interpretation: enhancing reproducibility and accessibility. Nat Rev Genet September 2012;13:667–71.

[18] American College of Medical Genetics and Genomics. American College of Medical Genetics and Genomics (ACMG) policy statement points to consider in the clinical application of genomic sequencing, http://www.acmg.net/StaticContent/PPG/Clinical_Application_of_Genomic_Sequencing.pdf.

[19] Wright C, Burton H, Hall A, Moorthie S, Pokorska-Bocci A, Sagoo G, et al. Next steps in the sequence. The implications of whole genome sequencing for health in the UK. PHG Foundation; 2011. http://www.phgfoundation.org.

[20] Burke W, Matheny Antommaria AH, Bennett R, Botkin J, et al. Recommendations for returning genomic incidental findings? We need to talk!. Genet Med November 2013;15(11):854–9.

[21] Selkirk CG, Weissman SM, Anderson A, Hulick PJ. Physicians' preparedness for integration of genomic and pharmacogenetic testing into practice within a major healthcare system. Genet Test Mol Biomarkers March 2013;17(3):219–25.

[22] Haga SB, Burke W, Ginsburg GS, Mills R, Agans R. Primary care physicians' knowledge of and experience with pharmacogenetic testing. Clin Genet October 2012;82(4):388–94.

[23] Haspel RL, Arnaout R, Briere L, Kantarci S, Marchand K, Tonellato P, et al. A call to action: training pathology residents in genomics and personalized medicine. Am J Clin Pathol 2010;133:832–4.

[24] Association for Molecular Pathology. Proposal to address CPT coding for genomic sequencing procedures. March 2013.

[25] Perkel JM. Sequence Analysis 101: a newbie's guide to crunching next-generation sequencing data. The Scientist 2011;25:60.

[26] Heger M. Technology no longer a hurdle, but clinical NGS faces reimbursement and biological knowledge issues. October 23, 2013. www.genomeweb.com.

[27] European Workshop on Genetic Testing Offer in Europe. EUR 25684—Joint Research Centre—Institute for Health and Consumer Protection.

[28] Root Ch. CPT codes for genomic sequencing procedures. Curr Future Reimburse January 22, 2015. webinar at www.knome.com.

[29] About Genomics England. How we work, www.genomicsengland.co.uk.

[30] Illumina. CEO on genome sequencing technology. FoxBusinness July 24, 2014;4(21). www.foxbusiness.com.

[31] UK to become world number one in DNA testing with plan to revolutionise fight against cancer and rare diseases. Posted on 1 August at: www.genomicsengland.co.uk.

[32] Heger M. Dutch study aims to demonstrate cost-effectiveness of reimbursing for exome sequencing Dx. October 19, 2011. www.genomeweb.com.

[33] Whole Genome Sequencing. Clinical impact and implications for health services. PHG Foundation; 2011.

[34] Heger M. With strong uptake and reimbursement, washU's GPS launches expanded cancer, cardiomyopathy panels. December 18, 2013. www.genomeweb.com.

FUTURE DIRECTIONS

Michal Okoniewski[1], Rafał Płoski[2], Marek Wiewiorka[3], Urszula Demkow[4]

[1]*Division Scientific IT Services, IT Services, ETH Zurich, Zurich, Switzerland;* [2]*Department of Medical Genetics, Centre of Biostructure, Medical University of Warsaw, Warsaw, Poland;* [3]*Institute of Computer Science, Warsaw University of Technology, Warsaw, Poland;* [4]*Department of Laboratory Diagnostics and Clinical Immunology of Developmental Age, Medical University of Warsaw, Warsaw, Poland*

CHAPTER OUTLINE

SEQUENCING PLATFORMS

At present the clinically used next generation sequencing (NGS) platforms are dominated by solutions from Solexa/Illumina and Ion Torrent (Chapter 1). However, it is clear that a need exists for further technical developments to overcome the limitations inherent in short read lengths, relatively low accuracy of sequencing, and high price. The objectives for NGS development have been articulated by the Advanced DNA Sequencing Technology program of the National Human Genome Research Institute (NHGRI)—an initiative to promote the development of methods permitting human whole-exome sequencing (WES) at a price of US $1000 or less, with an error rate below 1/10,000 and read length long enough to allow confident structural variants calling. It is believed that the full potential of NGS, especially in the clinical setting, will be realized only after these objectives have been reached (http://www.genome.gov/12513210).

NANOPORE SEQUENCING

The technology that is regarded as the most promising approach to meeting the NHGRI goals is known as "nanopore sequencing" [1]. The idea behind nanopore sequencing is to pull a single-stranded DNA molecule through a tiny hole in a membrane while monitoring the degree of obstruction of this hole by the passing molecule. The obstruction differs according to size, shape, and electric charge of passing nucleotides. The pulling force commonly employed is generated by the electric potential difference applied to the two sides of the membrane known as the *cis* and *trans* sides. The potential induces DNA strand translocation in the *cis→trans* direction in a process somewhat similar to electrophoresis. The degree of obstruction of the pore is continuously monitored by fluctuations in the strength of the electric current flowing through. Nanopore sequencing is a technology of single-molecule sequencing with the additional potential of detecting epigenetic modifications such as 5-methylcytosine and 5-hydroxymethylcytosine as well as abasic sites.

Although conceptually simple, nanopore sequencing poses considerable technical challenges. One obstacle is the need for precisely defined and uniform size of the pores—in both the lateral and the vertical dimensions (i.e., defining pore diameter and length, respectively). Generally, both dimensions should be as small as possible with the constraint on the diameter imposed by the necessity to allow passage of nucleotides present in a single-stranded nucleic acid. Too large a diameter makes changes in the current difficult to detect, as the obstruction is relatively small. Excessive length of a pore results in more than one nucleotide in the DNA/RNA strand contributing to the obstruction of current flow, which complicates the interpretation. One way to circumvent this limitation is to determine and subsequently interpret current signatures for all possible combinations of a maximal number of bases expected to influence the conductance of the pore (four to six bases or more). Another challenge in nanopore sequencing is slowing down the speed of migration of the DNA through the pore to allow accurate recording of current changes.

For the first time the principle of nanopore sequencing was demonstrated by Kasianowicz et al., who employed a lipid bilayer membrane with embedded molecules of α-hemolysin (α-HL)—a *Staphylococcus aureus* protein that spontaneously forms ion channels in lipid membranes [2]. Subsequently another pore-forming protein, the *Mycobacterium smegmatis* porin A (MspA), was shown to be useful for sequencing [3]. The properties of nanopores obtained using proteins can be considerably improved by site-directed mutagenesis as demonstrated for both α-HL [4] and MspA [5].

To reduce the speed of migration of the DNA through the nanopore, DNA polymerases have been employed. A single molecule of DNA polymerase properly positioned at the *cis* side of the nanopore will pull single-stranded DNA, forcing a relatively slow *trans→cis* movement if synthesis of the complementary strand is occurring at the *cis* side. The controlled movement of the DNA strand in the opposite direction is also possible. When there is no DNA synthesis but DNA at the *cis* side is double stranded the movement of its single-stranded portion along the potential gradient is slowed down by the resistance of the double-stranded part of the molecule to unzipping. The double passage of the same molecule across the nanopore yields redundant current readings, increasing accuracy [6].

In parallel to the development of nanopores based on lipid membranes with embedded channel-forming proteins, there have been efforts to design solid-state nanopores, which are very attractive because of their stability and ease of integration with electronic systems. The first solid-state nanopores were silicon based [7] but subsequently a range of other materials has been proposed as potentially suitable, of which particular attention has been received by graphene [8,9]. Graphene is an atomically thin membrane of graphite with excellent stability and electrical properties. However, before the full potential of graphene can be utilized the technology of sculpting pores in this material with virtually atomic precision needs to be developed [10]. In 2014, a subnanometer-thick monolayer of molybdenum disulfide (MoS_2) was reported as potentially superior to graphene regarding the prevention of undesired hydrophobic interactions with DNA [11].

An innovative approach to DNA sequencing, which builds on nanopore concepts and technology, relies on electronic tunneling (ET). The principle is that the four bases have different charge conductances, which can be detected when they are passing through the closely spaced (~1.5 nm) electrodes [12]. Quantum Biosystems is developing a platform in which DNA molecules are forced to pass between nanoelectrodes allowing conductance to be measured accurately enough to allow sequencing (www.quantumbiosystems.com).

The technology of ET can be extended into an interesting possibility of using a scanning tunneling microscope (STM) for DNA sequencing. It has been shown that an aluminum tip on an STM coated with any of the four DNA nucleosides can recognize a complementary base in DNA owing to hydrogen bond-mediated tunneling [13]. In 2012, a potential universal molecule was designed that could allow the use of a single tip (rather than four) for scanning a whole DNA sequence [14]. Further improvements in the STM approach have been proposed involving gold [15] or titanium nitride electrodes [16].

EMERGING COMMERCIAL SOLUTIONS

The first nanopore sequencing system expected to be widely available is the MinION system produced by Oxford Nanopore Technologies featuring 512 protein pores acting in parallel (https://nanoporetech.com). In 2014, the prototype of this instrument was offered to a limited number of labs on the basis of an early-access program. MinION is an attractive device similar in size and appearance to a USB memory stick and reportedly capable of read lengths up to 50 kb and read speeds of 100 bases/s. According to the published literature so far, the MinION was successfully used for sequencing a bacterial genome [17] and, in conjunction with Illumina sequencing, for identification of the position and structure of a bacterial antibiotic-resistance island [18]. Other papers describing nanopore sequencing on MinION have been published as preprints (https://nanoporetech.com). In parallel there is a noticeable interest in the development of software tools for analysis data from this platform [19–21]. In addition to the small MinION, Oxford Nanopore Technologies is developing larger machines using a similar technology—PromethION and GridION. Whereas the MinION of Oxford Nanopore Technologies appears to be close to marketing it remains to be seen if the company resolves issues linked to high error rate and robustness of experiments.

Genia Technologies (http://www.geniachip.com) develops nanopore sequencing performed in conjunction with tagged nucleotides. In this approach, as first described by Kumar et al. [22], tagged nucleotides are used for the synthesis of a complementary strand on a template of a single DNA molecule. The reaction is catalyzed by a polymerase tethered to the nanopore in such a way that when it incorporates a nucleotide into DNA its tag is released, entering a nanopore and producing a unique ionic current blockade [22]. Note that in this approach it is the tag, not the sequenced molecule that passes through the pore and is detected.

Noblegen Biosciences (http://www.noblegenbio.com) is pursuing nanopore sequencing combined with fluorescence detection. The approach relies on initial DNA expansion by circular DNA conversion with subsequent hybridization using probes labeled with molecular beacons and quenchers [23]. Although complicated conceptually, the approach has been claimed to be sensitive and amenable to parallelization.

An approach to nanopore sequencing pursued by Nabsys, Inc. (www.nabsys.com), relies on hybridization of a short oligonucleotide with a defined sequence to a long single-stranded DNA molecule to be sequenced. The oligonucleotide hybridizes only to stretches with complementary sequence, marking their position. The whole molecule is then passed through a nanopore allowing the mapping of the double-stranded regions. To accomplish sequencing of the whole molecule, each base is queried many times by overlapping probes of different sequence, hybridized in different pools.

Yet another company working on nanopore sequencing is Stratos Genomics (www.stratosgenomics.com), which aims to circumvent the limitations of this technology by increasing the size of the detected molecules in a process called SBX (sequencing by expansion). In the first step of SBX the DNA molecules to be analyzed are used as templates to synthesize Xpandomers, which then pass sequentially through a nanopore detector. Xpandomers are essentially high-signal reporter molecules optimized for nanopore sequencing, but the details of their chemistry are not disclosed.

Despite apparent attractiveness and widespread interest in nanopore sequencing, its future is by no means settled. Perhaps a sobering thought is that, although the technology dates back to as early as 1996 [2], it has still not lived up to expectations.

FUTURE DIRECTIONS OF CLINICAL GENOMICS DATA PROCESSING

It has become important that NGS and its medical applications need complementary developments in computing information technologies. This is evident not only in the size of data sets, but also in the growing complexity of the software tools to analyze the data and the sophistication of the data interpretation scenarios. The question of how information technology (IT) will be in tune with the development of clinical genomics in the near future can be answered simply: the IT tools for genomics will become a whole new generation compared to the current ones. They are driven by the technological developments in NGS, but also by the current progress in IT and the revolutionary impact of genome-based medicine on society. According to the comparison of Moore's law for computing and the sequencing equipment [24] the NGS area has been developing much faster. Thus in many applications such as DNA variant calling, RNA-seq analysis, and multilab and cohort studies, the computing hardware and software infrastructure can be a serious bottleneck in the analysis of the research data. This issue needs to be solved before NGS will be brought into widespread clinical use. Also the new technological developments on the side of NGS such as the growing capacity of the high-throughput sequencers and the advent of nanopore technologies [25] can cause a flood of genomic data, which can be handled only with new types and paradigms of data processing techniques. Genomic, precision medicine, and cloud computing are expected to be combined to bring about a synergy effect. This will result in the ability to process much larger data with a more fine-grained precision and in consequence an improved applicability and precision on the side of medical applications. Also the economical aspects are important. The falling prices of both NGS and cloud IT will make it affordable for an increasing number of clinical areas.

DATA SIZE ESTIMATES AND CURRENT PROCESSING LIMITATIONS

The new Illumina sequencers (as of the first quarter of 2015) are expected to produce almost half a terabase pair per day; the same is true for the Illumina XTen system. Also, with the popularity of nanopore sequencers, it is possible to imagine a data explosion that may produce streams of data comparable in size to current mobile telephony or other Internet-type [26] networks. As a result, the capacity of future sequencing databases will need to become orders of magnitude larger than the currently biggest ones (ENA 1.2×10^{15} bp, SRA 3.5×10^{15} bp) [27–30]. In theory, the stored data could be limited only to the genomic variants, for example, VCF or gVCF files of differences between the DNA of the patient and the fixed reference human genome [31,32]. However, the differences in variant calling software make it not practically feasible, and it is also made more difficult by technical artifacts of sequencers and data processing techniques [33–35]. The results of a clinical sequencing experiment need to be stored also in the form of raw sequencing reads, until a consensus of variant calling [36,37] is reached. As complete genome sequencing will become affordable and replace the commonly used exome enrichment [38], for a single 20× human genome just the FASTQ format takes over 1.2 terabytes of data, so a bit more than the hard drive of a typical current laptop computer. To manage efficiently the data of patients on a daily basis, there is a strong need for new storage and data management approaches. In addition, the manufacturers of NGS machines and related technologies, such as Illumina, LifeTech, and Qiagen, are extending their products into complete processing pipelines. They are offering not only presequencing library preparation kits and equipment, but also postsequencing software and services, as far as functional analysis of the DNA and RNA results.

DATA PROCESSING MODELS

The classic and most often used way to process data with a computer is to use a single CPU. In computers with multiple CPUs (called cores) in a processor, the processing can be split between them. The software needs to control the running of many threads of a program on many cores and coordinate putting together the single final result. Such a multithreading property, which enables the use of multiple cores for processing, makes up a large part of high-throughput data processing software, such as genome aligners, *de novo* assemblers, or formatting tools such as SAMtools [39]. The data processing that uses multiple computing machines at a time can be more efficient (see Figure 1). Still, it requires coordinating software that controls the running of the data processing on many machines in parallel. In this way, the same computing task can be done faster depending on the number of machines working together. This property of multicomputer systems is called scalability, as it enables accommodating the growing amounts of processed data in comparable time.

Multimachine data processing can be done within physical computing clusters or sets of virtual computers, called computing clouds. The virtualization of computing has been a fast-developing area of computer science [40,41]. The virtual machine is a computing unit running under the supervision of a standard operating system (e.g., Windows or Unix), but not located physically in a fixed way. Most often the access to a virtual machine is done by the network connection described by an identifier (IP number). This approach has become commercialized, and "computing clouds" are available as sets of virtual machines that can be set up for a specific period of time, for example, as long as a specific data analysis task will be performed. Physically, clouds are served by large computing clusters managed by a special multicomputer operating software, such as OpenStack (https://www.openstack.org/) or VMware (http://www.vmware.com/). The cloud technologies can be provided in several modes, commonly referred to as IaaS, PaaS, SaaS, and DaaS. IaaS [42], Infrastructure as a Service, is the crudest level of access to the cloud. The provider makes available just the virtual machine and the role of a client or user is to set up their own virtual networks. This approach is typically too general and complex

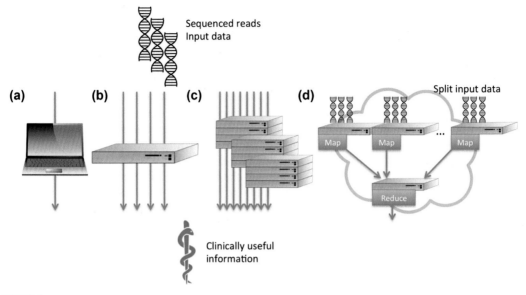

FIGURE 1

A schema of the data processing models. (a) Single machine. (b) Multithreaded. (c) High-performance computing cluster. (d) Map–Reduce cloud computing, in which the program code is sent to the nodes with distributed data.

to handle for clinical researchers, even those having bioinformatics support, as IaaS is basically for advanced IT system administrators. PaaS, Platform as a Service, brings the added value of analysis platforms and programming environments and thus can be used by experienced bioinformaticians. SaaS, Software as a Service, in addition encapsulates the internal cloud infrastructure in such a way that the end user has a specific software interface and does not need to comprehend the intricacies of IT implementation. In this case the domain knowledge is enough, so this level of cloud computing is accessible for clinicians. DaaS, Data as a Service, adds the ready data sets in the cloud to the software interface. In the case of clinical genomics these may be annotation databases or just sequencing data repositories.

HADOOP ECOSYSTEM TOOLS THAT CAN BE USED IN GENOMIC MEDICINE

Fortunately for the prospects of handling large clinical sequencing data [43], in parallel to the introduction of the NGS technologies in the area of data management and analysis, a fast development of the new parallel data analysis techniques can be observed. They are focused on the work being done for the projects belonging to the so-called Hadoop ecosystem. The main common quality of these software systems operates on the distributed data. In practice, they are often distributed on the data storage belonging to many machines in the cloud. The locality of the data is most often not transparent to the end user, while the distribution, replication, and control over the data is managed within the Hadoop File System (HDFS). Processing of the data is performed according to the Map–Reduce paradigm, in which first the "map" processing is performed on many machines in the cloud and subsequently the "reduce" step is performed, collecting and summarizing the data from the map step. This was historically the first approach to analyzing big data (mainly webpage links) in a distributed fashion introduced by Google [44]. Recent computing engines generalize this idea by representing jobs not just as a sequence of map and reduce stages but as a fairly more complicated directed acyclic (or cyclic) graph of specialized operators.

An example application of such a distributed analyses on genomics data may be performing a task similar to a "SAMtools pileup" [39,45], distributing the large number of alignments of sequencing reads split by genomic location into many machines, to calculate the "pileup" coverage function in parallel to all of them. Then the partial pileup results are sent to the "reduce" machine, which calculates, for example, the maximal coverage or finds genomic variants. In the distributed processing, one can say that because of the size and distribution of the data, the program is sent to them. In fact the program is sent to the data processing ("map") machines in the cloud. In the classic data processing paradigm, the data are sent to the processing machine. The distribution of data and processing can lead to high scalability and in consequence to orders of magnitude faster data operations, which will be an indispensable quality of the near-future genomic big data processing systems.

In the Hadoop ecosystem, particularly worth mentioning, because of its possible impact on near-future clinical genomic tools, is Apache Spark, another state-of-the-art case of a distributed computing framework. It is an open source software developed by AMPLab at the University of California at Berkeley. Apache Spark has been widely adopted as a successor to Apache Hadoop Map–Reduce and can be deployed on single machines, small clusters of a few nodes, up to large cluster deployment with thousands of nodes. Another very promising project is Apache Flink (formerly known as Stratosphere) [46]. Both computing engines exhibit many similarities (e.g., in-memory processing, graph job representation) but some features like pipelined execution or known form relational databases cost-based optimizer that adapt processing execution plan to data sets properties (like size, skewness, and cardinality) are unique to Apache Flink. Worth mentioning for the reason of their potential use in clinical genomics are the data warehousing solutions based on traditional massively parallel processor (MPP) architectures (such as Amazon Redshift), systems with MPP-like execution engines that integrate with the Hadoop ecosystem, such as distributed storage. Currently, the modern cloud-based softwares have found only partially direct applications in genomics; there is still a great potential for using it in specific applications and stages of NGS data analysis. The genomic-specific implementations and related tools that use the software described above are HadoopBAM [47], Seq-Pig [48], SparkSeq [45], BioPig [49], and Biodoop [50]. The ADAM Project [51] is an initiative to create a genomics processing engine and specialized format built using Apache Avro and Parquet. ADAM has currently several side projects that address RNA sequencing (RNAadam), DNA variant calling (Avocado and Guacamole), and other types of utilities for parallel processing of genomic data.

SparkSeq [45] is one of the first cloud-ready solutions for interactive analysis of NGS data. The particular development principle put the emphasis on single-nucleotide resolution of data processing and results. It was developed as a prototype that addresses many of the challenges present in parallel and distributed processing of NGS data. SparkSeq takes advantage of the Apache Spark computing framework as well as some other Hadoop ecosystem components. In addition, it adapts and implements many of the techniques and technologies that are gradually becoming de facto standards in big data ecosystems. To some extent it can be treated as an attempt to design a reference architecture, providing hints for cloud-based NGS data studies, and addressing challenges and many novel approaches.

DATA SECURITY AND BIOBANKING

Data security is one of the main concerns in sequencing data analyses performed in the cloud [52–54], as the cost of data production is still high, and the market value of the state-of-the-art biological data or patients' medical data is inestimable. This is why providing a solid security level for data both in motion and at rest seems to be a crucial factor for cloud-based NGS analytics adoption. A data-in-motion encryption has been for a long time available in Hadoop. It has been a part of the Hadoop RPC, HDFS data transfer, and Web user interfaces. For the data at rest in Hadoop, the encryption has been available since version 2.6. For transparent encryption a new abstraction to HDFS was introduced: the encryption zone. An encryption zone is a special directory, the contents of which will be transparently encrypted upon write and transparently decrypted upon read. Each encryption zone is associated with a single encryption zone key, which is specified when the zone is created. To enforce security policies at any level (i.e., files, folders, etc., stored in HDFS), and to manage fine-grained access control to higher level objects, one can take advantage of Apache Ranger. Finally, the way to protect access to the services provided by various Hadoop ecosystem components (such as WebHDFS, Stargate—HBase REST Gateway, or Hive JDBC) is to use the Apache Knox solution [55]. The area that will be developing dynamically with the growth of the volume of clinical sequencing data is digital biobanking [56,57]. With the proper security measures [58,59] and with the application of a scalable, most likely rooted in the Hadoop ecosystem, software solution it will be possible to store, organize, and prepare genomic data for analysis and use in the clinics. The digital biobanks can have various scales, from a single computer in smaller hospital units, through dedicated and detached computer clusters for whole hospital use, to semipublic, limited access cloud biobanking systems. The advantage of the modern scalable IT solution will be the fact that for every size of biobank implementation the same scalable software tools can be used.

CURRENTLY DEVELOPING SCENARIOS AND CASES

There are several open problems in clinical genomics data processing, which, however, can be potentially solved with the existing software tools in their state-of-the-art, distributed, and scalable form. Still, in many cases those open problems will require considerable work to practically solve. One of those problem is creating a functional system for storing genomic variants of patients. This would be a database that stores, indexes, and allows data-warehousing style slicing on the data currently stored as VCF (gVCF) files. Also, preparing the list of genome variants from the sequencing is not a trivial task, but a consensus in this area may be found after having applied larger computational power in the cloud. The higher capacities of data processing will enable us to explore more extensively the informational content of the sequencing data. In this area one can think of nucleotide-precision genomic signatures from RNA sequencing [60,61], precise mutation finding [62], or the use of machine learning to discover the hidden properties of sequencing data and filter valid biological phenomena from technology artifacts. Yet another practical use of cheaper sequencing and larger processing power will be the integration of various omics techniques on the genomic location and linking the outcome to the patient's phenotype. A scenario of running DNA genotyping, RNA sequencing, and methylation sequencing on the same patient samples is currently possible, but still the cost and workload are prohibitive. This will change at some point, and with dedicated integrative IT tools genomic medicine can get a multisided genomic insight into disease mechanisms. Such a software can produce multi-omics genomic signatures of diseases and their phenotypes or produce multi-omics decision rules to support precise personalized therapy. Ever-increasing NGS sequencer overall throughput is also changing the manner of implementing analytical pipelines [63]. They need to be not only scalable but also organized as semi- or fully automated assembly lines, using such approaches as data pipelining and streaming to minimize bottlenecks that may arise from unnecessary and inefficient processing stage synchronization. Automatic quality control of data and analytical algorithms should be imposed on as many computing phases as possible to ensure reliability and reproducibility of results.

EXPECTED CHALLENGES AND LIMITATIONS OF FUTURE DATA PROCESSING TOOLS AND MODELS

There is also a group of IT issues that are important for the progress of clinical genomics, but its development is currently hard to predict. Such factors are, for example, the price of sequencing itself and the computing cloud to process the data. If they even partly continue to follow Moore's law the progress will be natural. But if one of the technologies approaches a technological or economical barrier, then the development of all clinical genomics and personalized medicine may slow down for many years as technologically not feasible. Still, the current situation lets us be optimistic about the fast-paced growth of this area—both as a branch of science and in practical clinical applications. The area of data processing in clinical genomics brings about also many ethical psychological and social concerns, both in the use of big medical data sets and in the knowledge of genotypes associated with phenotypes of single patients and their families. Here the information technology on one side is creating new challenges, but on the other side can provide tools to solve them, for example, from the area of data security.

EDUCATION OF PHYSICIANS AND MEDICAL RESEARCHERS IN THE UNDERSTANDING AND USE OF GENOMIC DATA PROCESSING

In the process of generating synergies between modern genomics, personalized medicine, and information technology, there is also the critical aspect of the human factor. In genomic research, the bioinformatician was expected to bridge the gap between life science or medicine and the raw information technology. Those areas have different vocabularies of notions, different ways of solving problems, and presenting the results. The bioinformaticians had to learn both types of knowledge to create an interdisciplinary service useful to the life sciences and medicine. In genomic medicine the emphasis will be even more on the medical way of understanding patients, diagnoses, and therapies. IT will be complex, but even more just an instrument that serves personalized medicine goals. It is hard to expect IT engineers to learn medicine, as it requires long studies and practice. Perhaps it will be easier to create within medical universities a new group of professionals who will be genome medicine specialists. They will have enough medical knowledge to assist and support medical procedures and enough IT skills to take part in data processing. Perhaps some of them will be able to extend their studies to get a full medical diploma. Educating this group of genome–IT–medicine professionals can be achieved in a manner similar to the education of pharmacy specialists. Alternatively, genomic medicine can be yet another specialization after medical studies. In any case, the personalized genomic approach to medicine should soon become an integral part of medical education.

ELECTRONIC HEALTH RECORDS

An important challenge for the development of clinically useful NGS is the integration of the produced genomic data with other clinical information. Given the large volume of data the obvious direction is to use electronic health records (EHR), which have already proven valuable for individualized care [64]. However, before the genomic potential of EHR can be realized significant challenges must be met [65,66]. Tarczy-Hornoch et al. [67] surveyed informatics approaches to WES and whole-genome sequencing clinical reporting in EHR, showing a great diversity of attitudes to annotation tools and work flow, as well as to report generation and results interpretation. The NHGRI, together with the National Cancer Institute, funded the Clinical Sequencing Exploratory Research (CSER) Program, composed of six leading academic medical centers, to implement genomic data into patient care, in conjunction with ethical, legal, and psychological and social issues (http://grants.nih.gov/grants/guide/rfa-files/RFA-HG-10-017.html). Further, the CSER Electronic Medical Record Working Group was initiated to explore informatics issues related to annotation and prioritization of genomic variants as well as to the integration of genomic results into EHR and related clinical decision support approaches to decision support (http://www.genome.gov/27546194).

External and internal accumulated variant databases/knowledge bases (VDBKBs) available for EHR are a critical component of implementation of NGS testing results in clinical practice, while NGS platforms do not include tools for annotation of variants they report, nor are they integrated with VDBKBs. Tarczy-Hornoch et al. point to future directions to maximize the NGS impact on clinical medicine, such as collaborative creation of VDBKBs, standardization of applied ontology, and development of active decision support rules, all of them supporting the integration of reports into EHR [67]. Furthermore, Masys et al. identified some of the difficulties associated with the inclusion of big data infrastructure into EHR systems, like pipelines used for data processing and interpretation along with decision support tools [65].

While the genome of a patient is stable, the interpretation of the variants evolves, and the genomic findings may be reclassified over time. Automated e-mail alerts, sent to clinicians when a variant is reclassified and reannotated over time, especially if reflecting a possible change in patient management, are invaluable [67]. This approach is also consistent with the model presented by the Electronic Medical Records and Genomics consortium [64]. An external tool (GeneInsight, Alamut, or Cartagenia) can be integrated into the EHR to provide an alert at the time of electronic order entry [67]. The emerging discipline of translational bioinformatics focuses on such tools [68]. A challenge to putting actionable variants into the EHR is poor adherence to existing standards to represent genomic variants, such as the guideline proposed by the Human Genome Variation Society active genomic decision support (http://www.hgvs.org/mutnomen/recs/html) (in contrast to standardized coding systems such ICD10) (http://apps.who.int/classifications/icd10/browse/2010/en).

Frey et al. recognize that precise characterization of the phenotype to be processed in EHRs is a fundamental problem (deep phenotyping) [66]. Hripcsak and Albers further discuss the challenges of phenotyping in EHR such as incompleteness, inaccuracy, complexity, and bias. The discussion is followed by the description of possibilities for supporting phenotype development and consistency of vocabularies and ontologies, crucial to phenotype definition, improving EHR processes [69]. The Unified Medical Language System [70] provides a repository of vocabularies to be applied in medical records [66]. To support harmonization of phenotype categorization and establishment of unified semantics in human diseases, the Human Phenotype Ontology project was launched [71] to be further connected with the Online Mendelian Inheritance in Man database [66] and supported with software tools [72]. Another platform hosting information about rare variants and rare phenotypes are phenotype–genotype databases collecting phenotypic information and NGS findings, such as the Leiden Open Variation Database (www.lovd.nl).

Despite numerous difficulties in the implementation of NGS data into EHR systems, there are already some successful approaches such as the Electronic Medical Records and Genomics (eMERGE) network (http://emerge.mc.vanderbilt.edu). The eMERGE project was established in 2007 to promote genomic data using biorepositories linked to the EHR system, further complemented by the return of clinically relevant genomic findings in EHR, to be used by health care providers at a point of care. In addition, targeted resequencing of 84 pharmacogenes is currently being performed within the eMERGE network, and genotypes of pharmacogenetic relevance are being placed in the EHR to guide individualized drug therapy and develop best practices for integrating genomic data into the EHR [73].

CONCLUSION—VISION OF NEAR-FUTURE MEDICAL GENOMICS INFORMATION SYSTEMS

Having the current knowledge of the developments in NGS, its medical uses, and the relevant progress in information technology it is possible to imagine a near-future vision of ubiquitous medical software systems that will not only constantly support the "from-bench-to-bedside" translational medicine paradigm, but will also be available in a form of customized toolboxes for various stages of diagnosis and therapy. Such systems are not likely to diminish the role and authority of a physician, they should be treated instrumentally as information technology tools, a "lancet of the twenty-first century." The intricacies of database structure, cloud processing, and scalability should be encapsulated in a physician–user-friendly interface. In more complex cases these tools may require the assistance of the new group of genome medicine professionals, whose education programs at medical universities should start very soon. An important aspect of all future clinical applications of NGS technology will be the development of standardized guidelines for both laboratory "wet" procedures and bioinformatics analyses—a process that has just began [74].

REFERENCES

[1] Wang Y, Yang Q, Wang Z. The evolution of nanopore sequencing. Front Genet 2015;5.
[2] Kasianowicz J, Brandin E, Branton D, Deamer D. Characterization of individual polynucleotide molecules using a membrane channel. PNAS November 26, 1996;93(24):13770–3.
[3] Butler TZ, Pavlenok M, Derrington IM, Niederweis M, Gundlach JH. Single-molecule DNA detection with an engineered MspA protein nanopore. Proc Natl Acad Sci USA December 30, 2008;105(52):20647–52.
[4] Stoddart D, Heron AJ, Klingelhoefer J, Mikhailova E, Maglia G, Bayley H. Nucleobase recognition in ssDNA at the central constriction of the alpha-hemolysin pore. Nano Lett September 2010;10(9):3633–7.
[5] Derrington IM, Butler TZ, Collins MD, Manrao E, Pavlenok M, Niederweis M, et al. Nanopore DNA sequencing with MspA. Proc Natl Acad Sci USA September 14, 2010;107(37):16060–5.
[6] Manrao EA, Derrington IM, Laszlo AH, Langford KW, Hopper MK, Gillgren N, et al. Reading DNA at single-nucleotide resolution with a mutant MspA nanopore and phi29 DNA polymerase. Nat Biotechnol April 2012;30(4):349–53.

[7] Li J, Stein D, McMullan C, Branton D, Aziz MJ, Golovchenko JA. Ion-beam sculpting at nanometre length scales. Nature July 12, 2001;412(6843):166–9.

[8] Schneider GF, Kowalczyk SW, Calado VE, Pandraud G, Zandbergen HW, Vandersypen LMK, et al. DNA translocation through graphene nanopores. Nano Lett August 11, 2010;10(8):3163–7.

[9] Garaj S, Hubbard W, Reina A, Kong J, Branton D, Golovchenko JA. Graphene as a subnanometre trans-electrode membrane. Nature September 9, 2010;467(7312):190–3.

[10] Xu Q, Wu MY, Schneider GF, Houben L, Malladi SK, Dekker C, et al. Controllable atomic scale patterning of freestanding monolayer graphene at elevated temperature. ACS Nano February 26, 2013; 7(2):1566–72.

[11] Liu K, Feng J, Kis A, Radenovic A. Atomically thin molybdenum disulfide nanopores with high sensitivity for DNA translocation. ACS Nano March 25, 2014;8(3):2504–11.

[12] Zwolak M, Di Ventra M. Electronic signature of DNA nucleotides via transverse transport. Nano Lett March 2005;5(3):421–4.

[13] Ohshiro T, Umezawa Y. Complementary base-pair-facilitated electron tunneling for electrically pinpointing complementary nucleobases. Proc Natl Acad Sci USA January 3, 2006;103(1):10–4.

[14] Liang F, Li SQ, Lindsay S, Zhang PM. Synthesis, physicochemical properties, and hydrogen bonding of 4(5)-substituted 1-H-imidazole-2-carboxamide, a potential universal reader for DNA sequencing by recognition tunneling. Chemistry May 2012;18(19):5998–6007.

[15] Pathak B, Lofas H, Prasongkit J, Grigoriev A, Ahuja R, Scheicher RH. Double-functionalized nanopore-embedded gold electrodes for rapid DNA sequencing. Appl Phys Lett January 9, 2012;(2):100.

[16] Chen X. DNA sequencing with titanium nitride electrodes. Int J Quantum Chem October 15, 2013;113(20): 2295–305.

[17] Quick J, Quinlan AR, Loman NJ. A reference bacterial genome dataset generated on the MinION portable single-molecule nanopore sequencer. Gigascience 2014;3:22.

[18] Ashton PM, Nair S, Dallman T, Rubino S, Rabsch W, Mwaigwisya S, et al. MinION nanopore sequencing identifies the position and structure of a bacterial antibiotic resistance island. Nat Biotechnol December 8, 2014;33:296–300.

[19] Loman NJ, Quinlan AR. Poretools: a toolkit for analyzing nanopore sequence data. Bioinformatics December 1, 2014;30(23):3399–401.

[20] Watson M, Thomson M, Risse J, Talbot R, Santoyo-Lopez J, Gharbi K, et al. poRe: an R package for the visualization and analysis of nanopore sequencing data. Bioinformatics January 1, 2015;31(1):114–5.

[21] Jain M, Fiddes IT, Miga KH, Olsen HE, Paten B, Akeson M. Improved data analysis for the MinION nanopore sequencer. Nat Methods February 16, 2015;12(4):351–6.

[22] Kumar S, Tao CJ, Chien MC, Hellner B, Balijepalli A, Robertson JWF, et al. Peg-labeled nucleotides and nanopore detection for single molecule DNA sequencing by synthesis. Sci Rep September 21, 2012;2.

[23] McNally B, Singer A, Yu ZL, Sun YJ, Weng ZP, Meller A. Optical recognition of converted DNA nucleotides for single-molecule DNA sequencing using nanopore arrays. Nano Lett June 2010;10(6):2237–44.

[24] Wetterstrand KA. DNA sequencing costs: data from the NHGRI Genome Sequencing Program (GSP). National Human Genome Research Institute 2013.

[25] Branton D, Deamer DW, Marziali A, Bayley H, Benner SA, Butler T, et al. The potential and challenges of nanopore sequencing. Nat Biotechnol 2008;26(10):1146–53.

[26] Atzori L, Iera A, Morabito G. The internet of things: a survey. Comput Netw 2010;54(15):2787–805.

[27] Cochrane G, Akhtar R, Bonfield J, Bower L, Demiralp F, Faruque N, et al. Petabyte-scale innovations at the European Nucleotide Archive. Nucleic Acids Res 2009;37(Database issue):D19–25.

[28] Leinonen R, Sugawara H, Shumway M. The sequence read archive. Nucleic Acids Res 2010;39(Database issue):D19–21. http://dx.doi.org/10.1093/nar/gkq1019. gkq1019.

[29] Leinonen R, Akhtar R, Birney E, Bower L, Cerdeno T, Cheng Y, et al. The European nucleotide archive. Nucleic Acids Res 2010;39(Database issue):D28–31. http://dx.doi.org/10.1093/nar/gkq967. gkq967.

[30] Kodama Y, Shumway M, Leinonen R. The Sequence Read Archive: explosive growth of sequencing data. Nucleic Acids Res 2012;40(D1):D54–6.

[31] Danecek P, Auton A, Abecasis G, Albers CA, Banks E, DePristo MA, et al. The variant call format and VCFtools. Bioinformatics 2011;27(15):2156–8.

[32] Nielsen R, Paul JS, Albrechtsen A, Song YS. Genotype and SNP calling from next-generation sequencing data. Nat Rev Genet 2011;12(6):443–51.

[33] Ulahannan D, Kovac MB, Mulholland PJ, Cazier JB, Tomlinson I. Technical and implementation issues in using next-generation sequencing of cancers in clinical practice. Br J Cancer 2013;109(4):827–35.

[34] Deloukas P, Kanoni S, Willenborg C, Farrall M, Assimes TL, Thompson JR, et al. Large-scale association analysis identifies new risk loci for coronary artery disease. Nat Genet 2013;45(1):25–33.

[35] Harismendy O, Ng PC, Strausberg RL, Wang X, Stockwell TB, Beeson KY, et al. Evaluation of next generation sequencing platforms for population targeted sequencing studies. Genome Biol 2009;10(3):R32.

[36] DePristo MA, Banks E, Poplin R, Garimella KV, Maguire JR, Hartl C, et al. A framework for variation discovery and genotyping using next-generation DNA sequencing data. Nat Genet 2011;43(5):491–8.

[37] Duitama J, Quintero JC, Cruz DF, Quintero C, Hubmann G, Moreno MR, et al. An integrated framework for discovery and genotyping of genomic variants from high-throughput sequencing experiments. Nucleic Acids Res 2014;42(6):e44.

[38] Meienberg J, Zerjavic K, Keller I, Okoniewski M, Patrignani A, Ludin K, et al. New insights into the performance of human whole-exome capture platforms. Nucleic Acids Res 2015:gkv216.

[39] Li H, Handsaker B, Wysoker A, Fennell T, Ruan J, Homer N, et al. The Sequence Alignment/Map format and SAMtools. Bioinformatics 2009;25(16):2078–9.

[40] Foster I, Zhao Y, Raicu I, Lu S. Cloud computing and grid computing 360-degree compared. In: Grid computing environments workshop, 2008. GCE'08, IEEE. 2008. p. 1–10.

[41] Rosenthal A, Mork P, Li MH, Stanford J, Koester D, Reynolds P. Cloud computing: a new business paradigm for biomedical information sharing. J Biomed Inform 2010;43(2):342–53.

[42] Dai L, Gao X, Guo Y, Xiao J, Zhang Z, et al. Bioinformatics clouds for big data manipulation. Biol Direct 2012;7(1):43.

[43] Marx V. Biology: the big challenges of big data. Nature 2013;498(7453):255–60.

[44] Dean J, Ghemawat S. MapReduce: simplified data processing on large clusters. Commun ACM 2008;51(1):107–13.

[45] Wiewiorka M, Messina A, Pacholewska A, Maffioletti S, Gawrysiak P, Okoniewski MJ. SparkSeq: fast, scalable, cloud-ready tool for the interactive genomic data analysis with nucleotide precision. Bioinformatics 2014:343.

[46] Alexandrov A, Bergmann R, Ewen S, Freytag JC, Hueske F, Heise A, et al. The Stratosphere platform for big data analytics. VLDB J 2014;23(6):939–64.

[47] Niemenmaa M, Kallio A, Schumacher A, Klemela P, Korpelainen E, Heljanko K. Hadoop-BAM: directly manipulating next generation sequencing data in the cloud. Bioinformatics 2012;28(6):876–7.

[48] Schumacher A, Pireddu L, Niemenmaa M, Kallio A, Korpelainen E, Zanetti G, et al. SeqPig: simple and scalable scripting for large sequencing data sets in Hadoop. Bioinformatics 2014;30(1):119–20.

[49] Nordberg H, Bhatia K, Wang K, Wang Z. BioPig: a Hadoop-based analytic toolkit for large-scale sequence data. Bioinformatics 2013;29(23):3014–9.

[50] Leo S, Santoni F, Zanetti G. Biodoop: bioinformatics on Hadoop. In: Parallel processing workshops, 2009. ICPPW'09. International conference on IEEE. 2009. p. 415–22.

[51] Massie M, Nothaft F, Hartl C, Kozanitis C, Schumacher A, Joseph AD, et al. ADAM: genomics formats and processing patterns for cloud scale computing. Berkeley: EECS Department, University of California; Dec 2013.

[52] Lombardi F, Di Pietro R. Secure virtualization for cloud computing. J Netw Comput Appl 2011;34(4):1113–22.

[53] Dove ES, Joly Y, Anne M, Burton P, Chisholm R, Fortier I, et al. Genomic cloud computing: legal and ethical points to consider. Eur J Hum Genet 2014.

[54] Beck M, Haupt VJ, Roy J, Moennich J, kel R, Schroeder M, et al. GeneCloud: secure cloud computing for biomedical research. Trusted cloud computing. Springer; 2014. p. 3–14.

[55] Sharma PP, Navdeti CP. Securing big data Hadoop: a review of security issues, threats and solution. Int J Comput Sci Inform Technol 2014;5(2).

[56] Izzo M, Mortola F, Arnulfo G, Fato MM, Varesio L. A digital repository with an extensible data model for biobanking and genomic analysis management. BMC Genomics 2014;15(Suppl. 3):S3.

[57] Izzo M, Arnulfo G, Piastra MC, Tedone V, Varesio L, Fato MM. XTENS-A JSON-based digital repository for biomedical data management. Bioinformatics and biomedical engineering. Springer; 2015. p. 123–130.

[58] Hayden EC. Extreme cryptography paves way to personalized medicine. Nature 2015;519(7554).

[59] Lunshof JE, Chadwick R, Vorhaus DB, Church GM. From genetic privacy to open consent. Nat Rev Genet 2008;9(5):406–11.

[60] Anders S, McCarthy DJ, Chen Y, Okoniewski M, Smyth GK, Huber W, et al. Count-based differential expression analysis of RNA sequencing data using R and Bioconductor. Nat Protoc 2013;8(9):1765–86.

[61] Frazee AC, Sabunciyan S, Hansen KD, Irizarry RA, Leek JT. Differential expression analysis of RNA-seq data at single-base resolution. Biostatistics 2014;15(3):413–26.

[62] Okoniewski MJ, Meienberg J, Patrignani A, Szabelska A, Matyas G, Schlapbach R. Precise breakpoint localization of large genomic deletions using PacBio and Illumina next-generation sequencers. Biotechniques 2013;54(2):98.

[63] Merelli I, Pérez S, Gesing S, D'Agostino D. Managing, analysing, and integrating big data in medical bioinformatics: open problems and future perspectives. BioMed Res Int 2014;2014.

[64] Starren J, Williams MS, Bottinger EP. Crossing the omic Chasm a time for omic Ancillary systems. JAMA March 27, 2013;309(12):1237–8.

[65] Masys DR, Jarvik GP, Abernethy NF, Anderson NR, Papanicolaou GJ, Paltoo DN, et al. Technical desiderata for the integration of genomic data into Electronic Health Records. J Biomed Inform June 2012;45(3):419–22.

[66] Frey LJ, Lenert L, Lopez-Campos G. EHR big data deep phenotyping. Contribution of the IMIA genomic medicine working group. Yearb Med Inform 2014;9(1):206–11.

[67] Tarczy-Hornoch P, Amendola L, Aronson SJ, Garraway L, Gray S, Grundmeier RW, et al. A survey of informatics approaches to whole-exome and whole-genome clinical reporting in the electronic health record. Genet Med October 2013;15(10):824–32.

[68] Sarkar IN, Butte AJ, Lussier YA, Tarczy-Hornoch P, Ohno-Machado L. Translational bioinformatics: linking knowledge across biological and clinical realms. J Am Med Inform Assoc July 2011;18(4):354–7.

[69] Hripcsak G, Albers DJ. Next-generation phenotyping of electronic health records. J Am Med Inform Assoc January 2013;20(1):117–21.

[70] Bodenreider O. The unified medical language system (UMLS): integrating biomedical terminology. Nucleic Acids Res January 1, 2004;32:D267–70.

[71] Robinson PN, Mundlos S. The human phenotype ontology. Clin Genet June 2010;77(6):525–34.

[72] Deng Y, Gao L, Wang BB, Guo XL. HPOSim: an R package for phenotypic similarity measure and enrichment analysis based on the human phenotype ontology. PLoS ONE February 9, 2015;10(2).

[73] Kullo IJ, Haddad R, Prows CA, Holm I, Sanderson SC, Garrison NA, et al. Return of results in the genomic medicine projects of the eMERGE network. Front Genet 2014;5:50.

[74] Aziz N, Zhao Q, Bry L, Driscoll DK, Funke B, Gibson JS, et al. College of American Pathologists' laboratory standards for next-generation sequencing clinical tests. Arch Pathol Lab Med August 25, 2014;139(4):481–93.

ETHICAL AND PSYCHOSOCIAL ISSUES IN WHOLE-GENOME SEQUENCING FOR NEWBORNS

16

John D. Lantos

Children's Mercy Hospital Bioethics Center, University of Missouri – Kansas City, Kansas City, MO, USA

CHAPTER OUTLINE

INTRODUCTION

In 2010, the National Institutes of Health invited experts to propose a research agenda regarding newborn screening in the genomic era. Botkin made a distinction between newborn screening and screening of newborns: "The former is the current public health system used to identify a range of actionable childhood diseases, whereas the latter is far more encompassing, and could involve using sequencing (or other technologies) to identify adult-onset diseases and/or provide other information that would benefit patients throughout their lives."

Since 2010, another potential use of genomic testing of newborns has developed. Whole-genome sequencing (WGS) is now used not as a screening test but as a diagnostic test for children with illnesses that have been impossible to diagnose using standard diagnostic testing. The nature of WGS is such, however, that such diagnostic testing will screen for other genomic variants. So the diagnostic test is also, inevitably, a potential screening test. It is possible, though, to pay attention to and report only the diagnostic information.

WGS is associated with some well-known problems. It often yields results of uncertain clinical significance. Many researchers have developed algorithms to screen results and filter truly uninterpretable incidental findings from those that may have clinical significance. Solomon et al. report an approach taken at the National Human Genome Research Institute for sorting through findings:

"A working committee consisting of board-certified clinical geneticists, board-certified molecular geneticists, board-certified genetic counselors, bioethicists, and National Human Genome Research

Institute IRB members, as well as other genetic researchers, convened to discuss variants that met the above criteria. Finally, in several cases, experts in the study of individual genes and conditions were contacted when results remained equivocal."

This working committee based its recommendation on the following predetermined standards for disclosure:

1. The genetic change must be known or predicted to be of urgent clinical significance.
2. Knowledge of the finding must have a clear direct benefit that would be lost if diagnosis was made later; that is, knowledge of this risk factor would substantially alter medical or reproductive decision-making.
3. The potential benefit of knowing a genetic disorder exists clearly outweighs the potential risks of anxiety and subsequent medical testing that could result from this knowledge.
4. Unless they add substantial risk, risk factors for multifactorial disorders are not reported.
5. Recessive mutations will be reported only if (1) the carrier frequency for mutations in that specific gene is >1% (such that the disease incidence is more than 1/40,000); (2) the syndrome results in significant morbidity; or (3) early diagnosis and intervention would have significant benefit.

Using these criteria, they sought pathogenic genes in a pair of twins, one of whom had congenital anomalies and one of whom did not. Whole-exome sequencing identified 79,525 genetic variants in the twins. After filtering artifacts and excluding known single-nucleotide polymorphisms and variants not predicted to be pathogenic, the twins had 32 novel variants in 32 genes that were felt to be likely to be associated with human disease. Eighteen of these novel variants were associated with recessive disease and 18 were associated with dominantly manifesting conditions (variants in some genes were potentially associated with both recessive and dominant conditions). Only 1 variant ultimately met institutional review board-approved criteria for return of information to the research participants.

Such data suggest ways in which WGS might raise issues similar to those raised by other sorts of genetic testing in the past. They also suggest ways in which WGS might be different.

Ethical concerns about WGS mirror many of the concerns that have been raised over the years about other forms of testing. For example, concerns arose in the 1970s about screening newborns for sickle cell disease. Specifically, many pediatricians and bioethicists worried that such testing would lead to confusion or stigmatization without any compensatory benefit. In the 1980s and 1980s, similar concerns arose about newborn screening for cystic fibrosis. In both cases, controversy largely disappeared once it became apparent that screening and early identification led to better outcomes. With sickle cell, penicillin prophylaxis lowered infant mortality. With cystic fibrosis, early diagnosis was shown to be associated with better growth.

These examples illustrate a consistent feature of the debates about the testing of newborns for genetic disease. Initial concern focuses on the question of whether such testing will confer benefits on the children who are tested. The only way to find out if there is a benefit is to do the testing. Doing the testing often leads to benefits that were not the same ones that were initially anticipated. Few predicted that the benefits of newborn screening for cystic fibrosis would be primarily related to growth, rather than to pulmonary function.

Once it is shown that testing confers benefits, the ethical controversy disappears. There is widespread agreement that testing is appropriate if it provides immediate benefit to the child.

WGS has yet to prove itself in this regard. It is clearly a powerful diagnostic tool that has the potential to transform the way in which we diagnose disease and estimate prognosis. But the information generated by WGS is often difficult to interpret. This is partly due to the fact that most health professionals (and, of course, most patients or parents) are unfamiliar with this new sort of data. Green and colleagues at the National Human Genome Research Institute wrote, "Health care providers will need to be able to interpret genomic data. Patients will need to be able to understand the information being provided to them and to use that information to make decisions."

WGS has a very problematic signal-to-noise ratio. McKenna and colleagues highlighted this, noting, "… a large development gap between sequencing output and analysis results." Bieseker and colleagues recognize the problems of information overload: "A whole-genome or whole-exome result is overwhelming for both the clinician and the patient … (because) … variants from genome or exome range from those that are extremely likely to cause disease to those that are nearly certain to be benign, and every gradation between these two extremes."

At the same time, there are clearly some cases in which a signal can be discerned among the background noise. When that occurs, WGS may allow difficult diagnoses to be made in a timely way that would be impossible to make using any other diagnostic tools. Again, Bieseker, "What is also clear is that some of these variants can be not only highly predictive of disease but their return can enable life-saving treatment." And Green imagines, "The routine use of genomics for disease prevention."

In this regard, WGS is quite different from most prior forms of genetic testing. In general, those prior forms of testing tried to diagnose one disease at a time. The tests were either positive or negative (and the positives and negatives would be true or false). WGS, in contrast, tests for hundreds or thousands of diseases at once and also finds variants that may or may not be associated with disease. To highlight the ways in which WGS is different, I will contrast it with three other types of genetic testing that were done in the past using different technologies and deployed in different populations for different purposes. These three are: (1) mandated newborn screening programs; (2) carrier testing in populations known to be at risk for one specific disease; and (3) specimen collection for biobanking to be used in research. It is important to understand why and how WGS differs from such testing and why some of the ethical concerns raised by other testing programs may not be relevant.

DIFFERENCES BETWEEN WGS AND GENETIC TESTING IN OTHER CONTEXTS

NEWBORN METABOLIC SCREENING

Newborn screening is an unusual form of clinical testing. In most states, it is mandated and done without parental consent. (Some states allow parents to opt out, but most do not explicitly inform parents of this right.)

When newborn screening was first developed, the criteria were well defined and highly restrictive. Tests were designed "to identify infants with severe disorders that are relatively prevalent and treatable (or controllable)." The advent of new technology that made screening easier and less expensive led to the expansion of screening panels that can include diseases that are untreatable. Tests for such diseases were initiated before there was any empirical assessment of long-term outcomes, harms, and benefits.

Newborn screening is associated with a number of well-recognized problems that are inherent to all screening tests. Compared to clinical testing, it leads to a high rate of false positives and to identification of children who may actually have disease ("true positives") but whose disease is milder than anticipated.

WGS could be used for newborn screening. If so, it would raise issues similar to those raised by traditional newborn metabolic screening. Those issues would be even more complex because of the sheer volume of abnormalities that would be found. Instead, however, WGS can be used to test children who have been referred by their clinicians because they have symptoms of disease for which diagnosis has been elusive. Used in this way, WGS has been shown, in at least some cases, to be beneficial in finding efficacious treatments for critically ill children.

CARRIER TESTING IN CAREFULLY SELECTED HIGH-RISK POPULATIONS

Carrier testing has been conducted for a variety of conditions. Usually, people are selected for testing based upon their race, ethnicity, and family history. For example, screening for Tay-Sachs was originally carried out in Ashkenazi Jewish communities, screening for sickle cell in people of African descent, etc. There have been many studies of the psychosocial impact of such targeted testing for Tay-Sachs, muscular dystrophy, cystic fibrosis, and many other conditions. A meta-analysis of these studies revealed certain common themes. Such testing is predictably associated with anxiety and guilt in some people, euphoria or relief in others.

Currently, there is inconsistency in practices regarding communication about carrier status in different contexts. Many organizations oppose testing children for carrier status. Borry et al. reviewed 14 policy statements from 24 different groups. They note, "All the guidelines were in agreement that children preferably should not undergo carrier testing and that testing of children ideally should be deferred. All guidelines stated that it is in a child's best interest for him to decide whether to be tested at some stage later in life." When carrier status is not the goal but is an "incidental finding," the consensus dissolves. Four guidelines discussed the course of action in cases in which carrier status was discovered incidentally (e.g., during diagnostic testing, screening, or prenatal diagnosis, or in a research context). While guidelines from the British Medical Association and the American Academy of Pediatrics recommended that carrier status results obtained incidentally should be conveyed to parents, the American Medical Association and the German Society of Human Genetics recommended that this information should not be disclosed to parents or to other third parties. Miller and colleagues write, "The provision of carrier or predictive genetic testing is seen to infringe on the child's autonomy and right to confidentiality because it forecloses on the child's right to decide whether to seek this information and to whom it should be disclosed."

WGS can be used to screen for carrier status and, if so, would raise all the same issues as single-gene testing for autosomal recessive conditions. If used for diagnosis, however, then researchers would not necessarily need to either look for carrier status or return those results.

GENETIC TESTING OF STORED SAMPLES IN BIOREPOSITORIES

There are a number of well-recognized ethical concerns about large-scale genomic biorepositories, including (1) the initial consent process; (2) concerns about privacy and confidentiality; and (3) renewal of consent for further studies or at the time when a child turns 18.

Consent issues are complicated because, at the time when genetic material is first placed in a biobank, nobody knows what studies will be done on the material. In such situations, as Greely has noted, "It is may be effectively impossible for them to get true informed consent; they can only get a redefined, watered-down version of informed consent."

Concerns about privacy and confidentiality arise because, to be useful, biobanks must have a way to link genetic material to the person from whom it came and to provide detailed clinical information about that person that can be correlated with the genetic material. Most researchers have addressed this using computer technology that allow identifiers to be "scrubbed" from medical records, creating, essentially, a de-identified clinical record that can still be linked to the matched, but de-identified, genomic material.

Finally, the issue of when, if ever, to obtain renewed consent from people who were enrolled as minors but who have reached the age of majority is only beginning to be addressed. Some argue that parents should never be able to consent for the enrollment of minors. Others suggest that a robust process of recontact and re-consent at the age of majority will be sufficient. Given the rapid pace of change in the field, it is difficult to anticipate what we may or may not be doing 5, 10, or 18 years from now, in terms of both genomics and our ability to stay in touch with research subjects.

THE SPECIFIC ETHICAL ISSUES OF WHOLE-GENOME SEQUENCING FOR CLINICAL DIAGNOSIS

When WGS is used for clinical diagnosis, it avoids many of the problems that arise in other contexts. But it creates some new issues. Ormond and colleagues analyzed and summarized three sets of problems likely to arise in clinical WGS testing.

First, patients and parents need to understand the limitations of sequencing. For example, some methods do not reveal translocations, large duplications or deletions, copy number repeats, or expanding triplet repeats. When two variations are identified in the same gene, some whole-genome sequence analyses cannot be used to establish whether those variations are in copies of the gene on different chromosomes or in the same copy of the gene—a distinction that is crucial for recessive disorders.

Second, it is difficult to maintain up-to-date information on every known genetic disease. No centrally maintained repository of all rare and disease-associated variants exists as of this writing.

Third, the availability of WGS might result in a large increase in testing by cautious physicians. Such an increase not only would raise health costs, but would subject patients to the physical and psychological costs of increased testing.

It is hard to know how such issues will play out in the world of clinical medicine because WGS has not been used widely enough to yield the relevant outcome data. Furthermore, the technology for WGS is improving rapidly, as are the databases that allow more accurate interpretation of results. Thus, assessment is a moving target.

THINKING ABOUT HARMS AND BENEFITS

There are good reasons to imagine that clinicians and parents might perceive WGS testing as helpful, harmful, or simply irrelevant. It is likely that there will be cases in which they are uncertain whether the information from such testing was helpful or not.

WGS might be harmful if it provides false positive or false negative information. This may be true even if that information does not lead to bad clinical decisions. False information may lead to inappropriate complacency. It may increase parental anxiety. It may lead parents to overestimate (or underestimate) their own risk of disease or their risks of having another child with a similar illness.

WGS results might be beneficial if they lead to a more accurate diagnosis or prognosis, particularly if they lead to changes in therapy that improve outcomes. Improved outcomes could mean better treatments for treatable diseases or they could mean quicker recognition of an untreatable and fatal disease and a shift to palliative care. They might be beneficial even if they do not lead to changes in therapy for the baby who is tested but, instead, give parents information about future reproductive choices.

Results are likely to be perceived as of ambiguous or uncertain value if they do not clarify the diagnosis, the prognosis, or the appropriate treatment.

CONCLUSIONS

Looking for meaningful clinical information in the results of WGS is clearly challenging. Equally clearly, it is not impossible. There are numerous case reports of such testing being used successfully to diagnose conditions that had not been diagnosed using more standard tests.

While the controversies surrounding WGS are well known, the benefits are becoming equally apparent. WGS can be particularly useful in situations in which a child has a complex medical condition for which conventional testing has not yielded a diagnosis. In these cases, it is difficult for doctors to know how to treat a baby or to give parents an accurate prognosis. We do not know how often WGS will be useful in such situations.

To ensure that personalized genomic information and genome-based health interventions are safe and effective, we need to study their impact on individuals and their families and on doctors who are taking care of those individuals. As with earlier forms of genetic testing, this one will be assessable only if we try it and study the outcomes. Such early forays into the brave new world of genomic sequencing will truly be guided journeys into the unknown. As such, they will require humility from researchers, careful data collection to ensure that outcomes are measured, and a willingness to be surprised.

NEXT GENERATION SEQUENCING—ETHICAL AND SOCIAL ISSUES

17

Andrzej Kochański[1], Urszula Demkow[2]

[1]*Neuromuscular Unit, Mossakowski Medical Research Center, Polish Academy of Sciences, Warsaw, Poland;*
[2]*Department of Laboratory Diagnostics and Clinical Immunology of Developmental Age, Medical University of Warsaw, Warsaw, Poland*

CHAPTER OUTLINE

UNPREDICTABLE CONSEQUENCES OF THE NEXT GENERATION SEQUENCING-RELATED TECHNOLOGICAL REVOLUTION IN MEDICAL GENETICS

For centuries, the genetic background of familiar disorders remained a mystery, revealed only at the level of phenotype. Only obvious clinical abnormalities occurring across numerous generations have been observed. In 1753, a French philosopher, Pierre Luis de Maupertuis (1698–1759), described a family with a multigenerational occurrence of polydactyly and concluded that the probability of accidental occurrence of this phenomenon would be very low [1].

The genetic background of heritable disorders began to be only partly uncovered by cytogenetics, since the number of human chromosomes (karyotype) was established in 1956 by Joe-Hin Tjio and Albert Levan. Despite the groundbreaking discovery of the structure of DNA in 1953, the first method of genetic code sequencing, based on dideoxy-DNA, was not devised until 1977, by Fred Sanger [2]. Further advancements in sequencing were aided by progress in technology, allowing DNA samples to be isolated from various sources and single genes or bigger parts of the genome to be analyzed. The genes to be screened were selected by clinicians on the basis of the phenotype. Using such approach, from single gene to the phenotype, only one or a few genetic variants have been detected. The Human Genome Project and next generation sequencing (NGS) have further revolutionized the paradigm of clinical genetics. NGS-based technology opened up the possibility of analyzing hundreds or thousands of sequence variants, which may be categorized into four classes:

- classical pathogenic mutations of known clinical relevance,
- mutations of probable clinical relevance,
- DNA variants with unknown relevance,
- harmless polymorphisms.

The advent of NGS technologies created an unprecedented access to the genetic profile of a patient without providing an adequate means to interpret this information. A fundamental question and a matter of debate is to what extent information generated by NGS testing should be communicated to patients. There is a consensus that patients should be informed about findings of high clinical importance and that are at least partly correctable, thus balancing risks and benefits.

PROBLEM 1: THE RIGHT NOT TO KNOW

In recent years informed consent was focused on the right to know. As a matter of fact, for centuries in biomedicine, including medical genetics, the problem of the excess of data did not exist. Paradoxically, in the context of NGS, the right not to know seems to have reached a special and new value. According to some authors all the NGS testing results of clinical relevance should be disclosed to the patient [3]. Even, in a more radical form, it is the patient who decides to be or not to be informed about clinically relevant genomic testing results [4]. For example, the decision of the health care provider not to disclose the pathogenic mutation within the *BRCA1* gene collides in a direct way with the *Primum non nocere* rule. At present both women and men harboring a mutation within *BRCA1* can make informed decisions about their future, taking steps to reduce their cancer risk by being systematically submitted to various cancer prophylaxis programs. Several options are available for managing cancer risk in individuals who have a known harmful *BRCA1* or *BRCA2* mutation. These include enhanced screening or even prophylactic surgery. In addition, individuals who have a positive test result could participate in clinical research programs that may reduce deaths from breast and ovarian cancer.

On the other hand, a positive test result can bring anxiety and distress. Accordingly, the patients have the right not to know; thus both the European Convention of Human Rights and Biomedicine (Article 10.2) and the UNESCO Declaration of the Human Genome (Article 5c) confirm the right of every individual not to be informed about the results of their genetic tests. Distinct ethical dilemmas arise with respect to the reporting of findings unrelated to the indication for ordering the sequencing, but of medical value for patient health, for example, disclosure of pathogenic mutations resulting in late-onset neurodegenerative disorders. As of this writing, there is no consensus on how to manage such findings; ethical boards lack experience with this issue and vary widely on the extent of disclosure of the results to asymptomatic adults. Furthermore, genetic counseling before and after testing is imperative, because of the real and potential risks for psychological impact. The inability to accurately predict phenotype is a restraint of presymptomatic testing for many neurodegenerative disease genes. Nevertheless, some patients and family members may decide to pursue genetic testing to reduce uncertainty, or if not, their right not to know should be generally respected in the process of disclosure of NGS-generated data. Finally, it seems reasonable that the decision concerning the disclosure of genomic testing results should be made on the basis of both the opinion of the board of experts and the patient's own attitude. According to some authors the patient's "right" not to know is not an absolute right. The right not to know is conditioned by the risk of harm to other individuals [5]. Similarly, in the case of a direct threat to life (i.e., deleterious mutations responsible for potentially life-threatening arrhythmias) the patient's right not to know should not be overrated. The two important ethical values, that is, solidarity versus patient's autonomy, seem to collide sometimes in the understanding of the right not to know.

Given that the right not to know is not absolute, there is a question concerning patient autonomy in the decision-making process by the disclosure of NGS results. According to Andorno, the "right not to know cannot be presumed, but should be activated by the explicit will of the person" [5]. To conclude, the patient's right not to know with regard to the disclosure of NGS-generated data should be respected. However, in some circumstances (risk to other individuals and direct threat to life) the right not to know has a limited value. Given its conditioned value, the right to know should not be presumed, but rather activated in a particular situation. To respect preferences, the patient subjected to whole-exome sequencing (WES) or whole-genome sequencing (WGS) needs to undergo an extensive and complete pretest genetic counseling to ascertain his or her preference in terms of choosing whether to be tested and to obtain the test results of other disease-associated genes.

PROBLEM 2: INCIDENTAL/UNSOLICITED FINDINGS

The question of incidental findings (IFs), that is, findings not related nor relevant to the diagnostic indication for which the testing was ordered, did not arise with NGS technology [6]. In contrast to "old" genetic technology (i.e., karyotyping), in which IFs are very rare, the problem of unexpected positive findings by NGS is relatively common. In parallel to the term "incidental findings," the European Society of Human Genetics proposed the term "unsolicited findings" [7]. Even the testing of healthy individuals could result in an unsolicited detection of a pathogenic variant responsible for a late-onset disease. Some have argued that IFs should not be reported until there is strong evidence of benefit, while others have advocated that recognition of any known disease-associated gene could be clinically useful and should be reported. There is still insufficient evidence about benefits, risks, and costs of disclosing incidental findings to make evidence-based recommendations. In a study by Yang et al., 10% of 250 enrolled patients had medically actionable IFs [8]. In another study of 6503 participants, 0.7% of the European ancestry participants and 0.3% of the African ancestry participants had pathogenic/deleterious mutations within medically actionable genes [9]. The question has arisen as to whether a particular IF should be disclosed in the light of two dimensions, that is, medical and personal (right to know and right not to know). Rigter and colleagues proposed categorizing IFs according to their clinical relevance using the following criteria:

- Early- and late-onset disease,
- Level of risk,
- Burden of the disease,
- Options for treatment and prevention [4].

In fact, the disclosure of the NGS variants that require medical intervention seems to be reasonable in general; however, we cannot definitely exclude that even in the case of possible prevention, there may be psychological and social consequences. However, the American College of Medical Genetics and Genomics *Recommendations for Reporting of Incidental Findings in Clinical Exome and Genome Sequencing* proposes a minimal list of genetic conditions for which prevalence is high and early intervention is possible [17]. It is generally accepted that heath care providers should prevent harm by warning patients about certain incidental findings and this principle discards concerns about autonomy, just as it does elsewhere in medical practice. It is further recommended that the ordering clinician, before testing, should discuss with the patient the possibility of IFs. Patients have the right to decline NGS testing if they estimate the risks of possible IFs outweigh the potential benefits. Finally, the problem of the burden of the disease as a basis for disclosure of the results is also controversial. In fact the final phenotype cannot be completely predicted in an individual patient and in this case this criterion collides with the genetic determinism.

PROBLEM 3: GENETIC DETERMINISM AND DISCRIMINATION

The question of genetic determinism and discrimination of the individuals has been discussed in medical genetics for many years. While offering the promise of significant benefits, the information gained as a result of NGS testing also raises a number of sensitive human rights issues, including the possibility of discrimination and social stigma. Before the genomic era the deterministic model that genes alone define biology had become the prevailing paradigm. Contemporary genetics has had to redefine the meaning of genetic determinism. In every DNA sample analyzed with NGS technology hundreds or even thousands of sequence variants are identified. There is a question of how the combinations of the various NGS-identified variants determines complicated characteristics of physical and mental health, behavior, and intelligence. The false interpretation of NGS-generated data may result from an inappropriate genetic determinism based on the belief that physical and mental health is preprogrammed and depends on the composition of individual DNA [10]. Even though NGS has the potential to identify thousands of sequence variants or even unravel the whole genome, it is not possible to associate these findings with such complicated features as intelligence, behavior, or obesity. In practice, owing to the complex interactions between genes along with environmental conditions, the phenotypic consequences of an individual's genotype remain unpredictable [11]. On the other hand, individuals considering genetic testing are frequently concerned about discrimination as a result of predictive genetic information. In 2008 the Genetic Information Nondiscrimination Act (GINA) was launched. GINA is the first powerful act of legislation that protects the individual from genetic discrimination. GINA was signed to prevent health insurance companies and employers from discriminating against individuals based on family history of illness or results from genetic testing. Interestingly, in practice GINA has been used sporadically so far. Provocatively, GINA seems to increase the fear of discrimination, as 30% of respondents were more concerned about discrimination after reading this document. There is a possibility that GINA may provoke participants in translational medicine studies to decline and to raise questions of dignity and discrimination in particular [12]. In other words, the misuse of law and bioethics in medical genetics may slow the progress in biomedicine. In contrast to the euphoria accompanying the Human Genome Project in the late 1990s, the contemporary attitude of public opinion to achievements of genetics seems to be ambiguous.

PROBLEM 4: GENETICALLY-BASED SELECTION OF HUMAN EMBRYOS AND ASSISTED REPRODUCTIVE TECHNOLOGY

Preimplantation genetic diagnosis (PGD), introduced in the early 1990s, accompanying assisted reproductive technology is becoming an important technique for selection of early human embryos. In fact, the rate of pregnancy with PGD after array comparative genomic hybridization has increased from 41.7 to 69.1% [13].

An NGS-based approach for the detection of aneuploidy in early embryos has been reported [14]. As of this writing, the process of embryo selection is limited to a large-scale pathology, that is, aneuploidies. In a few years NGS may be expected to detect a wider spectrum of deleterious mutations residing in an early human embryo. In view of recurrent mutations occurring in hundreds of unrelated patients, the phenotype–genotype correlations aid in predicting the clinical course of a heritable disease. In contrast to postnatal genetic counseling, prenatal counseling is not accompanied by a clinical phenotype and may rely on only the prediction of the phenotype. Although, even in monogenic Mendelian traits, genetic counseling after PGD may provoke ethical questions, as even in Mendelian recessive conditions, new traits of inheritance, that is, three-allelic inheritance, could be established [15]. Furthermore, there is a question concerning the clinical relevance of novel, private mutations occurring within known and unknown genes. Thus, the limited efficiency of prenatal genetic counseling coexisting with the huge amount of data that may be generated by the NGS approach provokes a question concerning the spectrum of mutations that should be considered as deleterious. The current genetic counseling paradigm is not prepared for interpretation of the clinical relevance of the myriad of possible combinations of DNA sequence variants detected by NGS technology in human embryonal cells. Moreover, there is no agreement concerning the safety of the PGD procedure for early embryos (unknown long-term consequences of blastomere removal), which provokes additional ethical questions [16].

PROBLEM 5: NGS AND SOCIAL ISSUES

Genetic research is rapidly advancing, and understanding of the human genome can be beneficial to human health. On the other hand, the genetic information can also be misused. To address this issue, in 1990 the National Human Genome Research Institute launched the Ethical, Legal and Social Implications (ELSI) program as an integral part of the Human Genome Project (http://www.genome.gov/elsi/). The insights gained through ELSI research influence the development of guidelines, regulations, and legislation to protect against misuse of genetic information. Owing to the ELSI program, a variety of articles, books, newsletters, Web pages, TV programs, education projects, and meetings have further addressed this topic. NGS has created a lot of hopes and expectations and opened the door to the improvement of genetic diagnostics, but also has some limitations, including those listed below:

- The test results, which can reveal information about other family members in addition to the person who is tested, can have emotional, social, or financial consequences.
- lack of treatment strategies for many genetic disorders once they are diagnosed.
- Genetic testing cannot always predict the severity of the symptoms or whether the disorder will progress.
- There are still many genetic conditions for which tests have not yet been developed.
- Genetic testing results are not straightforward, and interpreting the vast majority of data generated by WES/WGS is challenging.
- Predictive testing for mutations associated with multifactorial diseases such as cancer is not definitive. Predictive tests deal in probabilities, not certainties.
- No test is 100% accurate. Accuracy varies depending on the disorder and the applied method.
- Lab errors may occur even though error reduction strategies are implemented.
- Not all deleterious variants may be detected and understood.
- Owing to the high cost of new medical technologies, NGS testing is not accessible to all, and patients may be required to pay some or all of the cost when the test is ordered.

CONCLUSIONS

NGS testing can be helpful to individuals and to some families but poses ethical, legal, and social questions. NGS has unrestricted potential to find and report incidental genomic findings of limited or unknown clinical utility. The ability to interpret the overwhelming majority of genetic variants is low; moreover, overreporting may be harmful. This fact has important ethical and legal implications and requires new ethical standards and policies. A particular emphasis should be put on confidentiality, informed consent, safety, dignity, rights, well-being, and patient privacy. A key limitation regarding the interpretation of incidental variants is that the full spectrum of genotype–phenotype correlations is not understood yet. This is the future of clinical genomics. The impact of NGS on social issues is not easy to predict since this technology is still reserved for a relatively small group of patients affected with heritable disorders. Some legal issues already forbid health insurers and employers to discriminate based on genetic information, but there are no recommendations about informing family members or intimates of a patient with a genetic condition. Emerging issues include genetic test results in pediatric populations, especially concerning adult-onset disorders. It is generally accepted that such testing should be postponed unless an early and effective intervention initiated in childhood may reduce future morbidity. There is also a fear that predictive genetic testing and carrier testing of a fetus for common and rare diseases and behavioral traits may lead to eugenic programs. In this context standard ethical considerations need to be reevaluated. Additionally, important ethical concerns raised by genetic testing and screening relate to accuracy and costs of the testing. Enormous and devastating consequences can result from receiving either a false positive or a false negative test result. A further question is the protection of NGS-generated data, which is crucial to avoid genetic discrimination against the patients.

REFERENCES

[1] Harper PS. "Before Mendel" in a short history of medical genetics. Oxford University Press; 2008. p. 13–52.
[2] Sanger F, Nicklen S, Coulson AR. DNA sequencing with chain-terminating inhibitors. Proc Natl Acad Sci USA December 1977;74(12):5463–7.
[3] Townsend A, Rousseau F, Friedman J, Adam S, Lohn Z, Birch P. Autonomy and the patient's right 'not to know' in clinical whole-genomic sequencing. Eur J Hum Genet 2014;22(1):6. http://dx.doi.org/10.1038/ejhg.2013.94. Epub May 15, 2013.
[4] Rigter T, Henneman L, Kristoffersson U, Hall A, Yntema HG, Borry P, et al. Reflecting on earlier experiences with unsolicited findings: points to consider for next-generation sequencing and informed consent in diagnostics. Hum Mutat October 2013;34(10):1322–8.
[5] Andorno R. The right not to know: an autonomy based approach. J Med Ethics 2004;30:435–40.
[6] Green RC, Berg JS, Grody WW, Kalia SS, Korf BR, Martin CL, American College of Medical Genetics and Genomics, et al. ACMG recommendations for reporting of incidental findings in clinical exome and genome sequencing. Genet Med 2013;15(7):565–74.
[7] Hastings R, de Wert G, Fowler B, Krawczak M, Vermeulen E, Bakker E, et al. The changing landscape of genetic testing and its impact on clinical and laboratory services and research in Europe. Eur J Hum Genet 2012;9:911–6.
[8] Yang Y, Muzny DM, Reid JG, Bainbridge MN, Willis A, Ward PA, et al. Clinical whole-exome sequencing for the diagnosis of Mendelian disorders. N Engl J Med October 17, 2013;369(16):1502–11.

[9] Amendola LM, Dorschner MO, Robertson PD, Salama JS, Hart R, Shirts BH, et al. Actionable exomic incidental findings in 6503 participants: challenges of variant classification. Genome Res 2015. January 30. pii:gr.183483.114. [Epub ahead of print].

[10] Frebourg T. The challenge for the next generation of medical geneticists. Hum Mutat 2014;35(8):909–11.

[11] Roubertoux PL, Carlier M. Good use and misuse of "genetic determinism". J Physiol Paris 2011;105(4–6):190–4.

[12] Green RC, Lautenbach D, McGuire AL. GINA, genetic discrimination, and genomic medicine. N Engl J Med 2015;372(5):397–9.

[13] Yang Z, Liu J, Collins GS, Salem SA, Liu X, Lyle SS, et al. Selection of single blastocysts for fresh transfer via standard morphology assessment alone and with array CGH for good prognosis IVF patients: results from a randomized pilot study. Mol Cytogenet 2012;5(1):24. http://dx.doi.org/10.1186/1755-8166-5-24.

[14] Fiorentino F, Bono S, Biricik A, Nuccitelli A, Cotroneo E, Cottone G, et al. Application of next-generation sequencing technology for comprehensive aneuploidy screening of blastocysts in clinical preimplantation genetic screening cycles. Hum Reprod 2014;29(12):2802–13.

[15] Katsanis N, Ansley SJ, Badano JL, Eichers ER, Lewis RA, Hoskins BE, et al. Triallelic inheritance in Bardet-Biedl syndrome, a Mendelian recessive disorder. Science 2001;293(5538):2256–9.

[16] Sampino S, Zacchini F, Swiergiel AH, Modlinski AJ, Loi P, Ptak GE. Effects of blastomere biopsy on postnatal growth and behavior in mice. Hum Reprod 2014;29(9):1875–83.

[17] https://www.acmg.net/docs/ACMG_Releases_Highly-Anticipated_Recommendations_on_Incidental_Findings_in_Clinical_Exome_and_Genome_Sequencing.pdf.

Index

Printed in the United States
By Bookmasters